# COMPLETE GUIDE TO DIGITAL CAMERAS

# COMPLETE GUIDE TO DIGITAL CAMERAS

BY MICHAEL D. MURIE

PROMPT®
PUBLICATIONS

**PROMPT© Publications** is an imprint of Howard W. Sams & Company, A Bell Atlantic Company, 2647 Waterfront Parkway, E. Dr., Indianapolis, IN 46214-2041.



**International Standard Book Number:** 0-7906-1175-9

**Library of Congress Catalog Number:** 98-68487

***Acquisitions Editor:*** Loretta Yates

***Editor:*** Pat Brady

***Assistant Editor:*** J.B. Hall

***Typesetting:*** Pat Brady

***Indexing:*** Loretta Yates

***Graphics Conversion:*** Brian Rees

***Cover Design:*** Christy Pierce

***Illustrations and Other Materials:*** Courtesy of the Author

***Front Cover Photo:*** Courtesy Olympus America Inc.

***Trademark Acknowledgments***:



PRINTED IN THE UNITED STATES OF AMERICA

9 8 7 6 5 4 3 2 1

For Skye

# ACKNOWLEDGMENTS

This book would not have been possible without the efforts of many people. In particular I would like to thank the staff at Prompt Publications; Editors Loretta Yates and Pat Brady, as well as J.B. Hall and Christy Pierce for their tireless work on behalf of this book.

A large number of companies provided assistance of one kind or another, including: Kimberly Reingold of Polaroid Corp, Wendy Applonie and Adam Anderson of Iomega Corp, Justin Johns of Lexmark International Inc, Scott Signore of Agfa-Gevaert Group, Kevin McLaughlin of Springboard Communications, Brian Solis, Juanita Josel and Matt Keowen of FlashPoint Technology Inc, Kari Garcia of Comm Strategies/MGVision, Brad Killam of Scitex America, Jim Datter of Surveyor Corp, Claudine Mangano of Sony, Brandon Talaich of SanDisk Corp, Lisa Grantham of Pictographics International Corp, Rania Alafoyiannis of Shandwick International, Bill Kircos of Intel, Bruce Ahern of Imaging Technologies Corporation, Jim Anders of Kaidan Inc, Rob Wiens of ACD Systems, Ltd, Markus Nigrin of Easytek Software GmbH, Laura Shook of Cerious Software, Inc, and Gale Livers of Ikelite.

And especially Joe Runde of Eastman Kodak and Mike Wong of Lexar Media.

I really appreciate the contribution of the photographers who were interviewed or provided images for the book and CD: Moose Peterson, Susan Murie, Barry Fogarty, Kent Borg, Jan-Erik Eriksson, Rommel Feria, Peter Cox, John "Jack" McLean, Walter Kimmel, Michael Chaney, Doug Knoerr, Rob and Robin Burr, Fran Traylor, Dave Burgess, Charlie Rohloff, Sandra Glaser, Michael Amick, Richard McDonald, Jim Wilhite, Jim Shaffer, Jr., Sue Williams, Roy Malitzke, Tom DiCesare and Rick Leger.

Finally, thanks to workmates, friends, and family for their help and support.

And Susan and Skye.

# ABOUT THE AUTHOR

Michael D. Murie is the Manager of Multimedia Development Harmonie Group, a healthcare application and multimedia development company in Boston, MA. Prior to that he was a freelance consultant and worked on a variety of projects including The Jack Kerouac ROMnibus CD-ROM. He is the author of the books Macintosh Multimedia Workshop and The Macintosh Multimedia Starter Kit and co-wrote with David Drucker The QuickTime Handbook. He has also written many reviews for MacWeek and New Media. In his spare time he attempts to keep the Multimedia Workshop website up to date: www.m2w.net

# CONTENTS

# INTRODUCTION

This book grew out of my personal interest and enthusiasm for digital cameras. For the past ten years I've been a multimedia developer; making interactive projects of all kinds. Most of these have a tremendous appetite for images and other media, and I'm always in a hurry to get the things assembled as quickly as possible. Being able to take photographs of objects, subjects, materials and scenery and get them in to the computer immediately makes the process that much easier, so using a digital camera just makes a whole lot of sense.

Digital cameras have come a long way in the last ten years. Now, for less than $1,000 you can buy a camera that takes photographs good enough to print at 8 x 10 sizes. Resolution and image quality has improved dramatically, and the functionality of the cameras is improving too. If you've been thinking of buying a digital camera, now is as good a time as any.

This book covers the digital camera world, from how cameras work and how to make a buying decision, through using the camera and downloading images to a computer. Since a digital camera really only makes sense if you have a computer, a lot of the book is devoted to dealing with images once you get them into your computer—from explaining resolution and color models through how to edit images, and the myriad of software applications available to manage, edit and massage your images. The chapter on printing covers the many issues involved in turning an image into hard copy form.

Since this is the age of the Internet, the book also covers the basics of Internet imaging; how to create a web page, how to upload images to a web server, and working with some of the popular auction websites. The chapter on webcams covers setting up webcams and digital conferencing.

The chapter "Making it Move" deals with formatting sequences of images as well as creating animated sequences for still images.

The last section of the book includes specifications for most of the digital cameras available or that have been released in the past ten years. Unfortunately, since the field is changing so rapidly—new cameras are being introduced every six months or less—it can be difficult to keep on top of the latest advances.

I hope that this book will provide you with the information you need to get the most out of your digital camera, as well as provide additional ideas for using your camera.

Michael D. Murie

# CHAPTER 1
# STUMBLING THROUGH DIGITAL IMAGING

...in which the author attempts to establish his digital camera credentials, while providing a brief history of the digital camera.

As a multimedia developer I often need to get photographs onto the computer, both to create rough draft prototypes, as well as for final artwork. Flatbed scanners have been one of the most reliable methods for getting artwork onto the computer. Over the past ten years—and particularly in the past two years—the price of flatbed scanners has fallen to the point where a reasonable scanner can be bought for less than two hundred dollars. But, flatbed scanners require that you first get the film developed and printed, which takes at least one hour. Also, the printing of the negatives can introduce unwanted color and exposure changes. Automated film printers use an automatic meter to calculate the exposure for the print. This can negate any manual exposure adjustments you made when taking the original photograph. Finally, the scanning process itself can further skew the accuracy of the colors in the image. So while scanners are reliable, cheap and easy to use, they aren't the quickest, most accurate way to capture an image and get it into the computer.

Slide scanners are another alternative. Again, you have to get the film developed, but you don't have to worry about the lab adjusting your prints when it prints them. While the price has come down, film scanners are still more expensive than flatbed scanners. Color accuracy—except for the expensive, several-hundred-dollar units—is also an issue.

About ten years ago Kodak introduced the Photo CD service, which scans film at high resolution and quality and places the images on a CD-ROM. For getting high quality, high resolution scans onto a computer it remains one of the best solutions for those who only need a few images at high resolution. Unfortunately, cost and convenience is a problem. The lowest price I've paid has been 52 cents per image, and that was with a week and a half delivery time. Service bureaus that offer same day Photo CD service tend to charge $2 - $3 per image. Kodak has recently introduced the Picture CD service, which offers good quality scans at a lower resolution the Photo CD. This might be an acceptable solution for many people.

## ENTER DIGITAL CAMERAS

But wouldn't it be great to take a picture and immediately transfer it to the computer? No developing, no wait. Well, that's the promise of digital cameras. Unfortunately, while they've always promised "no wait" for the picture, until recently the quality of the picture produced by the consumer units wasn't very good.

The first "digital camera" I used was a Sony Mavica made almost a decade ago. This was really a still video camera, not a digital camera. Though it resembled some of today's digital cameras, and used a CCD to capture an image, it stored its images on a mini-floppy disk, and the image was recorded in the form of an analog television signal. It only "output" analog video; its only connection was a video-out plug. You could connect the device to a television to watch the pictures, or if you had a computer you could use a frame grabber (video digitizer) and capture the image, much as you would from a video camera. In theory, it was more convenient for capturing stills than a video camera because it worked like a still camera (point and click) and, when digitizing on the computer, you didn't have to mess around trying to find the image on the video tape. In a sense this was true, however the quality of the images was very low because Sony's consumer unit captured in fields rather than a frame of video.

To understand the significance of this requires a bit of understanding of how television works. While a computer display is made up of pixels, arranged in a fixed grid, the common television standards (NTSC and PAL) are analog formats. In these video formats, the picture signal is made up of horizontal lines, which the scan gun draws one at a time on the display. A complete screen (called a frame) consists of several hundred horizontal lines, the exact number of which depends upon the equipment being used. At its maximum it's about 500 horizontal lines.

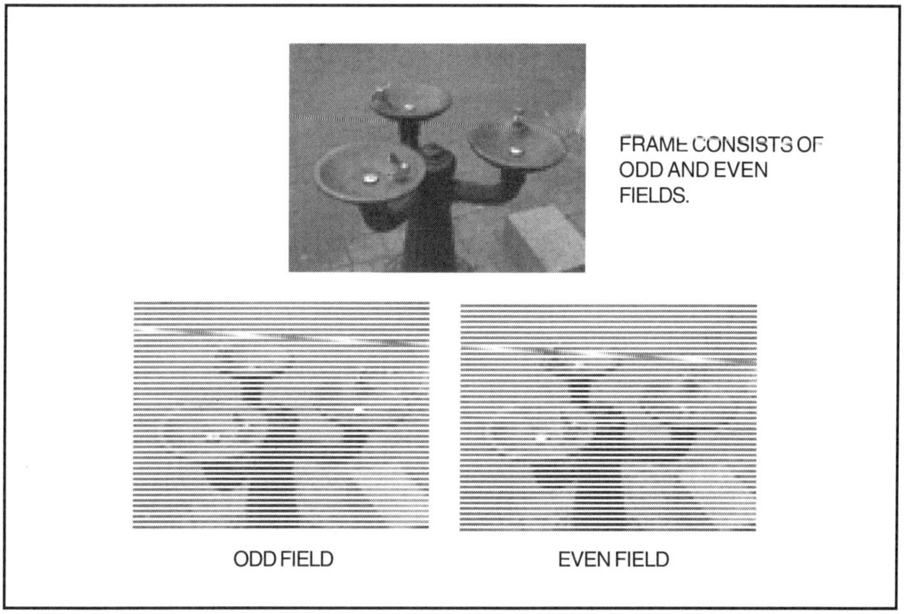


*Figure 1-1. Fields and frames in video.*

Since NTSC is an analog standard, conversion to a digital image is somewhat arbitrary in nature; however, it's generally converted to a 640 x 480 pixel computer image.

Because the scan gun took so long to draw the frames in the original televisions, a trick was introduced to improve the apparent screen redraw rate. Rather than draw each horizontal line one at a time from top to bottom, the gun draws every other line. It draws all the odd lines (1,3,5, etc.) then goes back and draws the even lines (2, 4, 6, etc.) Each set of odd or even lines is called a field. So while a frame is drawn in 1/30 of a second, a field is drawn in 1/60 of a second.

Going back to the digital still camera, on this consumer unit, the camera only captured a field instead of a frame. This meant the image had half the vertical resolution of a "regular" screen image. To make up for the missing lines, line doubling is used. This means that each horizontal line is repeated twice.

The next camera I got to play with was a Canon Still Video Camera, which was marketed as a semi-professional unit. It captured frames instead of fields, and was a more sophisticated camera than the consumer Sony unit. It was also a lot more expensive, and you had to buy a separate unit that attached to the SCSI port of your computer to digitize the images that were stored on the mini-floppy disk. Unfortunately, that was expensive too (about $2,000 back in 1990). The quality of the images was a step forward from the consumer unit, but still wasn't really good enough to justify the price.

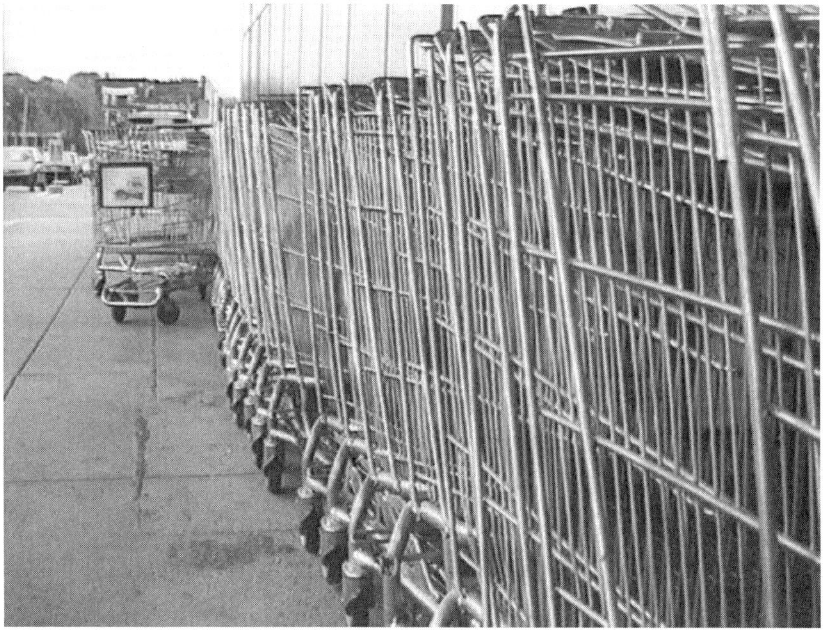

*Figure 1-2. Image captured by Canon still video camera. If you look closely you'll see there's very little detail in the wire of the carts.*

Around 1994 I got to play with the first low price (less than $1,000) digital camera; Apple's QuickTake 100. At the time I thought it was a cool piece of equipment, though it was still rather pricey. Ostensibly it produced 640 x 480 images, and it saved them in memory in the camera, not on a mini-floppy. Connecting the camera to a computer using a serial cable was easy; images were downloaded reasonably quickly (the pictures weren't really that large; only about 40K) and the camera was easy to use. Unfortunately, it could only hold a few images. Apple did a great job with the software; something that even today's camera manufacturers could learn a thing or two about.

All in all, it was a lot easier to get images into the computer from the QuickTake than from any other camera. Unfortunately, the image quality wasn't that great. At full screen, the images looked smudged—an artifact of the low-resolution CCD and the heavy JPEG compression used. I found that the images looked "okay" when scaled down to quarter size, 320 x 240, though even at that size, they exhibit color problems.

Apple did upgrade the QuickTake to the 150, which had slightly improved resolution and could take more pictures, but the quality was very similar to the original.

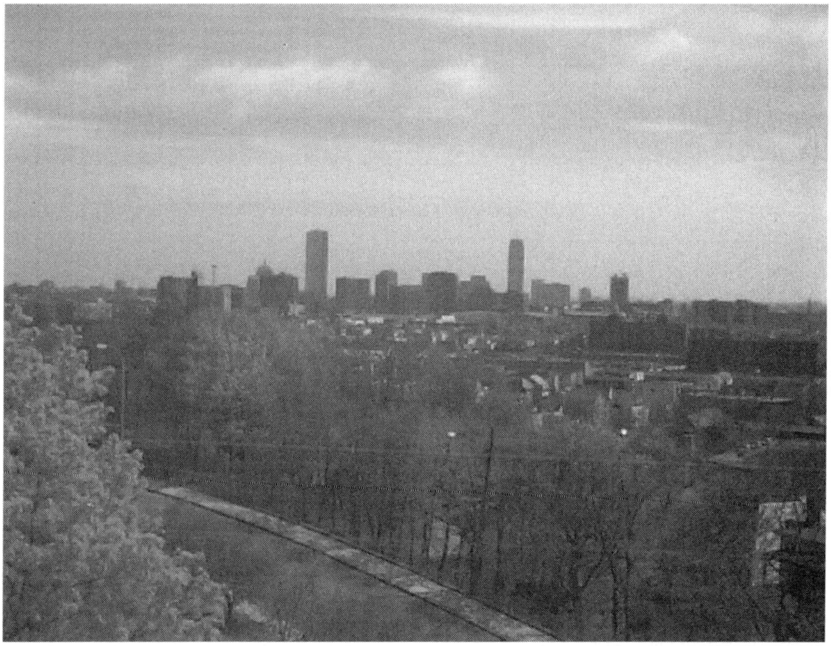

*Figure 1-3. QuickTake 100 image.*

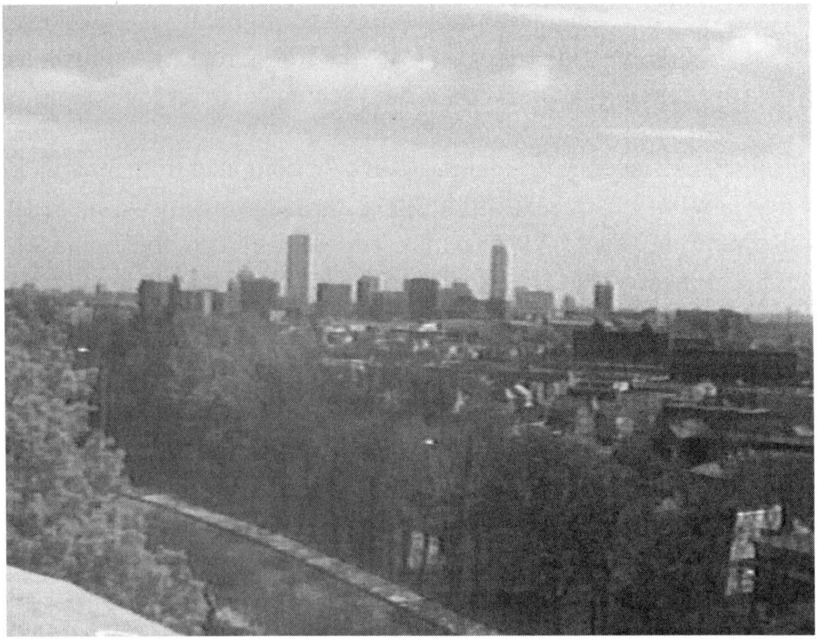

*Figure 1-4. Digitized image captured from video camera.*

*Figures 1-3* and *1-4* were both taken at the same time using a QuickTake 100 and an 8mm video camera.

While there's some color differences, there is very little difference in resolution between the image captured with the QuickTake 100 and an image captured using an 8mm video camera and digitized using a video frame grabber.

The next cameras I spent time with were the Kodak DC50 and Chinon ES-3000, two cameras which essentially were the same; the body and optics were manufactured by Chinon, but the DC50 had a different CCD. This camera had a lot to recommend it; the image quality was better (though still not really 640 x 480 in quality) and it featured a zoom lens. In addition to its internal memory, the camera accepted a memory card.

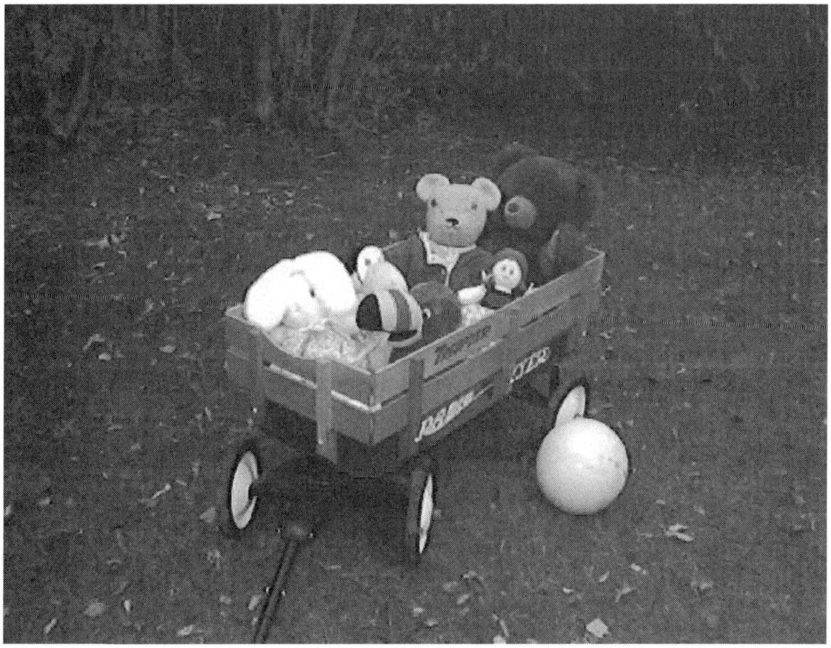

*Figure 1-5. Chinon ES3000 image.*

The zoom lens really added to the convenience of the camera, and the picture quality was much improved over the QuickTake. However, they cost several hundred dollars and the images still weren't quite good enough to be used at 640 x 480.

In 1996 Kodak introduced the first Megapixel consumer camera, the DC 120. This camera cheats a bit; the CCD has a resolution of 848 x 976, but the software scales this up to 1280 x 960. This results in rather blurry images when viewed at actual size. Still, the images were a great step up from the DC50. Finally, here was a camera that produced acceptable 640 x 480 quality images; I just rescaled the image using Photoshop. Conveniently, the image size was exactly double 640 x 480. The camera featured a zoom lens and a memory card, and though it was a completely new camera, it very much resembled the DC50 in form factor and user interface.

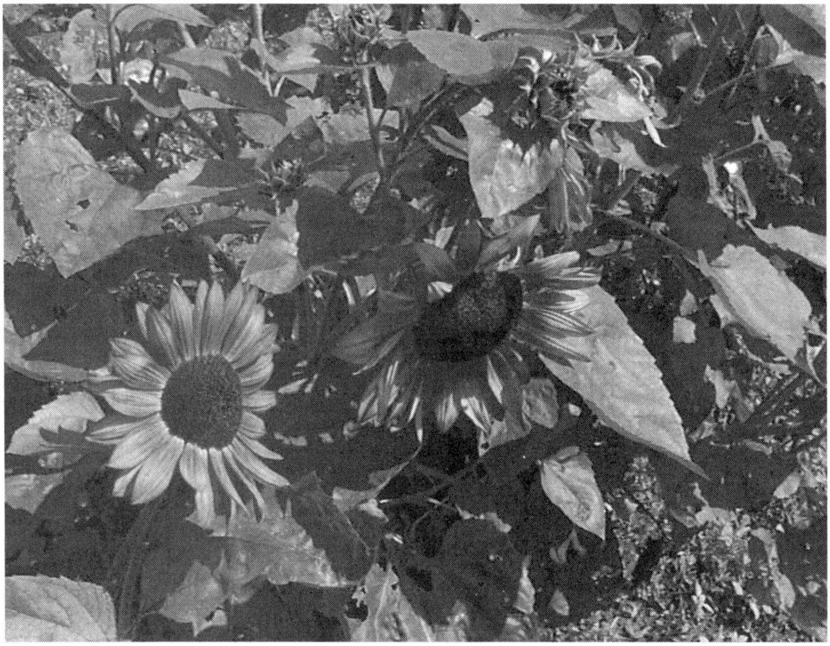

*Figure 1-6. Kodak DC120 image.*

I used the DC 120 for a couple of years, and was very happy with it. That was until I got to spend some time with Kodak's DC260 camera. The images from this camera, though not always perfect, are incredible at 640 x 480. At their actual size of 1536 x 1024, you can see a little noise in the image, but they are still usable. The other things I like about this camera are the color accuracy, the fact that it takes pictures almost immediately, and the speed with which you can transfer images using the USB interface. Finally, here is a camera which is what I wished I had when I first started playing with "digital cameras" all those years ago. Other manufacturers are offering cameras with similar capabilities.

Digital cameras have come a long way in the last ten years, and they still have a way to go before they can rival film cameras for price. Only the multi-thousand-dollar professional cameras currently available can match the image resolution you get with a Photo CD scan from a 35mm film camera (about 3,000 x 2,000 pixels). The latest generation of high-end consumer cameras produce images suitable for printing at 8 x 10 inches (with a resolution of 1600 x 1200).

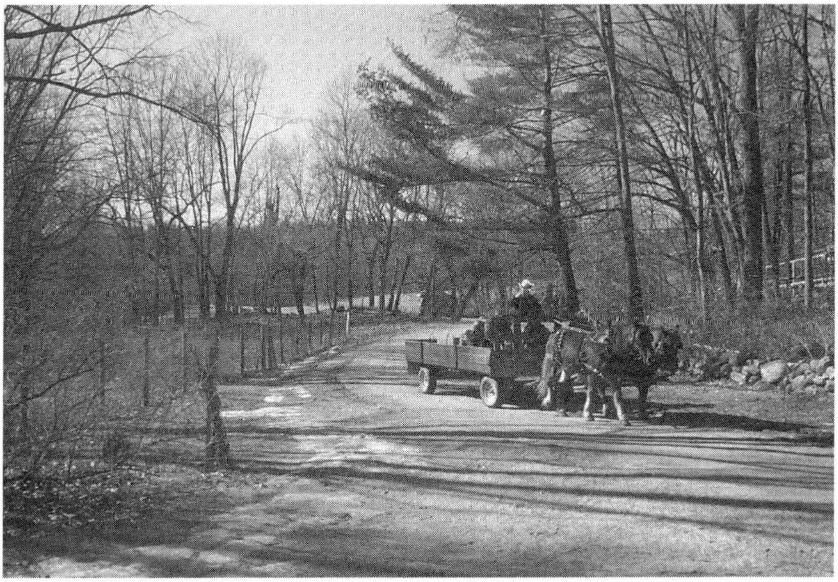

*Figure 1-7. Kodak DC260 image.*

It's a good time to be looking at digital cameras. Every few months new models come out. The quality keeps improving, yet the prices remain about the same (much like the computer industry). Should you buy a digital camera now? If you're reading this book then you're obviously interested in digital cameras. If you need or want to edit images on you computer; have a website or want to sell things on the Internet; want to email images to friends and family; or produce interactive or traditional media on the computer, then buy a digital camera.

If you don't do any of these things but enjoy taking pictures, then I want to encourage you to consider buying a digital camera. Despite the limitations of the current cameras, two things about them will change the way you use a camera. First, the ability to review the image immediately using an LCD display means that you'll know whether you got the picture you wanted before you go home. Second, since you can always delete the images you don't want, you'll experience a new freedom in picture taking.

Warning! It could be addictive. You may need to buy a bigger hard drive, but that can come later. First, let's look at how to go about buying a camera!

## CHAPTER 2

# HOW TO BUY A DIGITAL CAMERA

Like the desktop computer market, the field of digital cameras has been changing so dramatically over the period 1997-1999 that recommendations for which camera to buy become meaningless within a very short space of time. The models have been changing so quickly—new ones appearing every few weeks—that even magazine comparisons are out of date by the time they are printed.

For this reason, this book provides only a few specific recommendations for cameras because chances are that by the time you read this there will be newer cameras available, prices will have changed, and so will user expectations.

It is the intention of this chapter to help you choose amongst the features available and provide general recommendations. At the end of the chapter you'll find some checklists and guides to help you.

For those who think this is just an attempt to weasel out of giving recommendations, you'll even find some recommendations for currently available cameras at the beginning of Chapter 24. For more up-to-date recommendations, check the webpage: *www.m2w.net/digitalcamera.*

## ALTERNATIVES TO DIGITAL CAMERAS

If you've purchased this book then you're probably pretty committed to buying a digital camera. You might already have one and are just reading this chapter to confirm that you made the correct buying decision. In that case you should now skip to the section *Features to Consider*, (pg. 19).

But maybe you're standing in the book store, thumbing through the pages, wondering whether a digital camera is right for you, and trying to find some quick and free advice before you buy a camera. If that's the case, then I'd like to first say, "buy this book now!"

If you want to get digital images into your computer, whether for use on a web page, or for image editing, there are ways to do this without buying a digital camera. You might already have all the equipment you need. The first choice is to use a traditional film camera, then digitize the prints or negatives. The second choice is to use a video camera of some kind and digitize (capture) a frame of video.

## USING A TRADITIONAL CAMERA

If you have a good quality film camera, then you already have a potential input device. A flatbed scanner—units costing about $100 can produce very good results—can be used to scan the photographs. If you only have a few images or an occasional need, then this is the best way to go.

A good 35mm SLR camera will produce better photographs than any of the medium-priced digital cameras ($1000 or less) and can perform as well as even the more expensive studio cameras. The downside is that a film camera isn't as convenient; you have to get the film developed, which means a trip to the 1-hour photoprocessor, and accurate color reproduction can be difficult.

A Polaroid camera cuts the trip to the store, though the film is a little more expensive. Polaroid sells a small scanner, the PhotoMax Photo Scanner, which is perfect for this application; it scans images up to 4 x 6. If you're going this route, the Polaroid Spectra cameras are a better choice than the 600 models; the Spectra film and camera has a much wider latitude and produces a much better picture. Even so, partly because the prints are smaller than regular 4 x 6 prints, you'll find that the scans aren't as good as you can get with most film cameras.

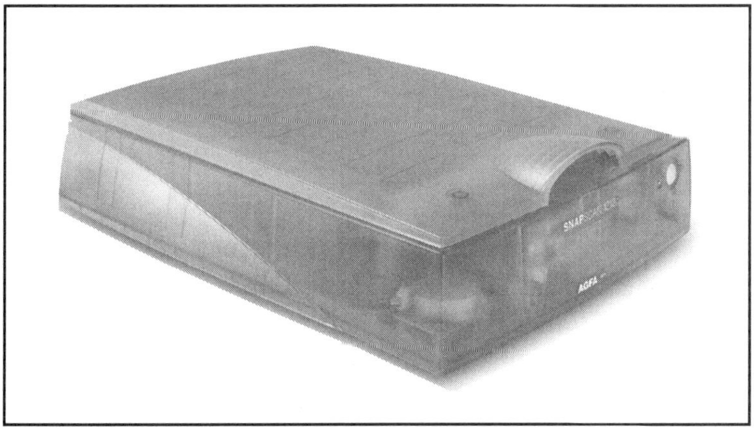


*Figure 2-1. Flatbed scanner. Easy to use, and inexpensive, a flatbed scanner makes sense for many uses. (photo courtesy Agfa Corporation)*

Don't have a scanner? Many photoprocessing outlets offer a service option that will scan the images and place them on a floppy disc. Quality of the scans isn't going to be great because they are small and heavily compressed in order to fit them all onto a single floppy. If you are going to be printing these images, then you'll probably want to use another method. For websites, scanning is probably acceptable.

If you need a good quality scan, Kodak's Photo CD and Picture CD provide high-quality scans on a CD. Price becomes a factor here; Photo CD is considered a professional format, and costs range from 52 cents an image up to $3 or more depending upon where you have the scan done and the number of images. But Photo CD will produce absolutely the best scan from your negative with a resolution of 3072 x 2048 pixels. Picture CD provides good quality scans (as good as you'll get from the $1,000 digital cameras) that are half the resolution of the Photo CD scans (1536 x 1024), but for only $10 per disk. Unfortunately, that's in addition to film processing, and if you have pictures made as well (some places require it) then the Picture CD can end up costing as much as Photo CDs.

Photo CDs are reasonably easy to use. Unfortunately, they are stored in a custom file format. Kodak has licensed this to most editing application developers, so you can usually open a Photo CD file with your editing application.

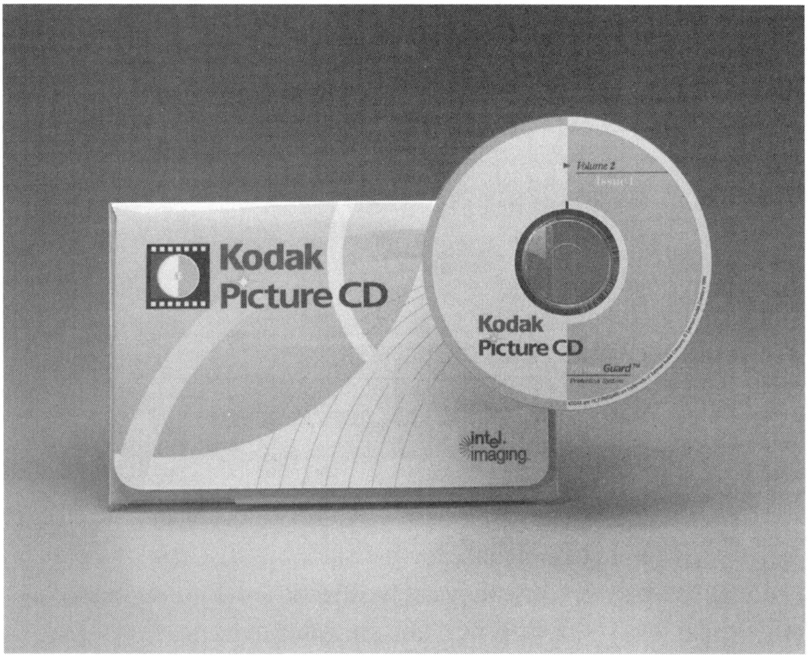


*Figure 2-2. Kodak's Picture CD format stores regular images digitized onto a CD-ROM.*

When you open the file, the application will let you choose between several different resolutions for the image. Once the file is opened you can save it in any other file format, but you can't save it back to Photo CD format.

Picture CD comes with special custom software for Windows computers which lets you view thumbnails of the images, and even perform basic editing functions. This is all provided in a magazine-like interface. Currently, the software doesn't work for Macintosh users, but all of the images are saved in standard JPEG format, so whether you have a Macintosh or a Windows computer you can open the files directly and skip the custom software.

Picture CD provides quality as good as you'll get from all but the most expensive digital cameras, while Photo CD rivals even those. The downside is that they aren't instantaneous. At a minimum it'll take about an hour to get a Picture

CD made, assuming you live close to a photo processor with the equipment necessary to make the disc. Otherwise, it could take anywhere from one day to a week to get the disc back. You also have the cost to consider; though you'll process many rolls of film for the cost of a digital camera.

## VIDEO

### HI8, 8MM, S-VHS AND VHS CAMCORDERS

If you have a camcorder, and you have a video digitizer (a device which "captures" frames of video and converts them into a digital image), then you can use these to capture still images on your computer. Fairly good quality digitizers are available for only a couple of hundred dollars. This may be a cheaper option than buying a digital camera, but the quality won't be nearly as good. If you only need a 320 x 240 pixel image then consider this option; otherwise, I'd look at something else.

Of course, if you don't already have a camcorder, the cost of one is as much as a digital camera, and a comparably priced digital camera will capture a much better image. Of course, a digital camera won't do as good a job recording motion!

### DV CAMCORDERS

The comparatively new DV camcorders provide a special still image capability which is very useful for capturing images. While the resolution is lower (about 640 x 480) than you'll get from the megapixel still cameras, the DV camera can capture images almost as quickly and easily as a still camera. Some of the cameras capture in a progressive mode, which is ideal for still images. Instead of capturing the frame as interlaced fields, a progressive scan CCD captures the horizontal lines one at a time from top to bottom. This eliminates the feathering that can be seen in interlaced captures when an object in moving. This is because it is in one place for the first field, and another for the second. Cameras that offer this feature include the Canon Optura, which Canon bills as having a Digital Photo Mode and Digital Motor Drive (special modes for capturing still images). The camera even looks more like a still camera than a video camera.

However, it's still a tape-based system; so finding an image may not be as quick and easy as using a still camera. And if your computer didn't come with one, you'll have to buy a FireWire card to be able to transfer the images to your computer. (See Chapter 5)

A DV camera will cost as much as—and usually more than—a digital still camera. For these reasons I recommend this solution when you are more interested in capturing video (motion) than still images. If you are looking to replace your camcorder, yet want to be able to capture a few still images using a "digital camera," then a DV camera might be the right choice. These are described later in Chapter 19. Otherwise, a digital camera for the same price will produce much better images, and is easier to use and more flexible.

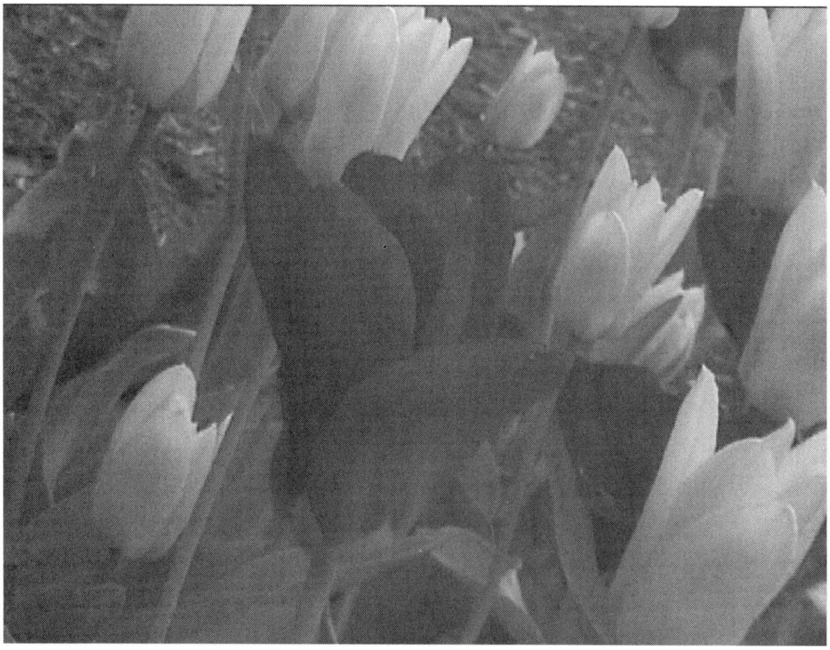

*Figure 2-3. Frame captured from DV camcorder. DV Camcorders provide good 640 x 480 resolution images useful for many purposes. They're best for video. (Photo courtesy Kent Borg)*

# FEATURES TO CONSIDER

This section reviews the major features that you should consider when evaluating a digital camera. The section after this, *Evaluating Cameras*, provides some typical uses of digital cameras, with suggestions for features to look for.

This section (and book) mainly covers the consumer and "prosumer" cameras which are priced at $1,000 or less. There are several cameras that are much more expensive than this, but the price of these units is out of reach of most users. For some background on these cameras see Chapter 22.

## RESOLUTION

Resolution is usually the first thing anyone considers when looking at a digital camera. But what does resolution mean? Resolution is a very misleading term because it is used very carelessly in many discussions. The term resolution should, strictly speaking, be used to describe the ability to resolve detail (changes in contrast between pixels). However, resolution is often loosely used as a term describing the total number of pixels in a captured image. While this figure is very important, it only has relevance when you are actually displaying or printing the printer—i.e., do you have enough pixels to display or print it at the size you want, without the image appearing to lack detail?

Obviously, the more pixels the better, but how many pixels do you need to have to be able to do what you want with the camera you just bought?

To answer that question you need to know what you plan to do with the photographs you will take. Is it to look at them on screen, or to print? To place on a web page, or use in a brochure?

The resolution of computer screens varies, particularly if you have a multiscan monitor and can switch between a number of view sizes. A 640 x 480 pixel image is "full screen" if your monitor is set to display at 640 x 480 pixels, but not if the monitor is set to 1600 x 1200.

Resolution is covered in more detail in Chapter 6 and Chapter 9.

Printing has a different set of requirements. But for a good average, assume you'll need a resolution of at least 150 pixels per inch. To get a good 4 x 6 print you'll need at least 600 x 900 pixels. To get an 8 x 10 you'll need at least 1200 x 1500.

A word of warning: Just because a camera says that it has a resolution of a certain size, don't assume that the images it produces at that size are acceptable for your purposes. You should always view samples of images that were captured with the camera. Many early cameras claimed to produce images at 640 x 480, but at that size the images were badly blurred and poor in quality. Only by scaling them down—to as little as 320 x 240—do you have an image that looks acceptable.

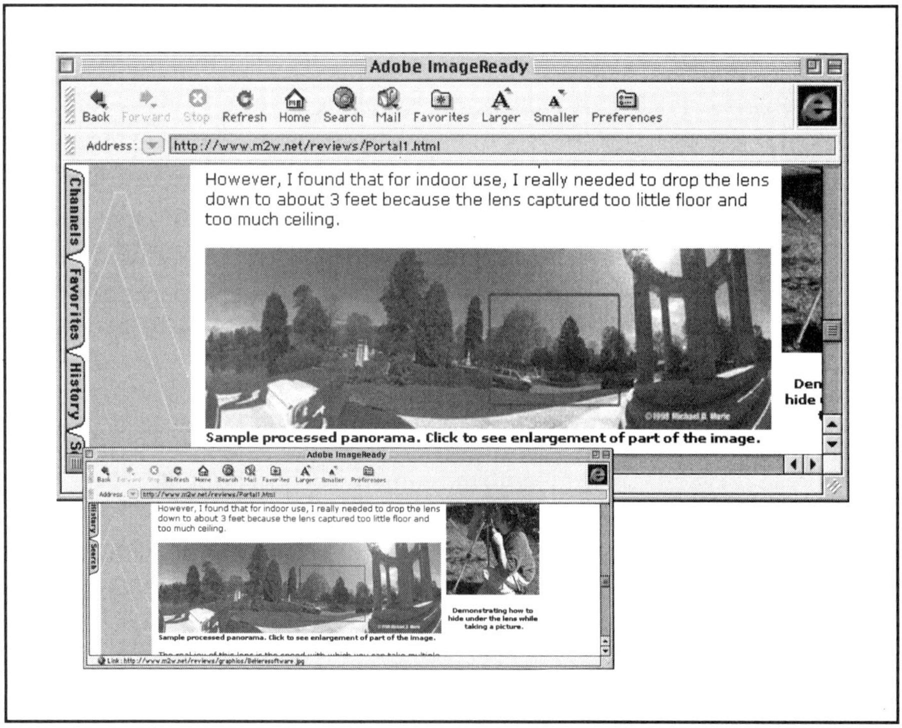


*Figure 2-4. The same web page viewed at different screen resolutions on the same monitor; the size of the graphic changes, though the relative size (i.e., compared to toolbars etc.) remains the same.*

The quality of the newest cameras has improved dramatically, but it's still a good idea to see sample images and not take the specification sheet at face value.

## MEMORY

All digital cameras, except for some studio cameras, webcams and video conferencing cameras, use either internal memory or a memory card to save the images after they are taken. Many early cameras contained only 1 or 2 megabytes of internal memory; this is important when you consider that even small files at high compression rates are 50K in size. The Apple QuickTake 150 allows you to take approximately 16 pictures—then you have to download the pictures and erase the memory before continuing.

Flash memory cards (sometimes called "digital film") have become the preferred method of storing images. There are two varieties of memory card in use today: CompactFlash and SmartMedia. In many respects these two formats are very similar; they are small plastic cards that contain memory chips.

While they are approximately the same size, they are physically different. The SmartMedia card is slightly longer, but much thinner than the CompactFlash card. Also, the SmartMedia card has a small cutout on one corner, while the CampactFlash card is completely rectangular.

| | **Length** | **Width** | **Thickness** |
|---|---|---|---|
| **CompactFlash** | 1.43 in (36.4 mm) | 1.69 in (42.8 mm) | 0.13 in (3.30 mm) |
| **SmartMedia** | 1.78 in (45 mm) | 1.46 in (37 mm) | 0.03 in (0.76 mm) |

They are different internally as well. SmartMedia does not include a controller, a chip that handles the transfer and storage of data on the card, while CompactFlash does. Because of this, the smaller (8 and 16MB) CompactFlash cards are more expensive than the same sized SmartMedia cards. But, because the memory management for SmartMedia cards is handled by the camera, some early cameras, built before larger sized SmartMedia cards were created, are unable to work with larger SmartMedia sizes.

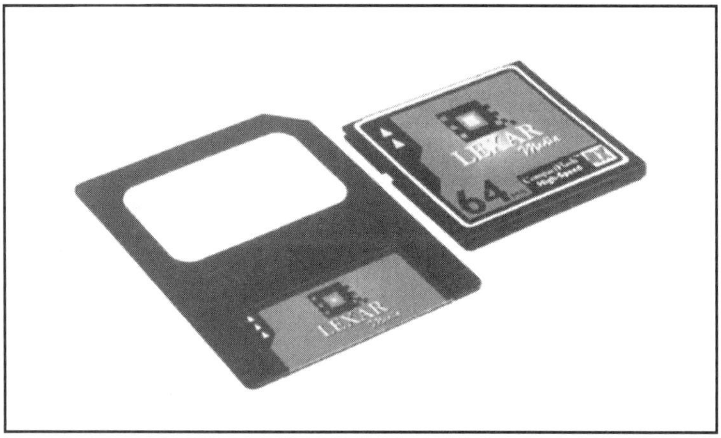


*Figure 2-5. CompactFlash and SmartMedia cards. (photo courtesy Lexar Media Inc.)*

At the moment, SmartMedia and CompactFlash manufacturers both claim that their formats will reach up to 1 Gigabyte in size, though no such card is currently available.

Currently, more manufacturers are using CompactFlash than SmartMedia; however, SmartMedia has shown up in some other devices, notably the Diamond Multimedia Rio MP3 Player.

Both cards can be inserted into a PCMCIA slot, providing you have the appropriate adapter. Also, some of the high-end cameras actually use PCMCIA cards as their storage device (see Chapter 10 and Chapter 22).

Today, while it appears that CompactFlash has a slight edge in market acceptance over SmartMedia, it's not enough to warrant ignoring a camera you are interested in. If you prefer the Olympus camera over the Kodak then go ahead and buy the Olympus. What's the worst that could happen? Five years from now you'll buy a new camera and you won't be able to use your old 32MB card in the camera. Of course, by then 1 Gigabyte cards will cost $50 and you won't want to use the 32MB card even if you could!

For more information about CompactFlash and SmartMedia cards, particularly ways to quickly (and in some cases very cheaply!) download images from these devices to your computer, see Chapter 5.

Sony has also introduced it's own memory card called the Memory Stick. Sony is adding this to cameras, digital video cameras, notebooks and desktop computers. Unfortunately, so far only Sony is supporting this format.

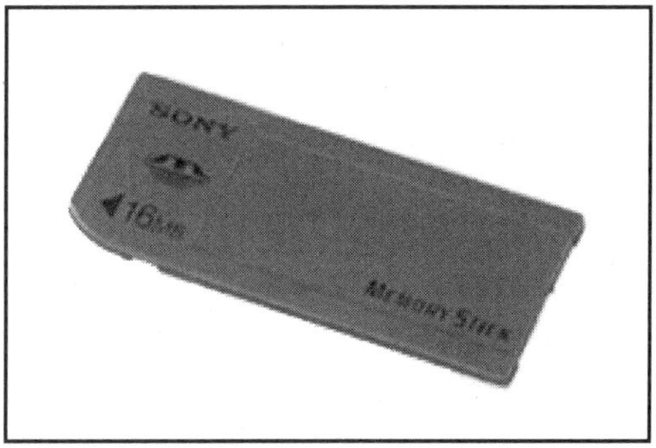


*Figure 2-6. Sony's Memory Stick card.*

## PCMCIA CARDS

Several large and expensive cameras, most notably most cameras in Kodak's professional digital models, use PCMCIA (Personal Computer Memory Card International Association) cards. This card format, also used by many notebooks, allow small hard disks (as well as other hardware) to be placed in a very small, removable format. The cards with hard drives can be expensive, but store a lot of information and provide fast random access to the information.

There are three PCMCIA formats; Type I, II and III. All three are the same width and height, but have different thicknesses, Type I are 3.3mm thick and are used for system memory products. Type II are 5mm thick and are used for flash cards, modem and network cards. Type III are 10.5mm thick and can be used for small disk drives. The standards are backward compatible; a Type I card will fit in a Type II slot, and Types I and II will fit in a Type III slot. But Type III will not fit in a Type II, and Types II and III will not fit in a Type I slot.

The PCMCIA format cards don't seem to be moving into the consumer and prosumer digital camera market; both SmartMedia and CompactFlash are about half the physical size of even the small Type I card.

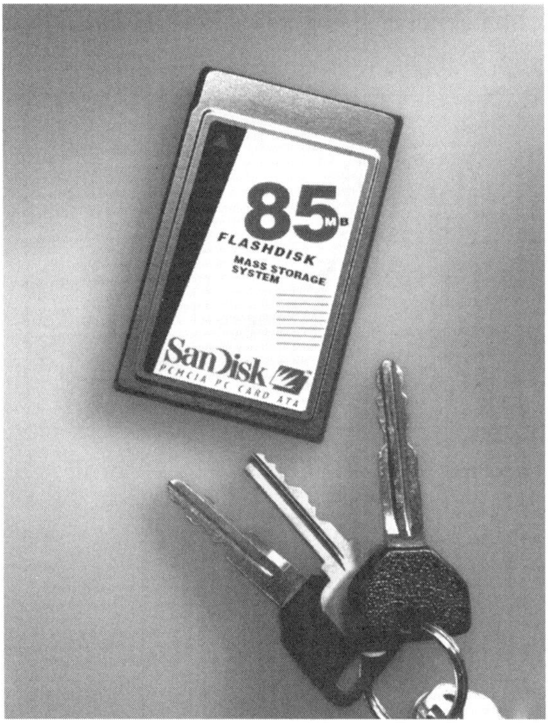


*Figure 2-7. An 85MB Type II PCMCIA card. (photo courtesy SanDisk Corporation)*

Apart from making it possible to use a hard disk that holds 300MB or more, the other advantage of PCMCIA cards is that they can be used in the many notebooks that come with PCMCIA slots. If your notebook contains such a connector, you can also use CompactFlash and SmartMedia cards, provided you buy an adapter (see Chapter 10). The CompactFlash adapter is very inexpensive because the controller in the CompactFlash card adheres to the ATA standard (ATA is part of the PCMCIA standard).

## DISK FORMATS

**Floppy Disk**. Sony uses floppy disks on many of their cameras, and by all accounts, the people who buy the Sony cameras like this feature a great deal. The floppy is formatted in the MS-DOS file structure (which can be read by both Windows and Macintosh computers). You can insert the floppy into your computer and copy the images; and floppies are so inexpensive it's almost like the "film" is free. Only buyers of newer Macintosh computers (which lack a floppy disk drive) will have a problem with this storage format.

This feature does, however, come at a sacrifice; Sony uses greater compression in order to store a reasonable number of images on a floppy disk. Even though floppies are cheap, it wouldn't be much fun if you could only store two images on a disk. Going away on a holiday would require packing 100 floppy disks! So it's understandable why they compress the images so much, but it means you'll get noticeably better images from other cameras that have the same resolution but use a flash card instead of floppy disks. Also, it can take a while to write large files to a floppy.

Still, most Sony users seem happy with the quality of the images their cameras produce. As with any camera, evaluate the quality of the images the camera produces before buying.

**MiniDisc (MD)**. Originally developed by Sony for recording digital audio, the MiniDisc stores about 140MB of data (that's uncompressed data; MiniDiscs use audio compression when storing audio files.) Sharp has marketed a camera that uses MiniDiscs in Japan, and was considering bringing the MD Data Camera to the United States in 1997 but abandoned the idea. The advantage of the MiniDisc is that it can store lots of images; Sharp claims up to 2000, though this would be heavily compressed and/or low-resolution images.

The biggest problem with the MiniDisc is that it has not caught on as a format for computer storage, so there are no computer drives available for loading the images into the computer. Using a serial connection to transfer images is not appealing, but if the camera had a USB connection it could be usable. Since Flash Memory cards are rapidly approaching the size of MiniDiscs, the window of opportunity for this technology may be closing.

**120MB HDD Floppy**. The large format floppy drives provide a lot of storage, almost as much as MiniDiscs. And there are computer drives available for these disks, though they have not replaced standard floppy drives, so there's a good chance that purchasers of the camera will also have to buy a new drive if they want a fast and easy way to transfer images. Still, it seems like a likely path for Sony to consider with their line of digital cameras.

**Iomega Clik! Drive**. This drive is examined in detail in the *Accessories* chapter because it includes a feature that lets you download images from any CompactFlash or SmartMedia card to the Clik! drive without being connected to a computer. It could be an attractive transfer medium for storing a large number of images while out in the field.

The Clik! drive has some advantages—it comes with a base that attaches to Windows PCs, and the 40MB drives are currently priced at about $10 each. However, 40MB may not seem very large a year from now. Also, flash memory cards should be more robust than the Clik! drive and mechanism.

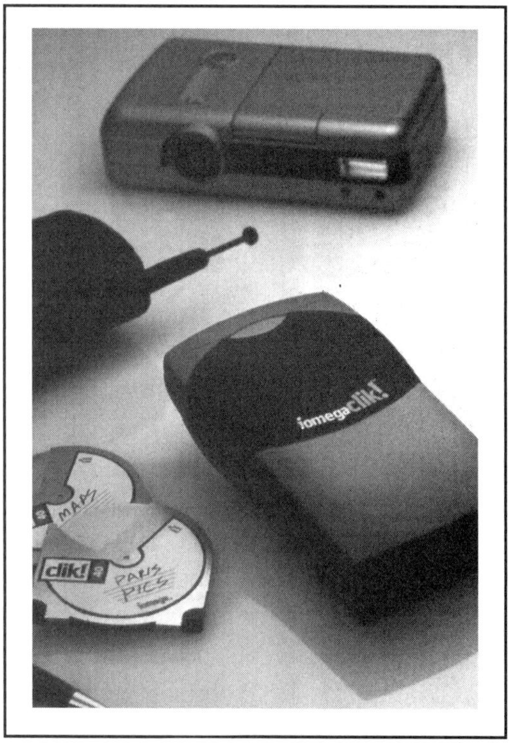


*Figure 2-8. Iomega Clik! drive. (courtesy Iomega Corporation)*

## TYPES OF MEMORY

To a certain extent digital cameras can be treated as islands. It's not important how the image is stored in the camera, or even what format it's in. The important questions are:

1. How many images can the camera store?

2. How easily can you add additional memory?

3. How easy and fast is it to transfer images to the computer in a format that is usable by your editing application?

If the answers are *plenty*, *very easy*, and *fast and easy*, then the only other issue is, how freely available is additional storage, and how expensive is it? Whether 90% of other cameras, or only 10%, use the same storage is of little interest to most users. Of course, if 90% of the market were using another format, this could make that format much cheaper.

Like all generalizations, I know that some people can see the desirability of exchanging cards with others. The advent of printers that accept memory cards (see Epson/Digital printer in Chapter 9) also suggest that the "camera is an island" view has some potential holes in it. Such are the joys of technology, and dealing with evolving standards and technology. Sometimes you just have to make your best guess and punt.

## LENS

In standard 35mm cameras, the quality of the lens is the focal point (forgive the pun!) of image quality. Since you can use the same film in any camera, the quality of the lens—as well as functionality and interface of the camera—become the overriding concern. Up until recently the quality of lenses in consumer digital cameras has been a minor concern. The quality of the image produced by the lens was so badly degraded by the low-resolution CCD, noise, and high JPEG compression, that worrying about the lens was largely irrelevant.

That is starting to change. Cameras such as the Olympus 620L, Nikon 950 and Kodak DC265 feature high-resolution CCDs, and very low, or even no-compression storage options.

Assuming that the CCD does a good job of capturing the image, optics quality does become a serious issue because most CCDs are small. While 35mm film is an inch and a half wide, most CCDs used measure less than one-half of an inch. This makes the lens smaller and more difficult to make. At the same time, the lenses are often built for one camera only. The cost of development cannot be amortized over many cameras.

What this means to you, the purchaser, is that it increases the cost of a camera with good optics. Also, you have to do some homework and look at the images to see what the camera is capable of producing.

Digital cameras can be very susceptible to barrel distortion (a rounding of horizontal lines), particularly at the wide angle settings. If you like taking wide-angle photographs of architectural objects (buildings, interiors) then this can be a problem. If you want to add a wide-angle lens adapter, then such problems will be magnified.

Most cameras feature a zoom lens (though a few of the cheapest have a single fixed lens). While photography purists claim that zoom lenses result in lazy photographers, there's no denying that zoom lenses are very useful. Also, since most digital cameras don't support interchangeable lenses, the one you buy is the one you are stuck with. There are add-on lenses that you can use to extend the range, either in wide-angle or telephoto. Most of these add-ons require that the front of the camera lens have a thread on the ring around the lens that the adapter can be screwed into. If the camera lacks a thread, or is of an unusual lens size, you might not be able to add attachments.

Cameras sometimes offer two flavors of zoom; optical and digital. The optical zoom moves the glass elements to adjust the focal length of the lens. (For more on focal lengths, read the beginning of Chapter 4.) Digital zooms use software to enlarge a portion of the image captured by the CCD. Optical zooms are preferable over digital zooms; the digital zoom images will be blurry because they are using interpolation to increase the size of the image (see Chapter 7,

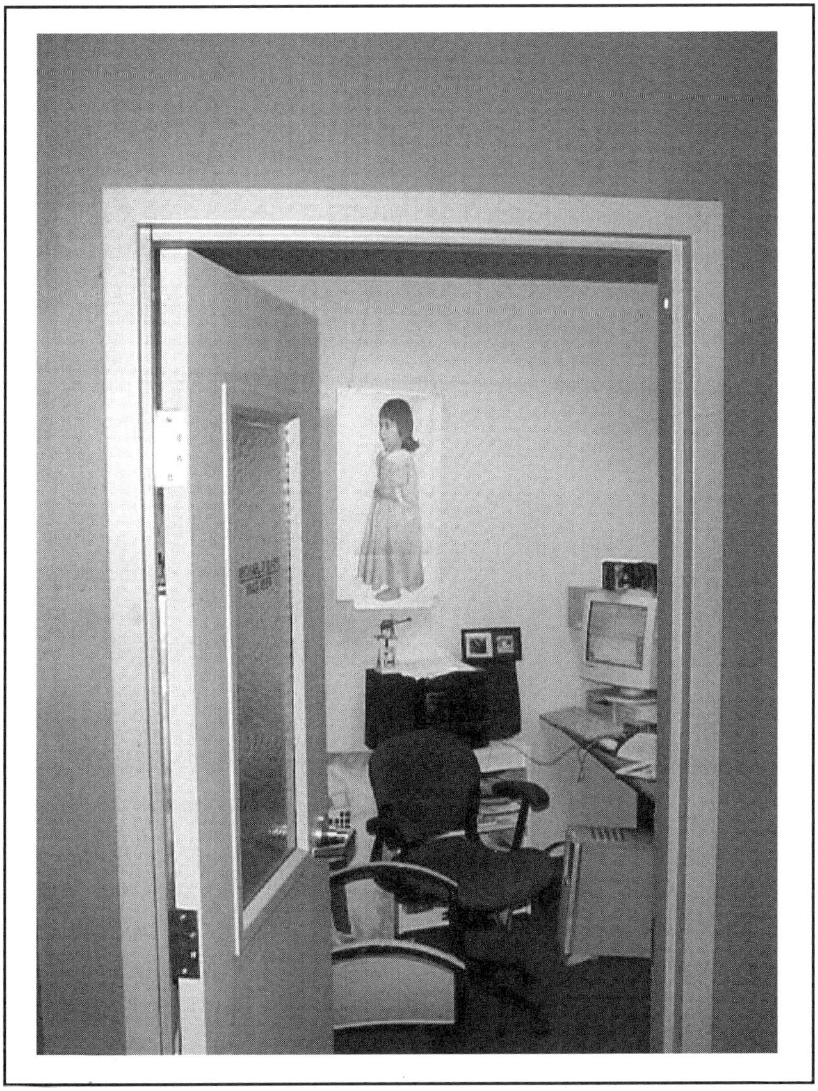

*Figure 2-9. Barrel distortion. What should be straight lines, the office door, curve slightly. Photo taken with the Kodak DC120.*

*Image Editing*). Most editing applications let you "zoom" (enlarge) an image after it's captured, so avoid the digital zoom, and if you must, enlarge the picture using your software editor.

Lens focal lengths (an indication of the field of view of the camera) are often quoted in terms of a 35mm camera lens simply because these sizes are more commonly understood. As an example, Nikon's CoolPix 950 has a f7mm to 21mm lens, which they claim is equivalent to a 35mm camera lens of a 38mm to 115mm zoom. Note that the conversion is not consistent, partly because different cameras use different sized CCDs (from 1/4 inch up to 1/2 inch) and thus have different diameter lenses. The Fuji DS-330 has a 9mm to 27mm lens, which is quoted as being equivalent to a 35mm to 105mm lens

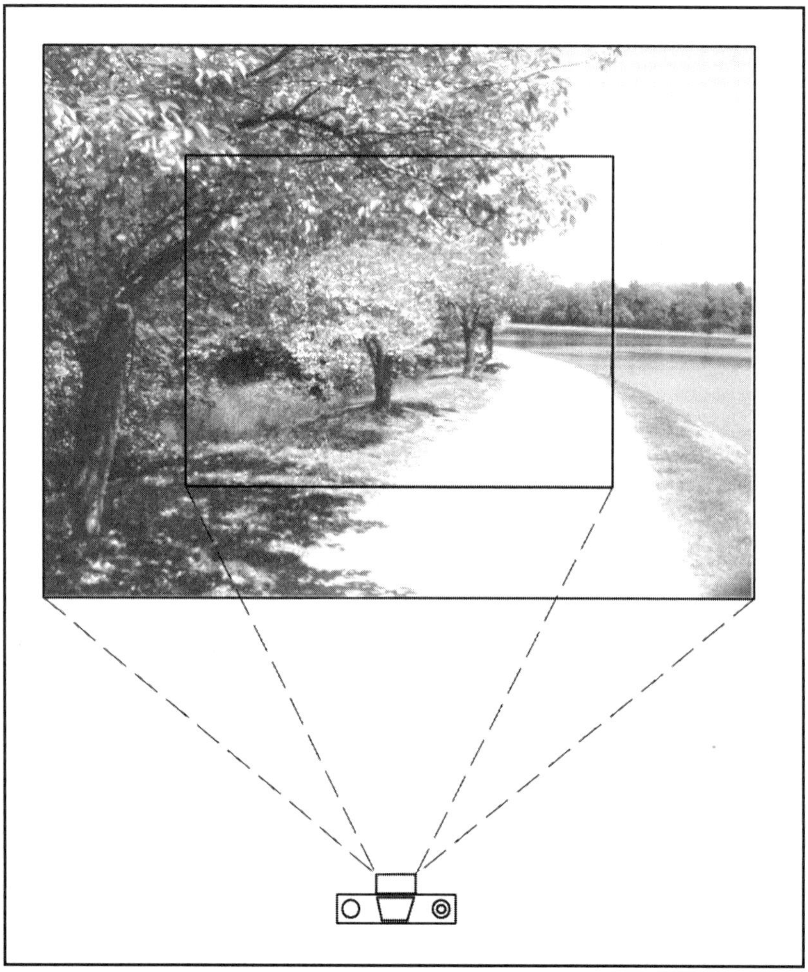

*Figure 2-10. Lens focal length.*

Close focus, or *macro* focus, is another feature to consider when evaluating the lens. Macro focus enables focusing on an object that's only a few inches from the front of the lens. Many lenses do not support this; they typically can only focus down to three feet or so from the front of the lens. While you can buy add-on adapter lenses which will allow you to focus closer to an object, those planning on taking lots of shots of small objects should look for a camera that features a macro focus feature.

## EXPOSURE, WHITE BALANCE AND FLASH

Exposure controls can greatly affect the quality of the picture you are able to produce. If the camera is fully automatic and provides no exposure adjustment, then this can make getting the best picture possible for a given situation almost impossible.

Cameras that let you choose exposure modes offer more flexibility.

Cameras determine the correct exposure—how long to open the shutter—using their own internal light meter. Exposure is explained in much greater detail in Chapter 4. There are three primary types of exposure metering: spot, center weighted, and matrix. Spot metering measures light from the center of the image. While limited, spot metering can be very effective if the camera has an exposure lock option. This lets you point the camera at something in the scene that you want the camera to expose for—you depress the shutter release halfway and the camera calculates the exposure. You can then point the camera somewhere else and the exposure remains locked on that setting.

Some cameras offer an exposure lock that works across multiple images; this can be very useful when taking multiple images to stitch together into a Panorama (see Chapter 15).

Center weighted metering measures light across a wider area of the image, but is biased toward the center of the image area. Finally, matrix metering measures light at many different points in the image to calculate the correct exposure. This is a much more sophisticated method of exposure measurement and has only recently found its way into sub-$1,000 cameras.

The camera metering may still not produce the right exposure. Exposure compensation is a feature that essentially lets you manually fool the camera's meter, telling it to increase or decrease the exposure time by a certain amount. The amount of compensation is measured in f-stops, and is indicated by + or - EV. So +1 EV is telling the camera to overexpose by 1 f-stop. Often the exposure compensation will let you adjust in fractions such as 1/3 EV over a range of several stops. If your camera doesn't have an exposure lock, then exposure compensation is vital. For those unfamiliar with photographic terms such as f-stop, see Chapter 4.

If you can't have both exposure compensation and exposure lock, which is the most useful? Personally, I find exposure lock the most useful, but I suspect that it is a matter of personal preference.

Finally, many cameras let you set the white balance manually. This can vastly improve the color accuracy of the images produced in different lighting conditions. All cameras feature some kind of auto white balance feature which assumes the color of white light in the scene (white light has a color temperature). Unfortunately, indoor lighting does not have the same color temperature as the sun, so photographs taken indoors will look slightly yellow or blue, depending upon the type of lighting present. By choosing a different setting, this color problem can be corrected. Some cameras feature specific settings for interior lights. The more expensive cameras let you manually adjust the white balance by pointing the camera at a white object that's present in the scene and telling the camera to measure and adjust for that value.

Most cameras include an internal flash. While internal flash units have many limitations (see Chapter 4), they are undeniably useful. If you think that you will want to experiment with off-camera flash units, then a PC synch jack is a useful feature to look for.

## FAST PICTURE TAKING AND BURST MODE

One problem that many digital cameras share is the time that it takes for the camera to take the picture, as well as the time it takes to save the picture after the shutter has been released.

Early cameras could be very frustrating because you would push the shutter and it would take a second or two before the camera would actually take the picture. Even newer cameras can have this problem. I found that the Kodak DC260 could be just as annoying if I tried taking a picture "off the cuff." At least with this camera there is a way to solve the problem; depressing the shutter halfway forces the camera to lock focus and get ready to take a picture. Then when you depress the shutter button fully the camera responds almost immediately. But you should check this before you buy a camera; it can be very frustrating trying to take photographs of fast moving subjects if the camera features an objectionable delay.

With a film camera, particularly if you're not using a flash, you can take photographs very fast (as fast as you can wind the film advance, or as fast as the powered film advance works). With a digital camera, you are hampered by the time it takes the camera to save the images to memory, which can be more than five seconds for some cameras.

The more expensive cameras have started to add "burst mode," which is extra memory in the camera designed to hold images while they are being processed. You can therefore take one picture and then take another almost immediately. How many you can take in rapid succession depends on the amount of memory available for this feature. The Kodak DC260 features a burst mode that lets the camera take a second picture almost immediately; however, to take a third you have to wait for the first one to be processed. The newer DC265 adds more memory to increase the burst mode to 6 images.

## FILE FORMAT

Camera manufacturers have been moving towards standardizing the JPEG compression and file format to save their images. The adoption of the CompactFlash memory card has hastened this change because these cards use the ATA file structure. A CompactFlash card, with an inexpensive adapter, can be inserted into a PCMCIA slot and the files read using any application that can open the JPEG file format.

But, there are cameras that use proprietary file formats. Kodak's early cameras (including the DC120) use a proprietary file format. The software that came with the camera will either save the files to the user's hard disk in that file format, or "export" the file to TIFF or JPEG. While the proprietary file format has advantages for the manufacturer, it usually is a problem for users because they have to use special software to open or convert the files.

## CONNECTION

If you're going to be using the camera frequently, don't overlook how you will transfer the files from the camera, or the memory card, to your computer. While nearly all cameras come with a serial connection, serial is not the fastest method of transfer, and can get quite tedious when transferring many megabytes of data. Many of the newer cameras feature USB as an option. USB is a lot faster than serial. If your computer doesn't have USB and you buy one of these cameras it would be worth considering getting a USB card.

For those who buy a camera that features a memory card, the best solution is a card reader that attaches to your computer (there are several different methods for attachment, including parallel port, SCSI and USB). If you plan to use the camera regularly, definitely include in your purchase price some kind of high-speed method of transferring data. It'll save you from going nuts. For more about the options, check Chapter 10 and Chapter 5.

## RANGEFINDER/LCD

The first camera I ever used was an old Kodak Instamatic. I remember being terribly frustrated because I would compose a photograph very carefully and the images I got back were nothing like what I had thought I had taken. My first SLR camera was a major revelation. Suddenly, all my shots were "perfect," or at least how I remember framing them. It wasn't me that was at fault, it was the camera!

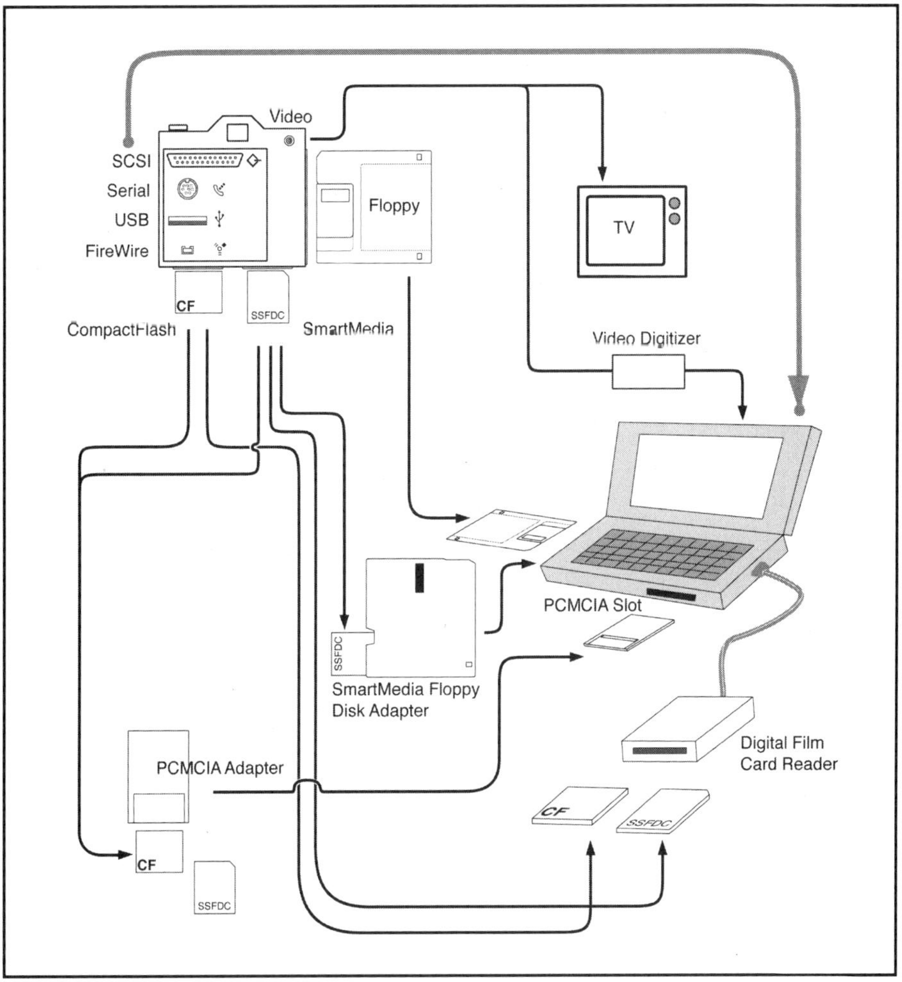


*Figure 2-11. Connection options.*

SLR (single lens reflex) cameras are the most popular design for professional photographers. They let the photographer "see through the lens" while focusing and composing the shot. In theory, this ensures accurate framing of the photograph, though not all SLR cameras show you all of the frame; many cameras don't display in the viewfinder 100% of what is captured on the film. But even if the mismatch is significant (say only 90% of the frame was displayed), you can always crop the unwanted part from the photograph. At least you don't have to worry about the reverse case—the camera excluding part of the image that you see through the viewfinder.

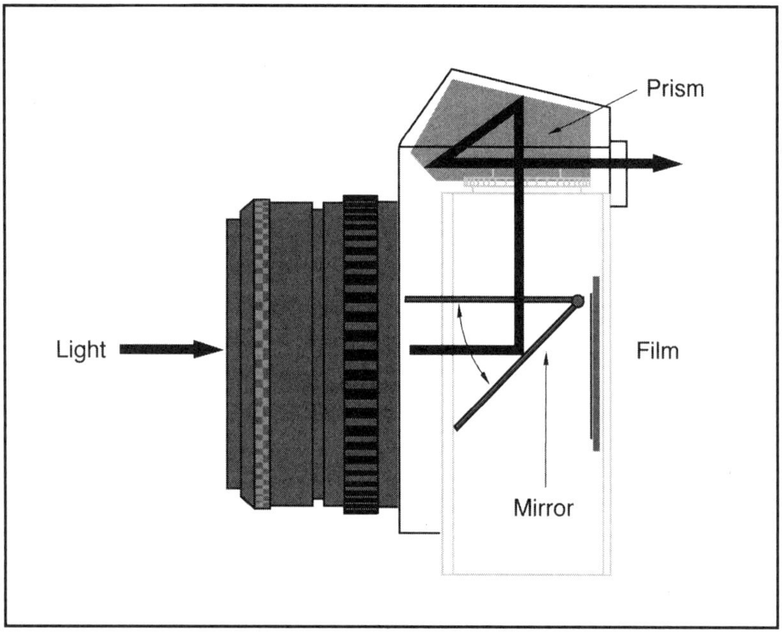


*Figure 2-12. How an SLR works.*

Unfortunately, few digital cameras are of the SLR type. Most use a *rangefinder*—a separate optical viewer that lets you "see" the approximate scene to be captured. These are fairly sophisticated in nature; for cameras with a zoom lens the rangefinder has to zoom too so that you see what the lens sees. While a rangefinder can do a very good job of approximating what the lens sees, they get progressively worse as the object gets closer to the camera due to parallax error; the viewfinder you are looking through is simply not on the same plane as the lens.

Depending upon how the rangefinder is aligned with the lens, this will result in the scene being photographed appearing off-center. Most cameras have the rangefinder to the left or right of the lens, but the Kodak DC260 viewfinder has the viewfinder above the plane of the lens. This results in the subject appearing higher in the frame than expected.

While you can compensate for this error, many digital cameras also feature an LCD panel that can be used to view either what the camera has just taken, or even what it is about to take. Using the LCD panel is the best bet for close-focus photography, but the LCD panel eats up battery power, and many LCDs

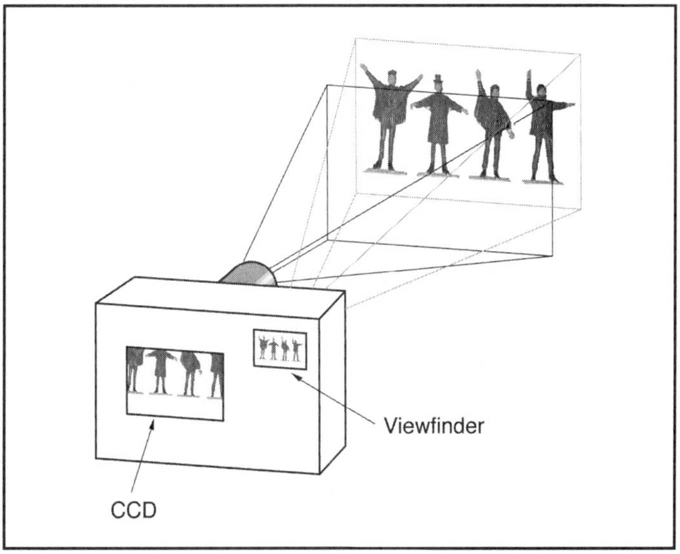


*Figure 2-13. Parallax error.*

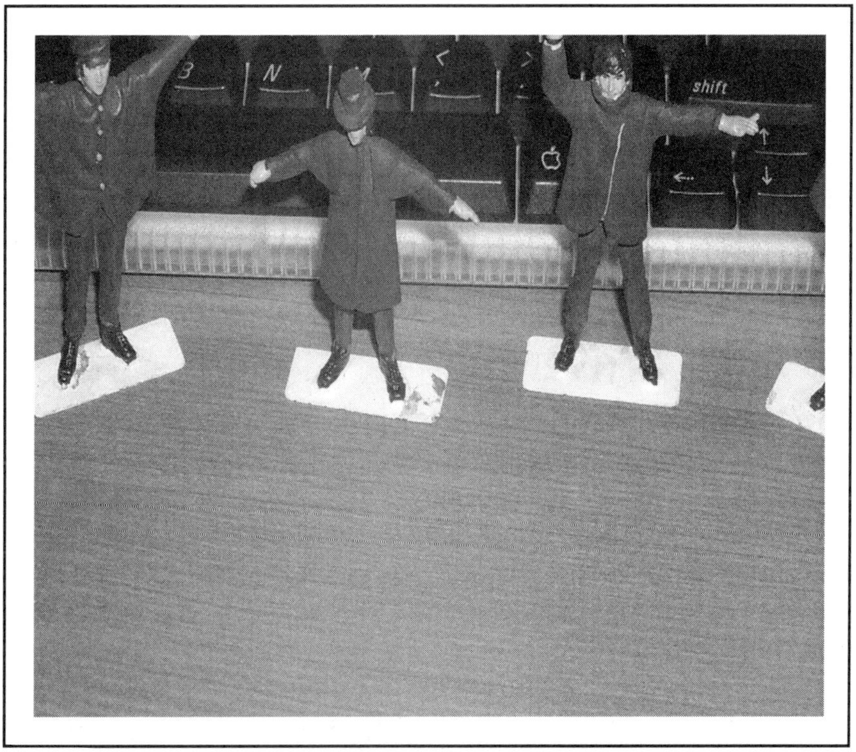


*Figure 2-14. Photograph taken with Kodak DC260. The figurines were in the center of the optical viewfinder.*

are hard to see in bright light (say a sunny day). A few cameras only have an LCD display panel, with no rangefinder. I've never used such a camera, so I can't offer a recommendation. I'm used to a viewfinder, but as long as the LCD was bright enough and could be used outside, it might be all you'd need.

Another factor to consider; the LCD panel on some cameras is updated "live" (almost like a video image), while others update very slowly, perhaps only 3 or 4 frames per second. While the latter is acceptable for many circumstances, if you are planning to use the LCD for composing a lot of shots, then the live update is much more pleasant to work with.

If the camera has a rangefinder, and you use glasses, you might want to get a camera with dioptic correction built-in. This allows you to adjust the viewfinder to suit your vision. This has two advantages; it's actually easier to see through and use a viewfinder if you're not wearing glasses, and you also eliminate one more source of tiny scratches on your glasses.

Whether you get a camera with an optical rangefinder is up to you, but I strongly recommend you gat a camera with an LCD display, even if it's just to review pictures after you take them. Not only is it useful for showing pictures to people after you take them, but it also provides a way to instantly make sure you got the photograph you wanted.

## BATTERIES

Camera manufacturers often generously include a set of batteries with your new camera purchase. Unfortunately, they often neglect to tell you that those batteries are probably not going to last you the first day. Digital cameras use a lot of power, and when you first get that camera home you're probably going to spend all day taking photographs and viewing them on the LCD panel, which sucks up even more power.

I went to a demo of the Nikon CoolPix 900 just after it came out, and the first thing the guy demo-ing the camera said was, "Right after buying this camera, buy some rechargeable batteries."

Some manufacturers have recognized this issue and started including rechargeable batteries. Even if your camera doesn't include rechargeable batteries, you can buy these as accessories from a number of sources. The chapter on accessories looks at the choices available, but in general I'd recommend buying a good quality recharger and two sets of NiMH (nickel- metal hydride) batteries if your camera doesn't come with these.

## CCD OR CMOS

Chapter 3 covers the CCD and CMOS technologies used in digital cameras. At the moment, for most users, CCDs seem to offer the best quality images in the widest number of choices. So far it's been difficult to really evaluate the CMOS technology; no camera has appeared with CMOS technology that rivals the best CCD cameras. And the CMOS cameras that have been released aren't noticeably cheaper than CCD cameras. So far there have no reports of differing operational life for CCDs and CMOS chips.

You should not base your purchasing choice on the sensor. All we can do, therefore, is evaluate the quality of the images a camera produces, and base purchase decisions on that.

## EASE OF USE

While taking a picture should be a pretty simple operation, as more features are added, they provide all kinds of options that the user has to interact with. Manufacturers choose to provide access to these features in a number of different ways; through buttons, multi-position knobs, LCD panels and even through scripting languages (see Chapter 20). The reality is that even if you've used several digital cameras, if you pick up a camera made by another manufacturer, or even a new camera made by the same manufacturer, you might have no idea how to use the thing.

If you want to take a picture with the Kodak DC120 all you have to do is turn the camera on, adjust the zoom lens, and press the shutter button. Nine out of ten people I gave the camera to couldn't figure out how to "make it work" (you had to slide the front cover) and many couldn't figure out which button was the shutter. The Kodak DC260 power ON and OFF button is right next to the shutter release. In the first two weeks of ownership, twice I turned the camera off when I meant to take a picture (which was a real annoyance because the camera takes several seconds to turn off and then power back up). The first time I picked up the Nikon CoolPix 900 I looked at all the little buttons and had no idea what to do.

Ease of use problems can extend to other areas as well. If you use the camera on a tripod, can you get to all of the switches and plugs easily? Can you remove and replace the flash card without removing the camera from the tripod? If the camera has manual functions, how easy is it to access them?

Many of these questions can't be answered without getting your hands on the camera. Some of them won't be answered until you've had the camera for a week or two. Ease of use is also a personal opinion in many cases; I might like the knobs and button layout of one camera while someone else prefers interacting with an LCD screen.

## COST

This issue was left for last, even though cost is really the first thing that needs to be considered. There is a wide range of cameras available at differing price points. Pick a number and you'll find a camera—of course, it may not be the best camera, or the right camera for what you want. The good news is that prices continue to go down, while image quality improves. Because of this, a digital camera may not be the best purchase for the casual photographer. A year from now a much better camera will be available at the same cost.

However, if you have a need for the camera, and are actually going to use it, then you should not wait—because if you wait you'll never buy a camera! It could be a while until the price/quality ratio stabilizes the way it has for film cameras.

Finally, if you are buying a camera, and plan to use it frequently, remember to factor in those extras; rechargeable batteries and perhaps a flash card reader. It will make using the camera much more pleasurable. And once you buy those, your ongoing operational costs will be marginal.

## EVALUATING CAMERAS

The list of features above is interesting, but you need to consider how to weight the factors; for example, is resolution more important than cost? How you are going to use the camera will determine which factors are most important. Here are some examples:

1. Want to print 8 x 10 images frequently?
   Resolution must be at least 1200 x 1500

2. Studio photography?
   Exposure control, PC sync

3. Objects and close-ups?
   Macro lens, live LCD

4. Multi-image panoramics?
   Exposure lock across images. Wide-angle lens (accepts adapter?)

5. Use with Macintosh iMac?
   USB connection

### OTHER SOURCES OF INFORMATION

Some of the best sources of information are other users. If you know someone who already has a camera, then talking to them, and using their camera (if they'll let you) is one of the best ways to find out what it's really like to have and use a digital camera.

Reviews appear in many national photography magazines. Unfortunately, I've found that the reviews often don't include the most recently released cameras. Comparison articles are often the best source of information because the same person reviewed all the cameras (hopefully). That way you see the cameras through one person's eyes. You still have to understand their biases—and everyone is biased in some way—but it can be easier than reading five different reviews by five different people.

There are some very useful sites on the Internet that carry reviews, comparison databases, user reports and even images that you can compare taken with different cameras:

*www.image-resource.com*

*www.dcresource.com*

*www.m2w.net*

## WHERE TO BUY

There are many places you can buy your camera, from camera and department stores to Internet stores. Which you choose is as much a matter of personal preference as anything. Certainly a store provides the ability to actually see the cameras and hold them in your hands. It's also convenient, should a problem develop in the first few days, to be able to take the camera straight back. On the other hand, you can sometimes get some great deals through mail-order.

If you've never bought a digital camera, then I recommend going to a store and trying out the cameras, assuming they have the one you are considering. If an electronic component is going to fail, it is most likely going to fail immediately. So, most problems should appear in the first few weeks. Since most stores will take back defective merchandise that is returned within the first month, it is worth considering buying through a store. Also, if you find the camera isn't what you expected, it is easier to return. If you get a defective camera through the mail it might require some work to get the camera returned or repaired.

Personally, I check the Internet to find the best price. There's a wide range of prices. I've found that some of the lowest priced vendors are also the worst for customer service; orders aren't always filled immediately and sometimes they sell something which, it turns out, they didn't actually have in stock. If you're willing to wait the couple of weeks it can take for the stars to align, then it can be worth the wait. Otherwise, you might want to consider using a slightly more expensive vendor that has a good reputation (ask friends for recommendations).

Personally I'd be very wary of buying used equipment through an auction site such as eBay. There are some resellers that auction and sell out-of-date discontinued equipment. As long as the unit is new and has at least a 90-day warranty, then that can be a good source for bargains. Unless you know what you are doing, avoid items with only a 30-day warranty, in "as-is" condition, in an open box, or refurbished equipment. If it's your first camera, you might be so excited when you get it that you might not notice one of the pixels in the CCD doesn't work, or the lens is out of alignment.

Don't forget, if the camera has been sitting in a warehouse for a year or so, the batteries included in the box may not be in great shape!

## WHAT TO DO WHEN YOU GET YOUR CAMERA HOME

Okay, so you've just bought a brand new digital camera, and you've probably already taken a few pictures to see how it works. Is there anything you should do to test the camera? Chances are, the biggest problems will appear immediately; the camera doesn't work, you can't connect to the camera, or something like that. One thing you should do is check the CCD for any bad pixels. To do this take two pictures; one of an all white piece of paper in a well-lit situation. The paper doesn't have to be in focus, just very well exposed. If you can, lock the exposure on another object before taking the picture so that the exposure isn't biased by the bright subject.

For the second picture, again exposure lock the camera and then place the lens cap (if you have one) over the front of the lens and take the picture.

Download the pictures and check for any pixels that are widely different from the others. If there are, repeat the test. If the pixel(s) are in the same position, then the pixel is defective; if it's random then it's noise in the CCD or electronics.

Part of the problem in performing any test is that digital cameras are inherently noisy; you'll nearly always see random patterns in the images. It's only when pixels are widely wrong—all the time—that you need to consider it a problem.

While you're looking at the images (particularly the black image) look and see whether there are any noticeable faults in the image. If there are no visible problems in the images then the camera is probably acceptable.

You should take test images, try all the features and buttons, and test the flash before you put the box in the basement.

## FIRST THINGS TO BUY AFTER YOU BUY THE CAMERA

The following things are the extras you should consider when buying a camera. If the camera only comes with alkaline batteries and a serial connection, I'd seriously recommend that you buy additional pieces with the camera. It will make the camera more usable and greatly improve your experience.

**1) Power**. If your camera doesn't come with both rechargeable batteries and an AC adapter, then these are the first accessories you should buy. If you are regularly using the camera (every day or so), get rechargeable batteries and skip the AC charger unless it's also used to charge the batteries, or if you will be using the camera primarily in a studio setting. NiMH batteries are considered the best choice for most circumstances, and get two sets so you can have one charging while using the others.

If you're averaging only one or two pictures a week, you might want to skip the rechargeable batteries and just get the AC adapter. Alkalines hold their charge best if use the camera infrequently, but the AC adapter will save the batteries while doing things like downloading images from the camera (which if you only have a serial connection will take quite a long time). It will also save the batteries when using the camera at home (or office) or experimenting with it.

**2) Connection**. This also depends upon what comes with the camera and how frequently you will use the camera. If it's only one or two pictures a week, the serial connection will be manageable. If you're taking ten pictures a day you'll quickly want to get something faster. (Read Chapter 5)

**3) More memory**. If your camera takes a memory card (and most do) and you take lots of pictures, you'll soon find that an extra memory card will be on your shopping list. Unless you know how many images you need to take at a time, or have money to throw around, I'd recommend waiting until you've had the camera a little while before going out and getting extra cards. More is definitely better, but if you only take ten pictures before downloading them to your computer, why spend money on more memory than you need? Flash cards are coming down in price, but they still cost more than $1 per megabyte.

# CHAPTER 3
# HOW A DIGITAL CAMERA WORKS

This chapter provides an introduction to the principles of digital cameras and their major components. It is not the intent of this chapter to provide guidance on how to build your own digital camera, but rather to provide information to help you better understand the differences between cameras and why features work the way they do.

If you are interested in building your own digital camera there is a section in Chapter 11, which covers some home-built digital cameras used for astronomy.

## THE BASIS OF IT ALL

Digital cameras use a light-sensitive chip to capture light entering the camera. The light striking the chip builds an electrical charge in the sensor, which generates an analog charge (a voltage between 0 and 1) which is then converted to a digital value which is saved as the digital image. The images, which are often saved in the standard file format JPEG, can then be transferred to the computer or a printer.

The light-sensitive chip may be either a CCD (Charge Coupled Device) or CMOS (Complementary Metal-Oxide Semiconductor). CCDs are currently used in more cameras than CMOS chips because they produce a better image. CCDs have been common in video cameras for many years. Early digital cameras used chips that were originally developed for video cameras, though this has started to change as the interest in, and market for, digital cameras has expanded.

CMOS chips for image capture are a more recent innovation. CMOS is produced using a different process that is similar to the production process used for memory chips. Because this process is cheaper, it's been argued that CMOS

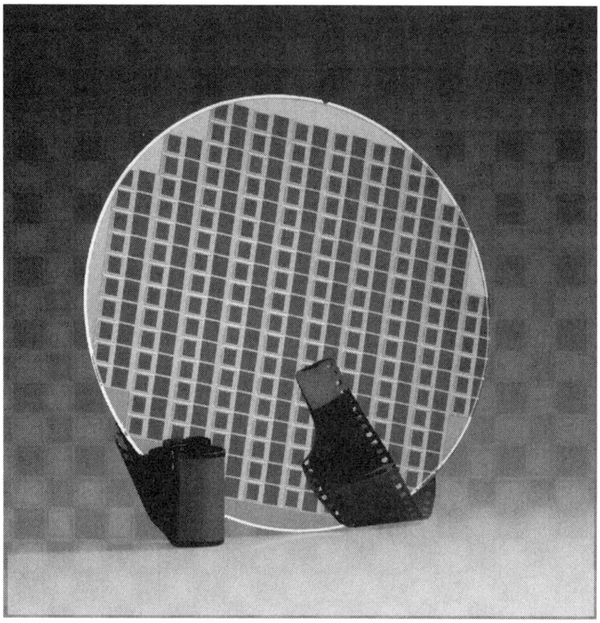

*Figure 3-1. CCDs on 8" silicon wafer. A roll of 35mm film shows the size of the chips. (Photo courtesy of International Business Machines, Inc.).*

is the future of digital cameras. CMOS chips are also more power efficient than CCDs. Unfortunately, they are less sensitive to light and produce more noise. So far, the image quality of the CMOS cameras that have been released has generally been poorer than the images produced by equivalent CCD cameras.

At the moment, a buying decision should not be based on the type of chip. Instead, rate a camera on resolution, picture quality and ease-of-use, and let the chips fall where they may.

Many early digital cameras—and many still in production today—are based upon sensors that were designed for video cameras. This accounts for the poor resolution of these cameras. A "standard" television picture is approximately the same resolution as a 640 x 480 image, so most consumer video cameras use CCDs which have only about 380K pixels. Even worse, a consumer video camera, particularly for VHS and 8mm formats, doesn't capture even that much detail. This is partly because television is a motion format; it appears to have more resolution because of the constant update of the fields and the eye averaging detail from the different fields. The quality of an individual frame looks much poorer when it is frozen than when it is seen "in motion" as constantly updating frames.

Digital camera manufacturers, many of whom also manufacture video cameras, are moving towards higher resolution digital cameras because if you want to print a good looking 8 x 10 inch print, or even a good 5 x 7 inch print, a 640 x 480 pixel image is not good enough. This has necessitated the development of higher resolution chips. Of course, this development does not continue completely separately from video cameras. Television is also moving toward higher resolutions as HDTV is adopted.

## THE HEART OF THE MATTER

The CCD or CMOS is a grid of light-sensitive receptors (called photosites) which react when light strikes them. The larger the size of the chip, generally, the higher the resolution of the image. The chip's size is measured diagonally. Typical sizes for digital cameras are 1/4, 1/3 and 1/2 inch. This is much smaller than a 35mm film frame, which explains the much smaller size of the lenses used by digital cameras.

Light falling on the photosite adds an electrical charge; the brighter the light the higher the charge. The photosite retains the charge even after the shutter has been closed, so that it can be passed to the reading electronics, which converts this charge (an analog value between 0 and 1) into a digital value that represents the brightness of the light falling on the pixel.

However, it is a little more complex than that. The photosites are light-sensitive, but not color-sensitive. Without additional filters and processing, the image produced by a digital camera would be black and white.

To produce a color image a color filter must be placed in front of the receptors. The color filter eliminates other colored light so that the photosite is only measuring the intensity of a single color; red, green or blue. There are many different ways to get color using a gray scale sensor.

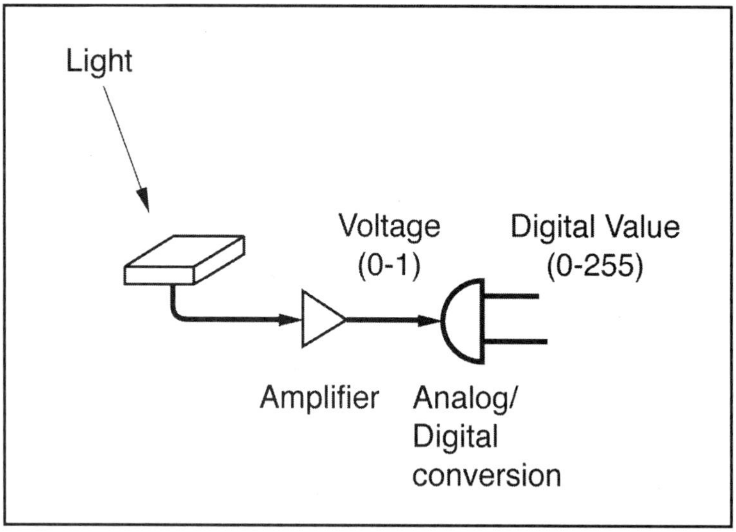


*Figure 3-2. Light strikes photosite, creates charge, charge is converted to digital value.*

The first method is to use three CCDs. Light coming in through the lens is split (using a prism) and directed toward the three CCDs. A red, green or blue filter is placed in front of each CCD so that each CCD is measuring the brightness levels for only a single color. The color image is created by combining the data from each CCD.

This method produces the most accurate color image because the color for each pixel is measured individually. However, splitting the light entering the camera reduces the amount of light going to an individual sensor, and reduces light sensitivity of the CCD. More importantly, the additional chips, circuitry and optics significantly increase the cost of the camera. This technique is often used in moderate to high-end video cameras, but so far has not found its way into consumer digital cameras.

The second, more common technique is to use a single CCD and place a grid of color filters in front of it. Each filter cell contains a single red, green or blue filter. The color filter eliminates other colored light so that the pixel is only measuring the intensity of a single color; red, green or blue. By arranging these filters side by side it is possible to estimate the color of a pixel by combining the color measurements of the surrounding pixels.

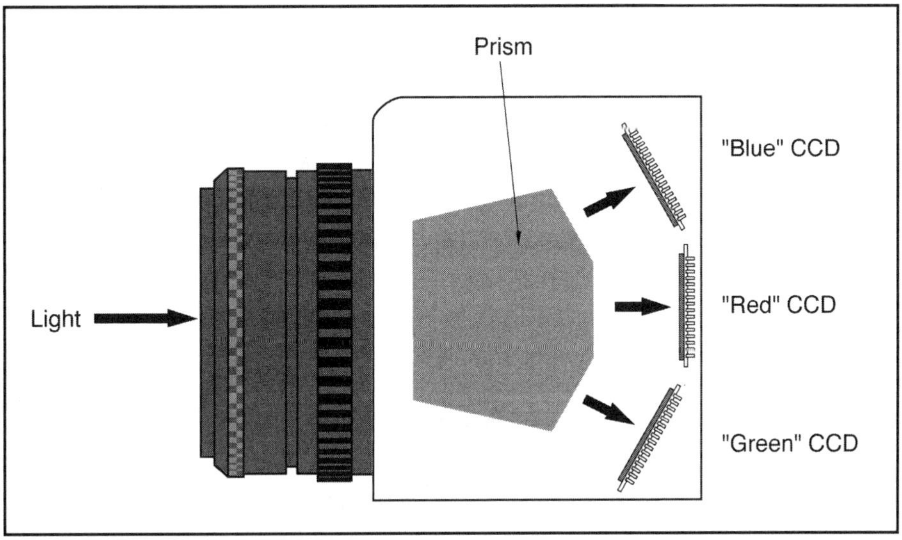


*Figure 3-3. A three-CCD camera.*

While the chroma (color) signal has less detail when captured using one chip, the luminance (brightness) is accurate across the image. This system produces images that are acceptable because the human eye has much poorer color resolution than it does sensitivity to changes in brightness.

In the diagram (*Figure 3-4*) you'll notice that the pattern of color filters does not have the same number of red, green and blue pixels. This pattern, which is G-R-G-B, is called the Byer pattern after Bryce Byer of Kodak, the inventor. It is biased toward producing accurate green color because the eye is more attuned to green than other colors. Half the photosites measure green, while a quarter measure red, and the other quarter measure blue.

Other patterns are sometimes used, and Canon has developed a filter pattern of cyan, yellow, green and magenta.

So while a single CCD camera may have 1 million pixels, and may produce an image with 1 million pixels, it may not accurately capture the color of any individual pixel in an image. And while it's true that the eye is not as sensitive to color changes as it is to changes in brightness (which is why this system works), this technique can result in very fine detail being reproduced incorrectly. Also, when there are sudden changes in color this technique can result in color fring-

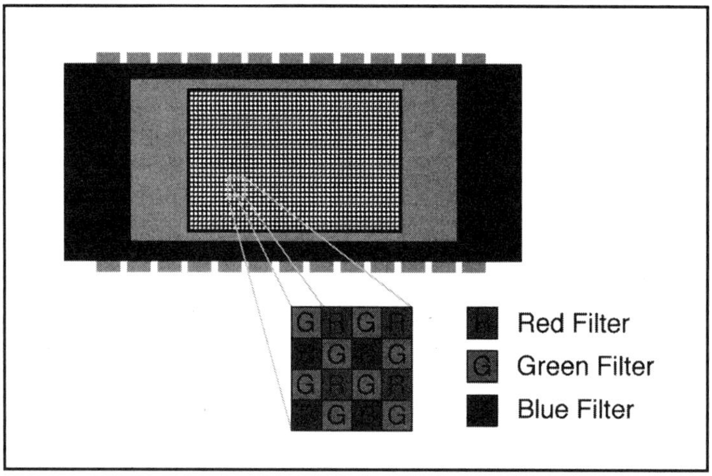


*Figure 3-4. Byer Filter pattern.*

ing; pixels at color boundaries that have incorrect colors. These errors are called *color aliasing*. The image may not be as accurate in color, particularly for small details, as the same resolution image created using a color scanner or film scanner.

## ARE ONE-CHIP DIGITAL CAMERAS GOOD ENOUGH?

In many circumstances a single-chip camera does such a good job that it's difficult to complain. Early consumer digital cameras were compromised in many areas. The chips were very low in resolution and had low light sensitivity. In-

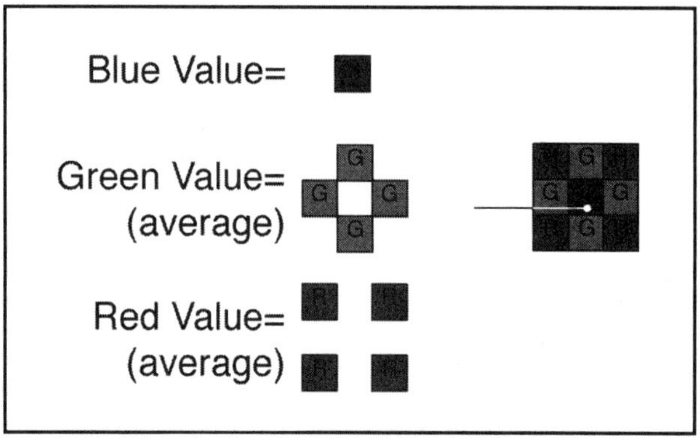


*Figure 3-5. Estimating the color of a pixel.*

creasing the sensitivity increased the noise produced by the chip (see below). The optics of the camera were often limited, and frequently contained a plastic lens. Finally, due to the expense of memory, the images were heavily compressed to fit into as few K (kilobytes) as possible. With all of these limitations, it was difficult to look at any one feature and complain about it as a limitation of the system.

Now, some of the new high-end consumer cameras, such as the Nikon CoolPix 950, feature glass optics and a non-compression option for saving images. Only now can you really start to determine whether the single-CCD approach seriously affects your images.

Is a three-CCD camera better? Certainly it should be if you compare a three-CCD and a single-CCD camera that has the same resolution. But if the three-CCD camera has a lower resolution than the single-chip CCD camera, then it's probably a good idea to stick with the single-chip camera. Sony's TRV-9000 DV (Digital Video) camera can be used to capture still images at a resolution of 640 x 480. It's a three-chip camera. But the images it produces aren't nearly as high in resolution as the Sony DSC-F55 digital camera (1600 x 1200), though the latter is a single-chip camera.

## A THIRD TECHNIQUE: LINEAR ARRAY

There are other ways for a digital camera to capture an image, though these methods aren't as flexible as the one- and three-chip cameras, and can really only be used for studio work. Linear array cameras use a single-element CCD which scans across the scene. A linear array is essentially a row of photosites. The array has three rows of sensors; one for red, green and blue. Working much like a flat-bed scanner, the linear array moves across the "film plane" of the digital camera, measuring the light entering the camera through the lens. Though it essentially produces the same results as a CCD camera, because the detector has to move to capture the image, and because it takes a period of time to do this, these cameras can only be used for studio work taking still objects (such as fruit or jewelry).

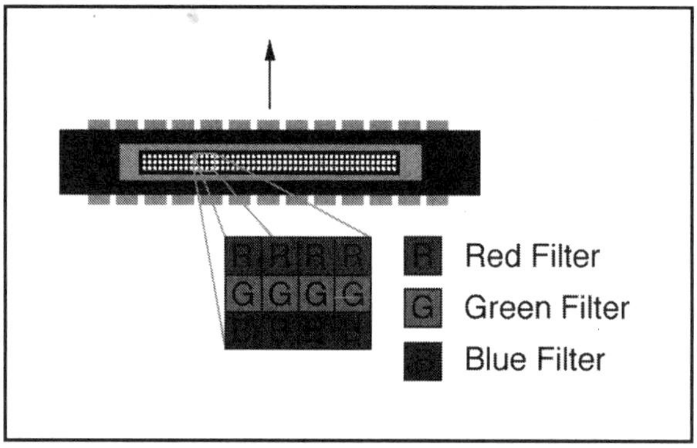


*Figure 3-6. How a linear array works.*

## FOURTH TECHNIQUE: THREE-SHOT CAMERA

A three-shot camera uses a single CCD, but places color filters in front of the sensor, one after the other, and captures three separate images. These are then combined to create the final image. While cameras based on this mechanism produce very accurate color, they can only be used for shooting objects that aren't moving.

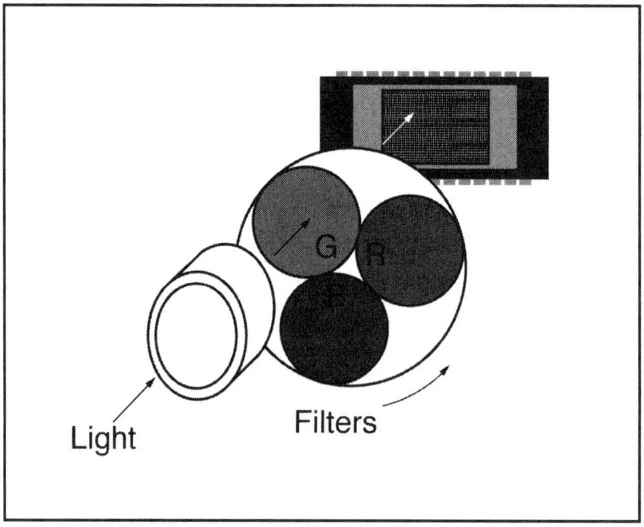


*Figure 3-7. Three-shot camera diagram.*

# THE PICTURE CAPTURE PROCESS

To take a picture, the CCD must first be discharged. This resets all of the photosites to zero. The camera tells the CCD to discharge the photosites, then the photosites are exposed and the charge builds up. Once the exposure time is reached, the levels of each photosite are read as a voltage. This voltage is passed through an analog-to-digital converter, which converts the analog scale to a digital value. In a sense, a digital camera isn't completely digital; the receptor is actually analog and it converts to a digital image.

CCDs are either *progressive* or *interlaced*. As described in Chapter 1, many CCDs were developed for video cameras which use interlacing; rather than scan all of the horizontal lines one after the other they read all of the odd lines, then all of the even lines. These two sets of lines (called fields) are exposed at slightly different points, so if an object is moving in the scene, a feathering effect (the object has moved slightly in one field compared to the other) can be seen.

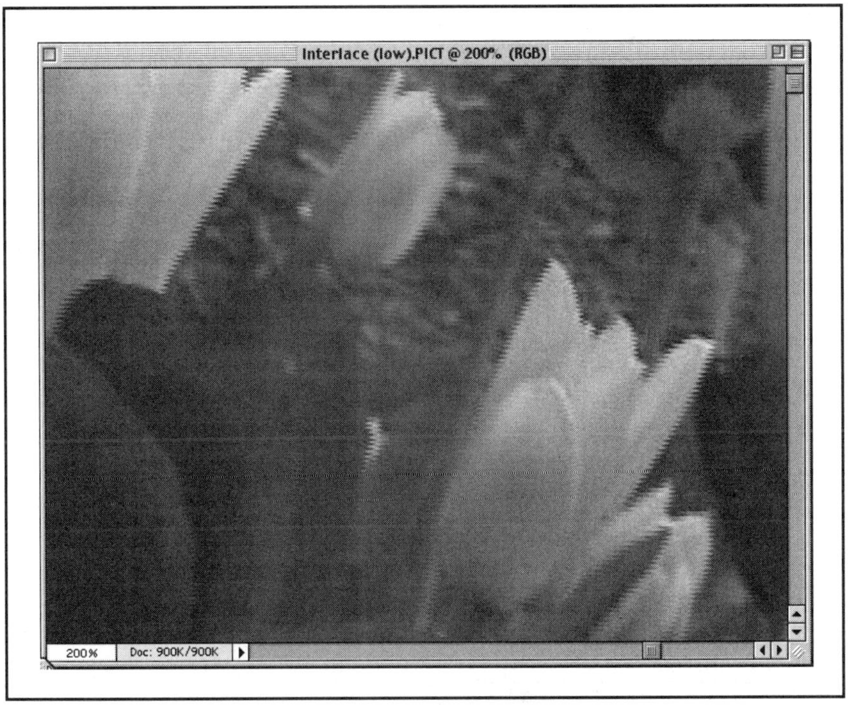


*Figure 3-8. Feathering in a video image.*

This feathering is not desirable for still cameras. Progressive chips read all of the lines at one time, so they don't suffer from this problem. Progressive CCDs are becoming more common in digital cameras.

The sensitivity of the CCD determines how much light is needed to generate an appropriate range of values in the image. Often the gain on the CCD is increased to make the CCD appear to be more sensitive to light than it actually is. Increasing gain means amplifying the voltages from the CCD to produce a "brighter" image. This also amplifies noise—random errors in the values returned by the individual photosites.

In theory, the photosites in a CCD or CMOS chip should produce a zero voltage when no light strikes the detector. This is rarely the case however, and photosites can produce a very slight voltage due to slight defects in manufacturing, variations in temperature and thermal effects. Assuming the camera is not defective, these effects should not be visible in your images. But you might be able to see these effects by forcing the camera to take a fairly long exposure while covering the lens with the lens cap. Using an editing application, you can take the black frame produced and amplify the contrast using the brightness and contrast controls in an editing application. But you should not be concerned about these effects unless they appear in the images you are taking.

*Figure 3-9. A black frame from a QuickTake 150.*

To create a digital image, an analog voltage must be converted to a numeric value. The range of this value determines the amount of detail stored from the signal. For example, if the voltage were translated into a number between 1 and 10, then the amount of detail stored would be much lower than if it was converted to a range between 1 and 100. If two pixels had a voltage of .9 and .91, then if stored in a ten value system they would both be stored as the same value (9). In the 100 step system they would be stored as different values. However, this conversion must be related to the sensitivity of the CCD itself; if the CCD is fairly insensitive to minor changes in light values, then increasing the accuracy of your reading from the CCD won't result in a more accurate image. Also, the most common image bit-depth on computers is 24-bit color, which assumes 8 bits per red, green and blue value per pixel. Some cameras actually capture at 10 bits (i.e., the analog-to-digital converter provides a 10-bit value), but then translates this to an 8-bit image. The extra information is not lost; it is used by the camera to produce a more accurate 8-bit image. (See discussion of gamma and gamma encoding in Chapter 8, to see how extra bits at capture can be useful for creating a more accurate image.)

Nearly all digital cameras exhibit noise (incorrect pixel values) at some time. This is accentuated when the images are under- or overexposed. Noise is always present in electronics. Even if the electronics were noiseless, simply the fact that random events are being measured results in noise; light photons striking the detectors is a random event and the level of two photosites measuring the same source will not be exactly the same. There is always going to be some noise, though the random nature of this noise is usually not noticeable. It is only when noise is accentuated—either by the electronics being used to transfer the charge on the CCD to the analog-to-digital converter, imperfections in the chip, or the artificial boosting of the signal—that the noise can become a problem (i.e., degrade the image). Larger CCDs should have less noise in the individual photosites because a larger photosite will have a larger signal and noise does not increase at the same rate as the signal. Unfortunately, larger CCDs can suffer from more noise introduced by the ancillary electronics.

Thermal variations (heat) within the camera electronics can actually impact the performance of the CCD. This will appear as noise in the image, but is often so small that is may be visible only when long exposures are made. For this reason, most cameras don't allow you to take very long exposures. Lowering the temperature of the CCD will reduce these problems, and some specialized cameras include special coding equipment.

## RESOLUTION

Once an image is captured, it is turned into digital information. Generally, the value from each photosite in the CCD is converted into a pixel in the image; however, there are exceptions to this.

In many cases not all of the pixels in a CCD are actually used to capture information. The manufacturer masks off, or doesn't make use of pixels, often because they want to produce an image that has a "standard" aspect ratio. So don't be surprised if you see in the specifications that a camera contains a CCD with 2 million pixels, but that number doesn't match the resolution of the images produced by the camera.

Some masking is necessary at the edges for the single CCD cameras because the color averaging technique can't produce accurate color for pixels at the absolute edge of the CCD.

The Nikon CoolPix 950 contains a 1/2 inch, 2.11 million pixel CCD, yet the resolution of the images it produces is 1,600 x 1,200 = 1,920,000. The CoolPix 950 is a very nice camera, and since it does have a 2 million pixel CCD it's understandable why the marketing department would want to talk that up, even though when you do the math the resolution of the images isn't quite 2 million!

Note too that digital cameras often have a different aspect ratio (the ratio between the horizontal and vertical dimensions) compared to standard print sizes. Computer screens are usually based on a 4:3 aspect ration (i.e., 640 x 480.) An 8 x 10 print has an aspect ratio of 5:4. Standard 35mm film measures 36mm x 24mm, which is a 3:2 aspect ratio (which matches a 4 x 6 print).

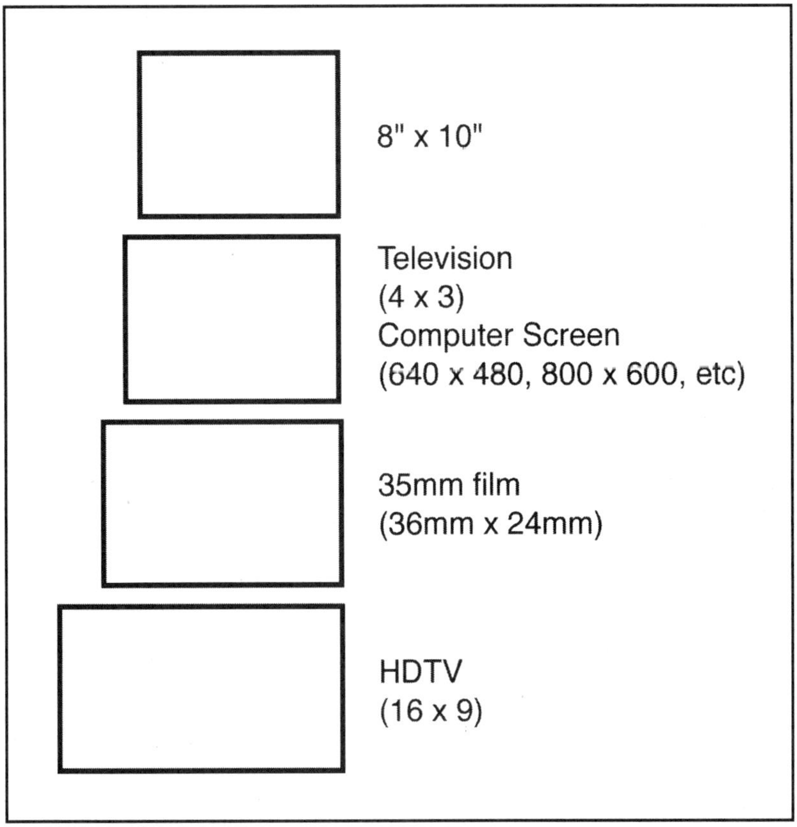


*Figure 3-10. Standard aspect ratios.*

## UPSAMPLING

While cameras throwing away pixels is understandable (note that dividing 1,600 and 1,200 by 2.5 gives 640 x 480, the standard VGA resolution), another practice is less desirable; upsampling. In upsampling, the original image is scaled to produce an image that is larger than the original resolution. This is done with software, and you can do it with most editing applications.

Some cameras perform the same trick. Kodak's DC120 takes images at an actual resolution of 848 x 976, but Kodak's software scales that to 1280 x 960 when you open the pictures. In its day the DC120 was a great camera (and still is if you get it at the right price!), and I can understand that providing images at 1280 x 960 is desirable for the end user because that number nicely divides into 640 x 480. But at the same time, the quality of the images is not as good as a "true" 1280 x 960 image.

When using the camera I always rescale the images down to much lower resolutions (usually 640 x 480), just because I never think the images are sharp enough. It would have been interesting to have the option of using the original files, but Kodak did not provide that option.

Chapter 6 discusses what resolution really means, but simply put, resolution is the ability to detect differences between the values of two adjacent points. Resolution can be measured using line charts, though a line chart doesn't tell the whole story; lenses have to handle many different wavelengths of light, and resolution is partly related to the contrast of different light levels produced by the lens. A lens can be "sharp" at some wavelengths, and less "sharp" at others. Since a line chart requires differentiating between two extremes in light (usually black and white), it may not reflect the way the lens works for other colors in the spectrum.

Before we leave the CCD, there's one more characteristic of these devices we should take note of. The CCD chip, even the highest-resolution ones, are smaller than standard 35mm film. This factor is important for those wondering whether a CCD can simply be inserted into the body of an existing 35mm camera, or if a digital camera could be made that works with "standard" 35mm lenses. At the moment this is difficult to do; the larger CCDs are much more difficult to manufacture, and are therefore more expensive.

And even the largest are still smaller than 35mm film. Kodak and Nikon do offer digital cameras based on 35mm SLR cameras, but they have some limitations. These cameras use standard 35mm lenses, but the focal length of the lens is effectively changed because the chip is smaller than the film (see Chapter 23). These are cameras that have been adapted for this use.

There is a company called Imagek (www.imagek.com) that said it was developing an insert that will work with standard film cameras, but there are many obstacles to this solution. Fitting the unit within the body of the camera, powering the unit, and synchronizing it with the cameras shutter are all problems that have to be overcome. It seems unlikely anyone will successfully produce such a product in the near future.

## LENS

The lens is a vital component of the system, as is the aperture and shutter. The interrelation between these parts of the camera is described in Chapter 4. Since most cameras have a fixed lens, you're kind of stuck with what you buy, though you can get add-on lenses—either wide-angle or teleconverters to either widen or lengthen your lens. A digital zoom is less desirable than an optical zoom. A digital zoom essentially does what upsampling does, though instead of taking the entire image from the CCD and enlarging that, it takes a portion of the image and enlarges it to the original image size, creating the illusion of a zoom effect.

## SAVING THE IMAGES

Once the image has been converted to digital values, it is saved in memory in the camera for storage until you are able to transfer the images to your computer. A few cameras lack memory for long-term storage, and must be connected to a computer so that the image can be transferred immediately to the computer.

In the case of cameras that save images to memory, almost all use some form of lossy compression such as JPEG to reduce the size of the images as much as possible. JPEG compression is described in some depth in Chapter 13.

If JPEG compression is used with a high level of compression, *artifacting* (incorrectly colored pixels) can appear in the image. Early cameras used very heavy compression, which resulted in very blocky images. However, the images produced by these cameras were somewhat noisy, so the JPEG compression partly hid this problem. If you have a noisy picture, throwing away detail in the picture isn't much of a concern because much of that "detail" will really be noise.

Also, a manufacturer might increase the gain on the CCD to increase the sensitivity of the camera. To reduce the effect of noise, they can use a smoothing algorithm which will soften the image, but reduce the apparent noise.

Pictures are usually stored in EPROM (Erasable PROgrammable Memory) chips. These chips retain data even when power is not being supplied to them. Some cameras have internal memory, while others use plug-in cards (*see Digital Camera Storage Options, pg. 64*). Some digital cameras use PCMCIA cards that contain small hard drives to store pictures.

## COMPARING DIGITAL CAMERAS TO FILM

Digital cameras can appear to be almost identical to a film camera in operation. Sure the sensitivity of CCDs is different, but even that is improving. There's no reason why a digital camera can't offer aperture and shutter controls, as well as exposure controls and all the other features found in a film camera (except for the rewind button!).

However, many of the low-priced cameras—and up until 1998 low-priced only meant under $1000—lacked manual controls. This is starting to change, but digital cameras in the sub-$1,000 category are only slowly adopting the exposure and manual adjustments found on a similarly priced film cameras.

Another noticeable difference is in the lenses. Most cameras have CCDs of only 1/2 inch or less. Because of this, the lenses are smaller in diameter than 35mm camera lenses. Interestingly, when manufacturers quote the lens zoom range they often quote it as an equivalent to a 35mm camera. Smaller lenses mean that the actual aperture (the width of the opening of the lens) for a given f-stop of the camera is much smaller for these lenses compared to 35mm cameras. The f-stop is calculated by dividing the focal length of the lens by the actual diameter of the opening. For a given f-stop a smaller lens will have a smaller actual opening.

This can also impact depth of field. Depth of field is the range within the image that is sharp. The smaller the f-stop for a lens, the larger the region of the picture zone that is in focus.

Many people don't concern themselves with the range of the image that is in focus, as long as the subject is sharp they are happy with the picture. However, depth of field can be used to create interesting effects; the figure in the foreground is sharp while the background is out of focus, for example. With smaller apertures it can be harder to do this because of the inherently greater depth of field produced by the smaller lenses.

The other things that film camera users will notice are the delay that often occurs before the camera will take a picture, and the sometimes long period of time before the camera is ready to take another picture.

**Why is there a delay in taking the picture?** When the shutter release is depressed the camera first must clear the CCD. It then does a white light balance to correct for color, sets the exposure, focuses (on autofocus cameras), and finally fires the flash (if needed) and takes the picture. Unlike a film camera, a digital camera uses white light balancing in its initial exposure calculation. This allows you to obtain correct color under most light. Unfortunately, all this together can result in a delay of as much as a second.

**Why does it take so long for the camera to be ready after taking the picture?** The main delay occurs in the saving of the image to the camera's memory. During this time the camera must convert the analog values in the photosites into digital values in the camera memory. It must then calculate the actual color of the entire image using color averaging. It then must compress the file and save the resulting file to the storage medium.

**Why don't cameras accurately tell you how many pictures are left?** It would be nice if the camera assigned an amount of space to an image and calculated the amount of compression necessary to create such a file. But this would require compressing the image multiple times, which would increase processing time. And since images with fine detail will, by definition, take up more space than other images, it is those images that will suffer the most from doing this. Better to choose a single compression ratio that produces reasonable results for all images, and stick with it!

For this reason, the cameras don't really know how many pictures will fit in the available space because it depends on what kind of images you take. If the photographs contain a lot of redundant data (all the pixels are the same color), then the images will compress much more than those with a lot of detail.

## DIGITAL CAMERA STORAGE OPTIONS

All digital cameras need to store the images that they capture in some kind of memory, or the image must be transferred immediately to the computer. If the image has to be transferred immediately to the computer then the camera cannot be used without a computer connected to it. In some situations that is acceptable, and the lack of memory cuts the cost of the cameras; such uses include webcams and video conferencing. For all other cameras, some kind of an internal memory chip, memory card, floppy disk, or some other storage is needed.

Here are some camera examples:

### Memory-Less Cameras

Example: **QuickCam**

This camera captures the image and immediately sends it to the computer. The camera will not store multiple images for later retrieval.

Uses: Webcams and video conferencing. Capture of images while the camera is connected to the computer.

### Cameras with Internal Memory Only

Example: **QuickTake 150**

Apple's QuickTake includes enough memory to store 16 high-resolution images, or 32 low-resolution ones. Once the memory is full you must either delete the images or download them to the computer (and delete them) before capturing more images. Low resolution and limited amount of memory hamper this camera.

Uses: Now largely replaced by newer, higher-resolution cameras. Might be a good cameras for kids to learn with (and better quality than the Barbie Camera, and the Gameboy Camera!).

**Cameras with Internal Memory that also Accept External Memory Cards**

Example: **Kodak DC 120**

The Kodak DC 120 has internal memory that can store about 16 high-resolution images. The camera also includes a card slot for inserting a memory card. The camera will then capture images to that card. The card slot makes it possible to purchase multiple cards and keep taking pictures by switching the cards.

Uses: While it made sense at the time, now that memory cards have come down in price and are widely available, the small amount of internal memory is largely irrelevant.

**Cameras that Only Use Memory Cards**

Example: **Kodak DC260**

Because of the limitations of internal memory, manufacturers have moved away from including any for image storage, and simply use external cards. This camera has high resolution, many user options, and produces excellent images. Has been replaced by the DC265.

Uses: Excellent general-purpose camera.

**Cameras which Use External Disk Drive**

Example: **Sony MD81**

Sony has made a point of using the floppy disk, a very common disk format that can be read by most computers (unless your computer lacks a floppy disk drive). The advantage is that this medium is relatively inexpensive. However, the size of files is limited by the size of the floppy disk; for example, only two or three images could be stored on a disk if images were compressed at the same rate Kodak's DC260 uses on its memory card.

While many people have decried Sony's use of floppy disks, it has been very popular with the buyers of the camera.

Uses: Very good general-purpose camera.

## CONCLUSION

While digital cameras work very differently than film cameras, the differences in quality and in functionality of the camera is lessening as the digital camera market matures. While you should spend some time becoming familiar with how the camera works, and in what situations the CCD produces the best results, the camera actually hides all of the functions of the CCD and you just have to point and shoot!

CHAPTER 4

# WORKING WITH A DIGITAL CAMERA

Digital cameras share many similarities with traditional cameras, yet they offer many new features and the number of choices is daunting. This section provides a short introduction to the digital camera process, and points you to places in the book where these features are described in depth.

If you're new to cameras, don't know your aperture from your f-stop, then the next couple of pages should provide enough of a guide to the principles of cameras to get you going. If you are already comfortable with camera principles, then jump ahead to the section "Picture Taking."

## THE FIVE-MINUTE CAMERA COURSE

Cameras capture an image on light-sensitive material; in the case of regular cameras it's film, in the case of a digital camera, it's a grid of light-sensitive pixels which produce a voltage based on the amount of light that strikes them.

All light-sensitive material has a measurable sensitivity which, in simplest terms, is the amount of light that is needed to create a properly exposed image. For film this is usually measured as an ISO rating, and most digital camera manufacturers will quote an ISO equivalent for their cameras. Digital cameras are usually "slow," meaning that the ISO is low and it takes more light to get an acceptable image. The chart below shows ISO ratings. Most digital cameras rate at about 80 or 100, though some allow you to switch to a higher ISO rating; this often increases the amount of noise in the resulting images.

| SLOW | MEDIUM | FAST | VERY FAST |
|---|---|---|---|
| <=100 | 200 | 400 | 400+ |

*Figure 4-1. ISO Range. The terms slow, medium and fast are relative terms based on the films that are currently available.*

Cameras have two ways of controlling the amount of light that enters the camera—by adjusting the aperture, and shutter speed. The aperture is the size of the opening of the lens. The wider the aperture, the more light that can enter. The shutter opens and closes to allow the light to enter. The longer it is open, the more light can enter.

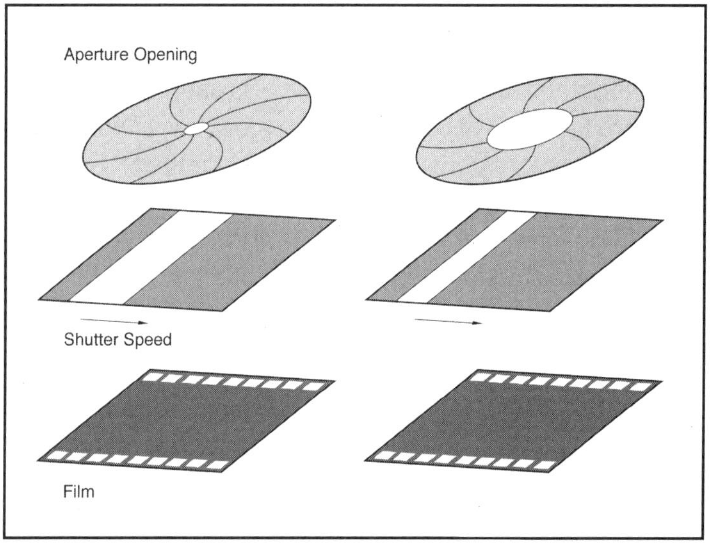


*Figure 4-2. Aperture and shutter speed.*

Given that only a specific amount of light is needed to expose the film correctly, the two work together; if you open the aperture, then you have to shorten the shutter time and vice versa. If you think of the CCD as a bucket you have to fill with water and the aperture as a variable width pipe, then the shutter speed is the length of time the tap is turned on and off to fill the bucket. Since the object is to provide only enough water to fill the bucket, changing the width of the pipe changes the length of time the tap is opened.

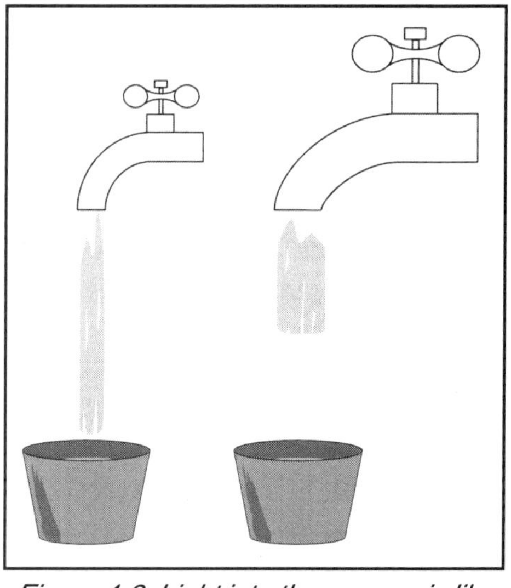

*Figure 4-3. Light into the camera is like water into a bucket.*

Since shutter speed and aperture are interrelated, the aperture and shutter speeds are calibrated so that the steps in one match the other (see *Figure 4-4*). Adjusting the shutter speed up means adjusting the aperture down, and vice versa. The chart shows the setting for 100 ISO film for a particular scene; if the speed of the film were changed then the aperture changes too (it's like you've changed the size of your bucket), but the interrelation between shutter speed and aperture remain. In other words, for 400 ISO film the shutter speed at the 5.6 aperture changes to 500, and the shutter speed for f8 changes to 250.

Apertures are often referred to as "stops," and the term "stopping down" means opening the shutter. Changing the exposure by "a stop" means changing to the next f-stop. Aperture is measured in terms of f-stop, which is the ratio of the aperture opening to the size of the lens.

| | | | | | | |
|---|---|---|---|---|---|---|
| **Apertures** | 2.2 | 4 | 5.6 | 8 | 11 | 16 |
| **Shutter Speed** | 500 | 250 | 125 | 60 | 30 | 15 |

*Figure 4-4. Example aperture and shutter speed combinations..*

## IT'S NOT THAT SIMPLE

This first discussion talked about "a properly exposed image," as though there is a constant amount of light that's needed to expose a piece of film. This is not strictly true. While in general there is a set amount of light that's needed to expose an average scene, different subjects might require different amounts of light to create an acceptable image. This is because any film, and CCD, has only a certain amount of exposure latitude. This is the range of light values over which differences are detectable. No light-sensitive device can accurately capture everything from the darkest blacks to the whitest whites. The light-sensitive material just doesn't have enough sensitivity to handle that. So, assuming that the camera can only accurately reproduce a certain range of values, we try to adjust the range to better capture what's in the subject. Because of this we might underexpose or overexpose an image to capture particular details in the scene.

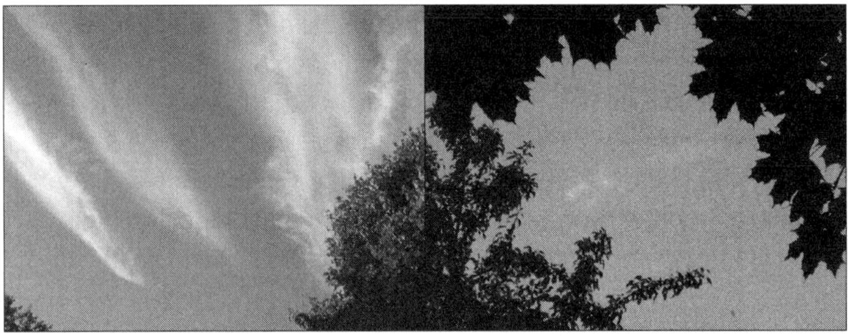

*Figure 4-5. Exposure range. Two images taken at the same time. The image on the left is correctly exposed. The image on the right is underexposed because the camera meter was fooled by the mix of dark and light regions.*

Say you're taking a picture of a black cat in snow. Your camera suggests a certain setting for taking the picture, which averages the black and white. The resulting picture has little detail in the snow or the cat. To see detail in the cat, we might overexpose the image, which brings out the hairs in the cat's fur but washes out the snow. Alternatively, if we want to capture the patterns in the snow, we might underexpose the scene. Many cameras include exposure compensation, which lets you adjust the exposure one or two stops.

## WHY CHANGE APERTURE INSTEAD OF SHUTTER SPEED?

Many first- and second-generation digital cameras are fully automatic; you have no control over aperture or shutter speed. Just point and shoot. It's only with more recent cameras that the ability to manually control aperture and shutter have become available. Being able to control these settings lets you control how your images look. Assuming that you have a range of aperture and shutter speeds to choose from, then you can switch between these to produce different effects.

Adjusting shutter speed lets you freeze or blur action; using a fast shutter speed will freeze a high jumper in mid-air, while using a slow shutter speed will blur a waterfall into a jelly mass. Aperture, on the other hand, has a much subtler, but just as interesting effect. The wider the aperture, the shorter the amount of the scene that is in focus. When the lens is wide open you can focus on the subject in front, and the background will be out of focus. But using a smaller aperture (higher number) will widen the range of the image that is sharp.

Unfortunately, since the actual size diameter of the lens used in digital cameras is comparatively small, the depth of focus is greater. This means that a greater range of the scene will be in focus and it will be harder to create depth-of-field effects.

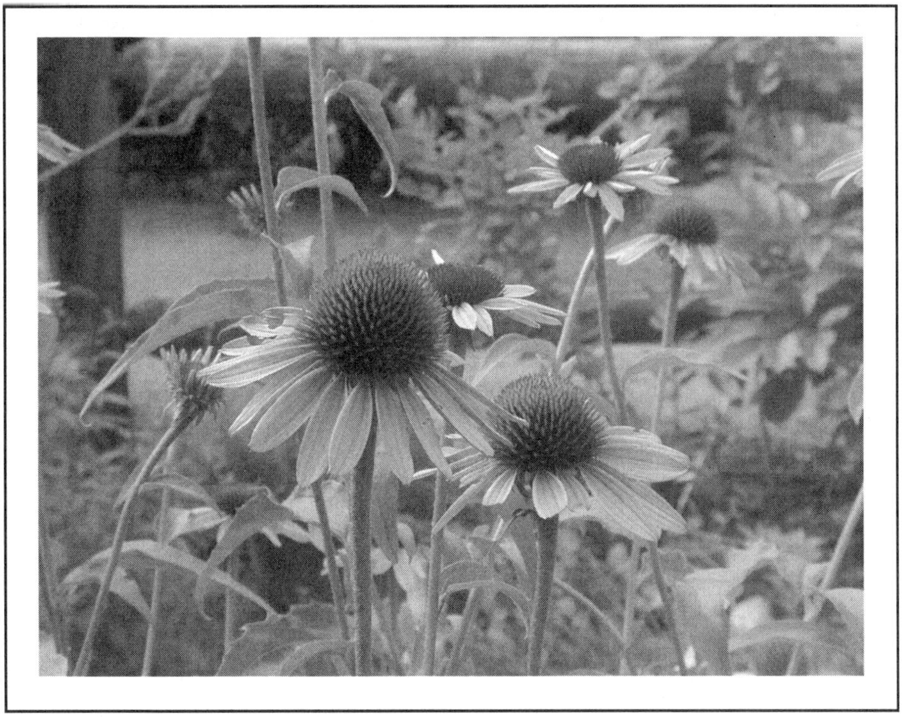

*Figure 4-6. Focus range (depth-of-field).*

Of course, if you're at the limits of the exposure range of your camera, say in a coal mine or in Antarctica, then you might have very few aperture and shutter choices. In a coal mine all you can do is bring in lights, but in Antarctica you can try neutral density filters, which reduce the amount of light entering the camera lens.

## ZOOM LENSES

Lenses are usually measured in terms of focal length in millimeters (mm). The focal length indicates the angle of view that the lens captures. For 35mm film cameras, a 50mm lens is considered "standard," with lenses less than 50mm being considered "wide." Lenses at the other end of the scale are referred to as being "long."

A zoom lens is great. It lets you get closer to your subject without getting off the couch. However, changing focal length not only brings you closer to you subject, but it also changes the relationship between foreground subject and background subject. If you framed a picture with the foreground subject taking up half the scene and took two pictures with different focal lengths, you would notice that the wider focal length allowed more background than the zoom setting.

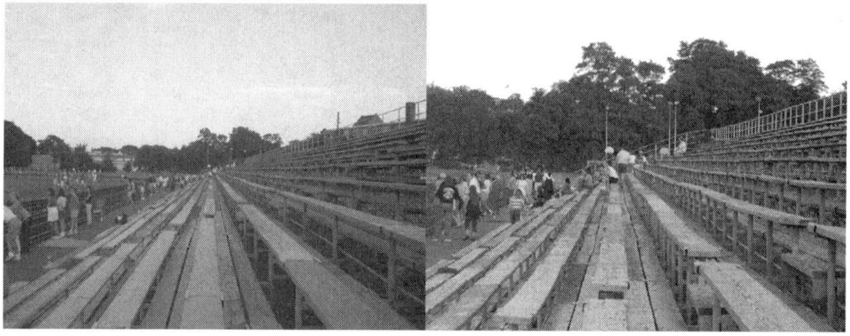

*Figure 4-7. Changing focal length changes angle of view.*

Finally, before we look at using a digital camera to take pictures, here are some tips for the new photographer:

1. Always take your photographs with the sun behind you; that way your subjects will be squinting badly and look like idiots.

2. Hold your breath while pressing the shutter; you never know when that birdie might pop out. Besides, you'll actually hold the camera steadier.

3. Always have everyone say "cheese" before you press the shutter; it makes people look like they're posing for a photograph.

4. Before you press the shutter look at what's behind your subject; are tree limbs growing out of Aunt Emily's head? Does she look better that way?

5. Always put the subject in the center of the photograph. Right in the center. Always.

6. Cameras only work when held horizontally at eye level. Using the camera in any other orientation, at any other angle or height, will void the warranty.

7. Some great pictures are taken of people getting ready to have their picture taken.

8. Just because the flash has no effect once your subject is more than 20 feet away doesn't mean you can't annoy everyone by using your flash at the opera.

9. Bigger and closer is better. That's why they made zoom lenses; so you don't have to walk.

10. Don't ask a guy on a motorcycle to take your picture using your camera. Trust Me.

## RULE OF THIRDS

One last tip before you go out and buy a photography book; I was just joking about putting the subject in the center. I don't know where it came from, but the rule of thirds has appeared in just about every photography book Kodak has published. The rule suggests that rather than placing the point of interest in the center of the subject, you should divide the scene into thirds (both horizontally and vertically) and place points of interest on the intersections of those lines. Like all rules, it's made to be broken, but try avoiding putting everything in the center of the picture. It makes for better pictures.

Now on with the show!

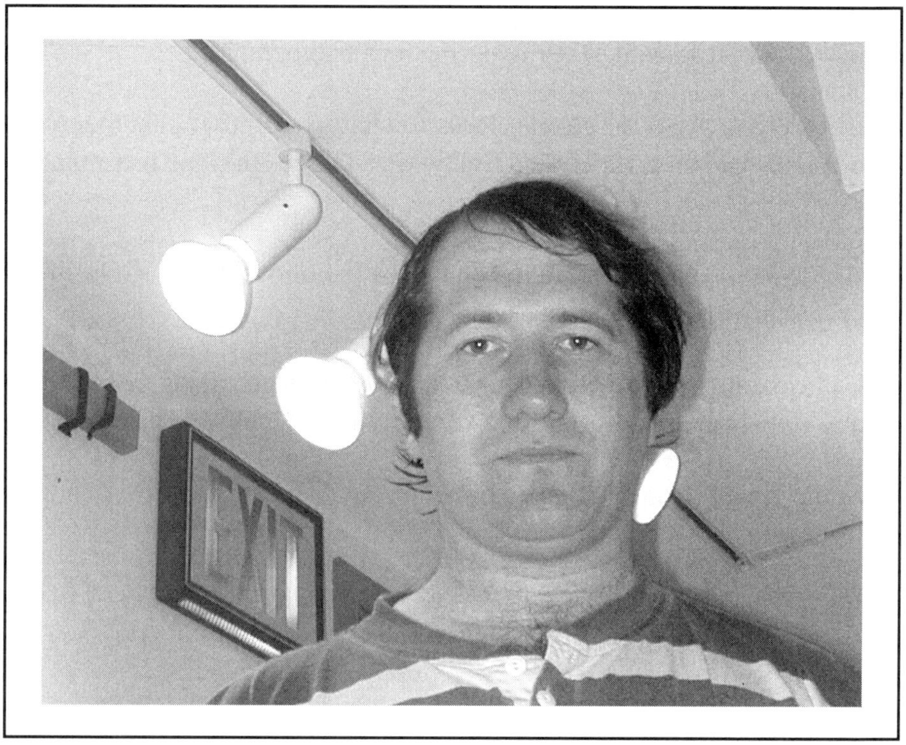


*Figure 4-8. What's that in the background?*

## PICTURE TAKING

Digital cameras provide an optical viewfinder and/or an LCD panel to preview the scene you are going to capture. Like a regular camera, digital cameras have aperture and shutter adjustments which adjust the amount of light entering the camera. Many digital cameras don't provide control over these settings, and some use neutral density filters (a dark lens) rather than an aperture adjustment to regulate the amount of light entering the lens.

The first thing you need to do is turn the camera on. This is not as silly as it sounds; some cameras take several seconds to wake up, power the flash, and get ready to take a picture. Cameras also turn themselves off automatically after a certain time of inactivity.

Once on, most cameras are in an automatic mode—just point and shoot and wait for the camera to take a picture. Many digital cameras have a noticeable delay between the time you depress the shutter release and the time the picture is taken. With long delays it can get very difficult to capture spontaneous moments; something happens, you press the button and the moment passes and the shutter fires. If you practice you can anticipate moments, but it's not always easy.

Fortunately, the problem is lessening; newer cameras have much shorter delays. The Kodak DC260, for example, has a noticeable delay if you simply point and click the shutter. If you point and half depress the shutter button, the camera focuses and gets itself ready to take a picture. When you press the shutter all the way, it takes the picture almost immediately.

It's possible to measure how long a delay there is between pressing the shutter and capturing a picture. Karen Kenworthy, a programmer and Contributing Editor at *Windows Magazine* has written a small utility called the Digital Eyes Camera Timer. The program is very simple. In a large window it counts down to zero and displays the words "NOW," then displays large numbers in tenths of a second. You simply start the countdown, look through the viewfinder and click the shutter when the word "now" appears. The number captured in the picture that you take indicates the shutter lag.

You can download the utility from the website:

*http://www.image-acquire.com/articles/timing.shtml*

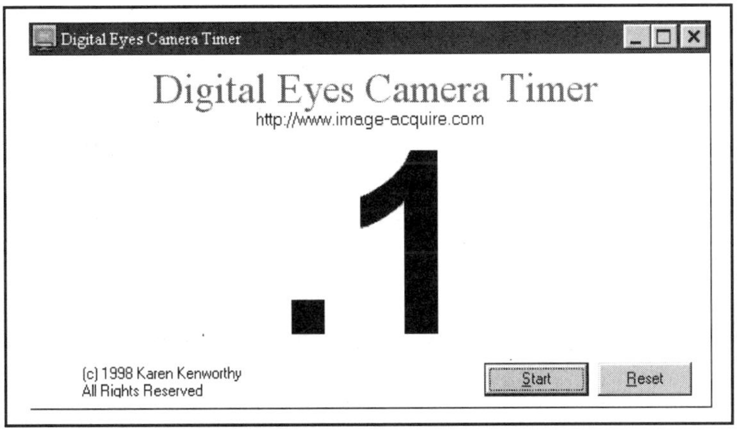


*Figure 4-9. Digital Eyes Camera Timer.*

## ADJUST SETTINGS

While the inexpensive cameras are fully automatic, more cameras are offering manual controls for adjusting white balance, exposure adjustment, metering, and manually choosing the aperture and shutter speed. Unfortunately, these options are often hidden away behind menus accessible via the LCD, or require pressing odd combinations of buttons. Despite this, it's worth spending the time to read the manual and learn how to adjust these settings for those occasions when the sully automatic setting doesn't work. Adjusting white balance, where appropriate, can greatly improve the color accuracy of your images. Similarly, experimenting with exposure compensation can produce better results. The joy of digital cameras is that providing they have a display, you can review the image and see whether it's correctly exposed.

Unfortunately, there are limitations to the LCD display itself. However, it will provide some indication of the quality of the image. Learning to use the manual settings takes time, but is worth the effort.

## WHAT RESOLUTION & COMPRESSION SHOULD I USE?

Cameras provide a range of options when capturing images. Some capture images in only one size, but offer difference compression options. With JPEG compression (see discussion of JPEG in Chapter 13) using a lower quality (higher compression rating) enables more images to be stored in the same amount of memory. The total resolution (number of pixels) of the images remains the same, but the file size (and the level of detail in the image) changes. Which setting to use? This can be a problem, because while the difference between the low- and high-compression options can be very small, it can still be noticeable in fine details (hair, etc.). Thankfully, the price of memory has fallen dramatically, and most cameras use memory cards, so it's getting easier to simply use the highest setting and switch memory cards. I recommend using the highest compression setting you can, within the confines of how many pictures you have to take in one setting and the amount of memory you have available.

Several cameras offer different resolution options as well as different compression. This results in a file with a different number of pixels (see Chapter 6), so less data is actually being stored. Again, this is typically done to increase the number of images you can store in a given amount of memory. The less data being saved, the faster it can be saved, but again, you are saving less resolution. Again, follow the advice above and choose as large a resolution as possible.

Finally, some of the high-end cameras are offering noncompressed files. This creates a very large file. The chief disadvantage of doing this is that it cuts down dramatically on the number of images that you can take. This is much more significant than the difference between the compression and resolution settings. Unless you can afford lots of memory cards, you will probably only use this option occasionally.

## WAITING TO TAKE ANOTHER PHOTOGRAPH

Once you take a photograph you'll have to wait to take the next one—unless the camera has a *burst mode*. The length of time you have to wait depends upon several factors; the size of the image you are capturing, how much charge is left in the batteries (the less power the longer it takes to recharge the flash, for example), and if your camera uses flash cards, the speed of the flash card. While there are some things you can't change, you can at least keep the in-between picture time as short as possible by: having a second set of batteries with you (you should use rechargeable batteries) or using AC power if you can; using noncompressed file formats unless you absolutely have to; and buying memory cards with fast access speeds.

## REDUCING POWER DRAIN

Hopefully you'll follow the advice provided and buy at least two sets of rechargeable batteries. If they're always with you, then you should have enough power in most shooting situations. If you find that's not the case, maybe you need to buy an extra set of batteries. There are also techniques you can use to conserve battery consumption; digital cameras are ravenous beasts!

Battery saving techniques:

1. Always charge your batteries!
2. Have a spare set of alkaline batteries for emergencies (they don't lose their charge).
3. Don't use the LCD unless you have to (it draws more power).
4. Turn the camera off when not using it.
5. Don't use the flash unless you have to (an external flash may be a better way to manage your battery consumption).
6. Taking several pictures in rapid succession will drain the batteries faster.
7. Rechargeable batteries lose their charge over time, even if not used.
8. Cold reduces the output of batteries.

## FLASH MEMORY PERFORMANCE

Flash memory card performance is quoted as several different numbers, but most commonly:

**Startup Time:** Quoted as "Sleep to Write" and "Sleep to Read," the amount of time it takes for the card to start accepting or returning data when it is told to wake up. Measured in milliseconds.

**Data Transfer Rate:** Rate at which data is transferred to or from the card. This will often be quoted as a "burst" figure, which is the highest possible rate and not necessarily the sustained rate that the card will achieve.

While these numbers can be almost meaningless—often because they are measured under ideal conditions and not typical use—they can be taken as a guide. In general, the higher the transfer rate the less time it will take to save an image.

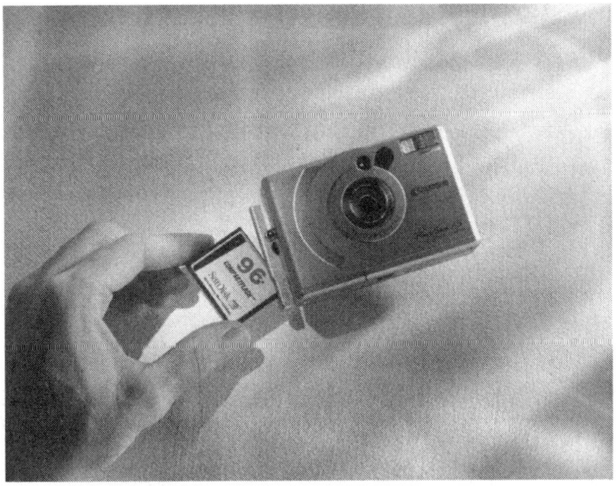

*Figure 4-10. CompactFlash card.*

It's been reported by some users that larger cards can be a disadvantage; that it takes longer to write to the large MB cards compared to earlier models. Since I've not seen any actual tests or documented reports I can't confirm this. Certainly it's not inconceivable; hard disks can exhibit this problem as they become "full" with files—it takes longer to write files because the disk space can become discontinuous (the file has to be broken up into chunks and written in different places on the disk). Flash memory shouldn't have this problem because you usually write files one after the other, and access time from one point to another in memory should be a lot faster than with a disk drive. I suspect that the problems are caused by different speeds of the cards being used, not by the increased size of the cards.

## "IF ALL THE WORLD'S A STAGE, I WANT BETTER LIGHTING"
## —WOODY ALLEN

Perhaps the biggest limitation of digital cameras (after limited resolution) is the lack of sensitivity of the CCDs. While this is improving, most CCDs are still not as sensitive as film. Speed improvements sometimes come at the expense of noise; manufacturers increase the gain (amplify the signal coming from the CCD) to make up for a lack of light sensitivity. This increases the noise in the images. The weak, built-in flash is just as bad. Photographs taken at night turn out with bright, washed-out foreground subjects with dark backgrounds.

While you can't do much about the limitations of the camera, you can work to give it every chance to produce the best image that it can. This usually comes down to one thing; adding more light. The more light in the scene, the better the image produced.

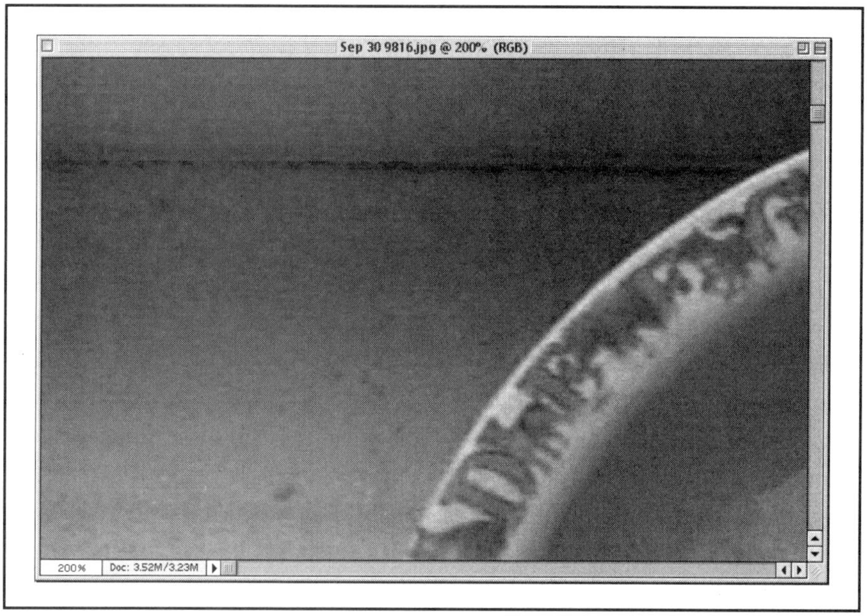


*Figure 4-11. Image noise.*

There are several ways to improve lighting; some of which involve hardly any work. Others involve a bit of thought and extra equipment. Here are some tips:

1. **Use natural light whenever possible.** Sunlight is the most common, and inexpensive accessory. Open blinds, and move your subjects into sunny rooms. If all else fails, go outside!

2. **Use a bounce card.** Once you're outside, you can "paint" with the sunlight by using a large white card to bounce the light onto the face of your subject, or otherwise increase the light.

3. **Use separate or additional flash units.** The built-in flash is small and limiting, and it creates red-eye and badly washed-out faces. Using an external flash (see below) can greatly improve your pictures.

If you are taking photographs inside, and you choose to turn off the camera's flash, be aware that the lighting used inside—fluorescent and incandescent—have very different color temperatures compared to sunlight. While it appears that these lights emit "white" light, they really don't, and this becomes readily apparent in your photographs.

Incandescents tend to have a yellow/orange cast, which some people consider to be "warm" and is actually desirable in some circumstances. Fluorescents, on the other hand, tend toward blue/green which provides a rather ghoulish cast to people's skin. Using a flash will minimize this; otherwise, you can use a color filter in front of the camera lens. Camera stores sell filters specifically for this. Using a filter does cut the amount of light entering the lens, which can impact your exposure time. Alternatively, you can use image editing software to adjust the color cast in an image.

## USING FLASH

External flash units are available in a variety of sizes. There are two connections used for flash units; a hot shoe and a PC sync jack. While hot shoes (a slot on the top of the camera) are uncommon in digital cameras, a PC sync jack is being included in several cameras. The flash unit is connected to the jack using a cord that's usually included with the flash. Even if you don't have a sync jack, it's still possible to use an additional flash with your camera. There are a variety of slave units as well as slave controls. These are light-sensitive mechanisms which fire the external flash when they detect light from the camera flash. These only work if your camera shutter fires when the first flash is fired; they won't work with the red-eye reduction feature which fires the cameras flash multiple times before taking the actual picture. Some cameras fire a short burst to calculate exposure before taking the pictures. Those cameras will also not work with a slave unit.

You don't have to point the flash at your subject; much more pleasing results can be accomplished by bouncing the flash off another surface (like the wall). You point the flash at the wall and let the light bounce off that wall onto the subject. It softens the light, making it appear more natural.

Experiment! The best thing about digital cameras is that you can take lots of pictures to practice at very little cost.

If you only use one flash and bounce light off a wall, you might find that the faces of your subjects have long shadows across them. To reduce this you need to direct some light from the flash into the faces of the subjects. You can do that by taping a small piece of card to the top of the flash which directs some light onto the subject. Some large flash units have a second small flash that points forward (while the main unit can be pointed in any direction). This fill flash accomplishes the same purpose as the card, but without unsightly masking tape! Alternatively, you can use a second smaller slave flash.

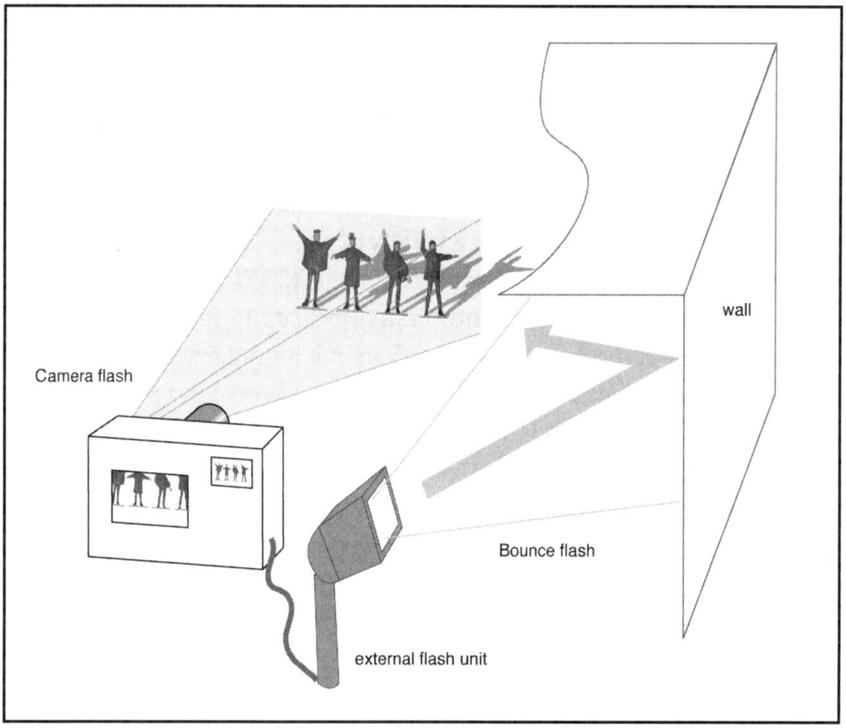


*Figure 4-12. Bouncing light and fill light diagram.*

Finally, some built-in flash units don't work well when you are very close to your subject; too much light is released and the subject appears washed out. Moving away, reducing the flash power setting (an option on a few cameras), or even using an external flash that you can move away from the subject, are probably the best solutions.

If all else fails, a flash diffuser (a piece of transparent paper placed over the front of the flash) may reduce the flash enough—yet allow enough light through—to produce good results. If you go this route, use a piece of paper that can be bent around the flash and taped with a good half inch or more of space between the surface of the flash and the paper. If you tape something onto the flash it's possible that the paper will interfere with the cooling of the unit. Excess heat could permanently damage the unit. WARNING: You could damage the flash. Refer to your manual before doing this.

## BATTERIES AND AC ADAPTERS

Described in some depth in Chapter 10, if you don't already have them, rechargeable batteries are almost a necessity for those who use cameras frequently. NiMH are preferred over NiCD.

If you use the camera in the studio, an AC adapter is a more convenient option.

## LENS ATTACHMENTS AND FILTERS

Filters and special effects lenses are available for SLR cameras. While it's also possible to use them with a digital camera, attaching them can be a problem. Also, because most cameras are rangefinder cameras, you can't see what the filter is doing except by using the LCD panel. This can be a particular problem for polarizing filters where the effect varies, depending upon the angle of the lens.

There are a range of close-up, wide-angle and other lenses that have been released for digital cameras. *Close-up lenses* let you focus on a subject that's close to the lens if your camera does not have a macro setting.

*Wide-angle* attachments enlarge the field of view of the lens. You should make sure that the attachment will work with your camera. A wide-angle lens, if it isn't the correct size for the lens, can result in vignetting—the image being cut off at the corners. Wide-angles will magnify any problems in the camera lens, such as barrel distortion (the bending of lines that should appear straight).

*Fisheye lenses* are wide-angle, which allow a very wide field of view, but greatly distort the image; no straight lines are straight. Use them only for special effects.

*Teleconverters* extend the zoom range so that far away objects appear closer.

As with all optics, the quality of the lens will affect the quality of your results.

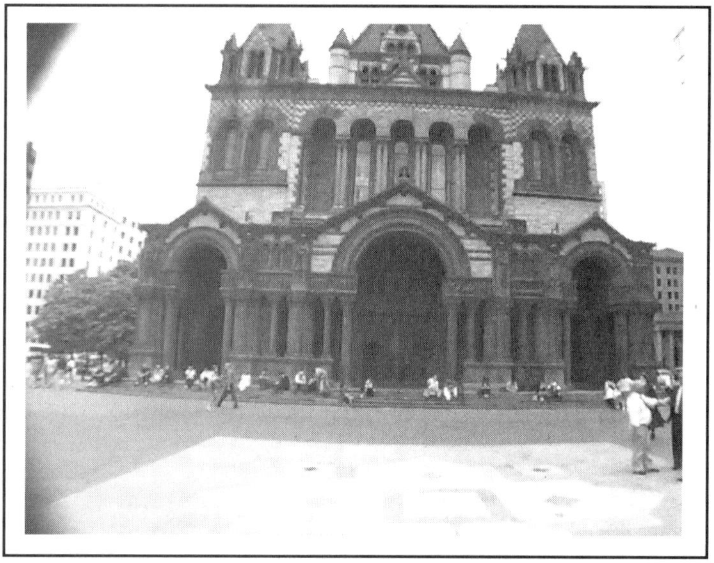

*Figure 4-13. Vignetting seen in an image taken with a wide-angle adapter.*

## ADJUSTING FOR PARALLAX

As described in Chapter 3, most digital camera use a viewfinder, and suffer from parallax (inaccuracy in what the rangefinder shows versus what the lens sees) when photographing subjects that are close to the lens. Usually less than three or four feet—the rangefinder should be acceptable accurate beyond that range. While it's possible to learn how to compensate for this error, it's much easier to use the LCD panel of the camera (assuming the camera has one).

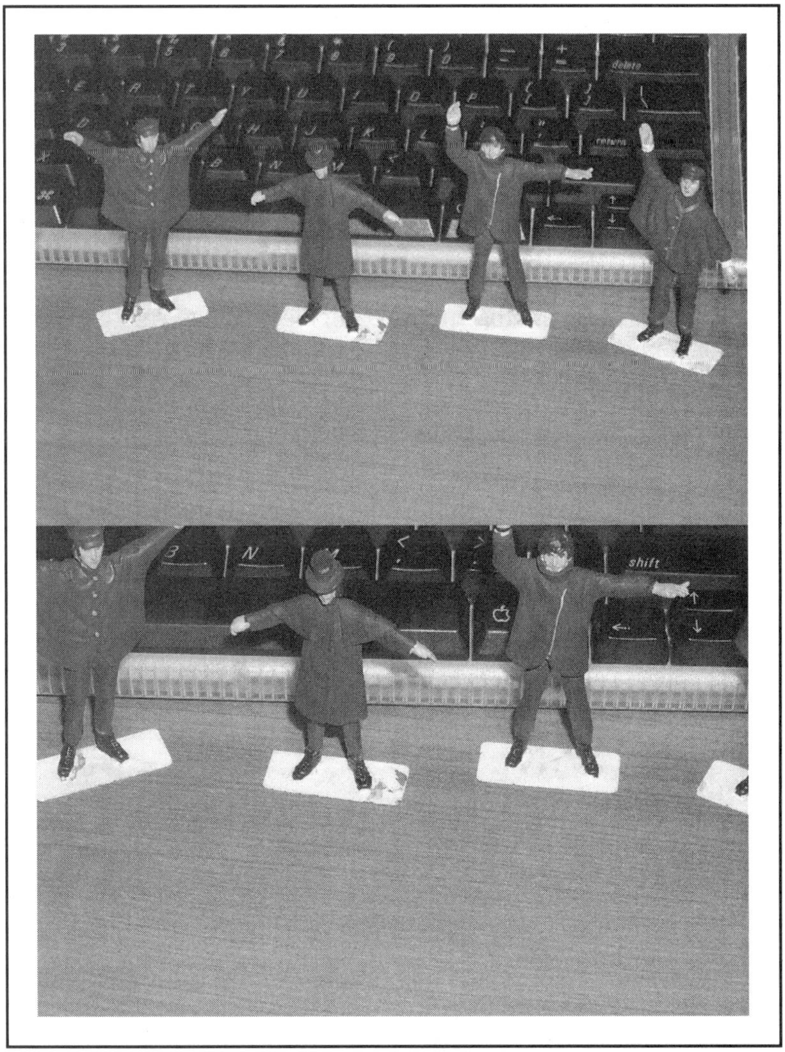

*Figure 4-14. Parallax error.*

## CLOSE-UP PHOTOGRAPHY

If you plan to do a lot of close-up photography, a camera that has a macro function (the ability to focus on objects that are close to the lens) should be factored into your buying decision. You will get much better results than you will with a close-up lens. However, if the camera you have doesn't have a macro function, then a close-up lens is your only solution.

You'll still have problems with parallax error and will need to use the LCD to properly frame and manually position the camera to focus it. You may also have to experiment with your flash (see the earlier section on flash).

## KEEPING THE CAMERA DRY

Water and condensation are the enemy of all electronics. And while regular cameras have added more and more electronics over the years, a digital camera is still more susceptible to water (though maybe a film camera is more susceptible to sand). If you don't have a camera bag, it's still possible to construct impromptu protection using a plastic freezer bag with a zipper lock. These are reasonably strong, seal very tightly and are easy to open and close. Don't be shy, use them for all occasions!

If you are going to be using the camera in very high humidity or damp situations, you might even want to consider buying a water-tight shell like those used for underwater photography (see Chapter 11).

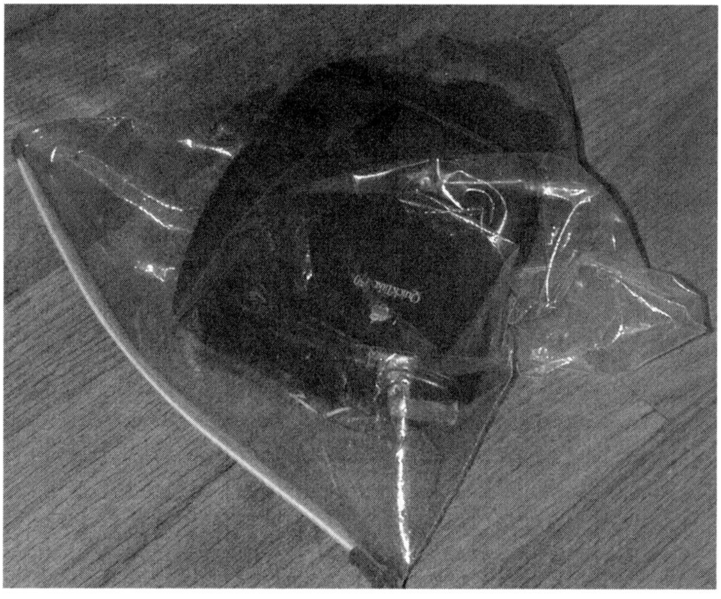

*Figure 4-15. Camera bag.*

## DIGITAL CAMERAS AND THE COLD

Can you use digital cameras in extreme cold? Manufacturers generally will quote a temperature range within which the camera will work satisfactorily. You should protect the camera from extremes in heat and cold. As long as it's well insulated (say in a coat pocket) there should be no trouble using the camera in lower temperatures. Just bring it out, take a picture and return the camera to your coat pocket. Batteries are the item most likely to be affected by cold; if your batteries are weak, they will be even weaker when cold. In theory, the CCD of the camera should actually work better as it gets colder because the background noise level will be lowered (see Chapter 3). However, it's unlikely that you will actually notice any variation in image quality.

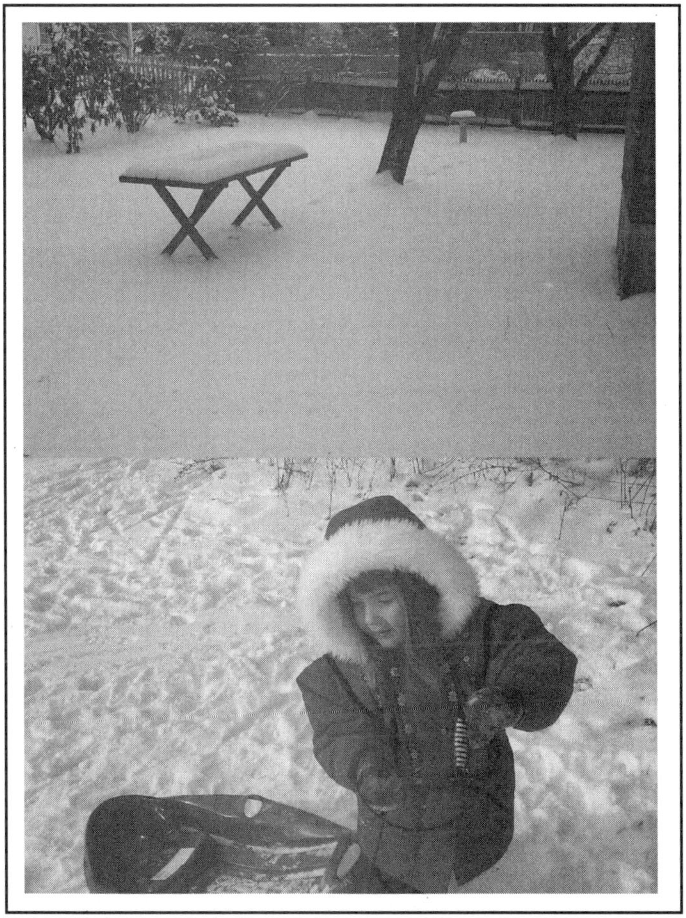

*Figure 4-16. Snow pictures.*

Of more danger is the effects of condensation caused by rapid changes in temperature. Condensation, while a minor annoyance when it forms on the lens, can cause major problems with the internal electronics of the camera; therefore, try to avoid rapid changes in temperature, and keep the camera in a bag with desiccant or silica gel which absorb moisture. After being out in the cold, keep the camera in your coat pocket or camera bag so that it slowly returns to room temperature. Thermal expansion and contraction might also be a problem, but this would be at extreme temperatures. Don't leave the camera in a car during the day in the summer.

LCD displays can be very sensitive to cold temperature, slowing down or not working at all.

## USING A DIGITAL CAMERA FOR OPTICAL CHARACTER RECOGNITION

Optical Character Recognition (OCR) is software that reads a scanned image and tries to find text within that file. It then converts that text into an editable text file that can be saved in a word processing format. If you have an old document that you would like to get into the computer so that you can edit it, then OCR offers the promise of reducing typing time. The OCR software does the work for you.

The effectiveness of this software is very much dependent upon the quality of the image being interpreted, as well as the quality of the original (a very clean, typewritten page will be more accurately recognized than a scan from an old newspaper).

Digital Cameras can be used to capture files to be recognized by OCR, though there are better ways to do it. A flatbed scanner, which you can get very cheaply, will do a better job than most digital cameras. If you can't use a flatbed scanner (you can't take one to the site, or the object is fragile and cannot be placed on the scanner), then you have three options:

**1.** If the type is large, use a digital camera.

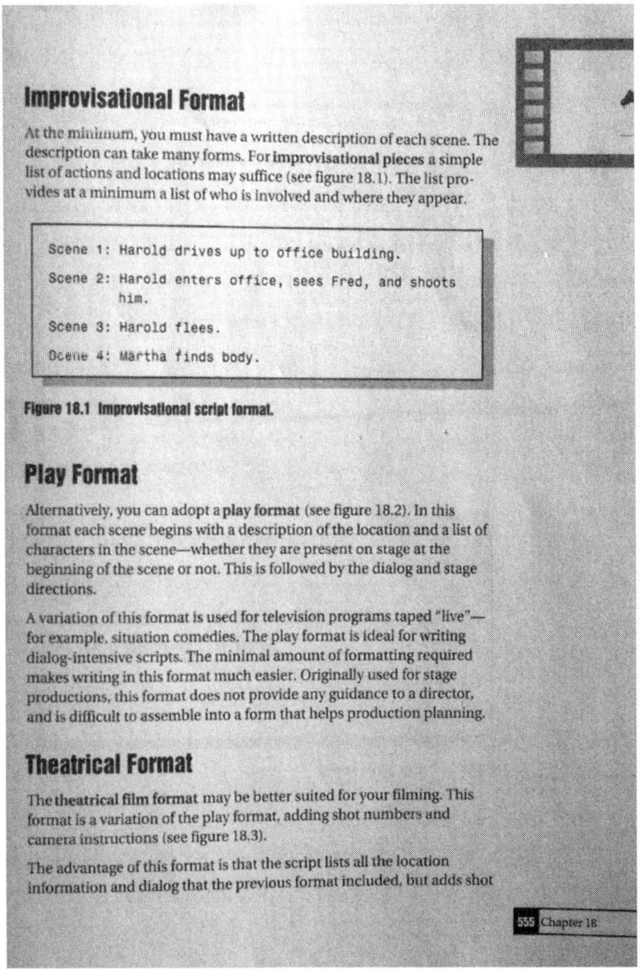

## Improvisational Format

At the minimum, you must have a written description of each scene. The description can take many forms. For **improvisational pieces** a simple list of actions and locations may suffice (see figure 18.1). The list provides at a minimum a list of who is involved and where they appear.

```
Scene 1: Harold drives up to office building.
Scene 2: Harold enters office, sees Fred, and shoots
         him.
Scene 3: Harold flees.
Scene 4: Martha finds body.
```

**Figure 18.1 Improvisational script format.**

## Play Format

Alternatively, you can adopt a **play format** (see figure 18.2). In this format each scene begins with a description of the location and a list of characters in the scene—whether they are present on stage at the beginning of the scene or not. This is followed by the dialog and stage directions.

A variation of this format is used for television programs taped "live"—for example, situation comedies. The play format is ideal for writing dialog-intensive scripts. The minimal amount of formatting required makes writing in this format much easier. Originally used for stage productions, this format does not provide any guidance to a director, and is difficult to assemble into a form that helps production planning.

## Theatrical Format

The **theatrical film format** may be better suited for your filming. This format is a variation of the play format, adding shot numbers and camera instructions (see figure 18.3).

The advantage of this format is that the script lists all the location information and dialog that the previous format included, but adds shot

555 Chapter 18

*Figure 4-17. Document photographed using Kodak DCS 420.*

2. If it's 10 or 12 point type, then you'll need either a high-resolution camera (you need about 300 dpi) or you'll have to take multiple shots of the object and then stitch them together using an editing application.

3. Use a regular film camera and then digitize the film or have Photo CD made.

Whichever technique you use, you should first perform a test to see what level of accuracy you will get with your documents and the equipment you are going to use (relying on statistics and samples provided by others is virtually meaningless).

The lower the accuracy of the OCR software, the less useful it is. If you're getting only 50% accuracy you'll spend as much time correcting the document as you would if you actually typed it in (and that's not counting the time you spent doing the capture and OCR in the first place). Clearly you need to have an accuracy much higher, in the 90% range, to make it worthwhile.

### MAINTENANCE AND TROUBLESHOOTING

While all cameras are different, and each can develop unique problems, Chapter 23 provides some general advice for dealing with problems you might encounter.

## CONCLUSION

Now that you've captured your images it's time to download them to the camera for editing and storage. You can read more about that in the next chapter.

Later chapters cover manipulating images using image editing software, printing, using images for web pages and in panoramic VR technologies such as QuickTime VR.

# CHAPTER 5
# CONNECTING TO A COMPUTER

Once you've taken your photographs it's time to transfer them from the camera to the computer. While there used to be only one way to do this— through the serial port of the computer—there are now several different ways to get your images to the computer. Your camera probably came with a serial cable, but it can take a long time to transfer several megabytes of images using a serial connection. By spending only a little money you might be able to greatly speed up the process and make your picture taking (and transfers) more pleasurable.

> Until recently you had to download you images from the camera before you could do anything with them. That's started to change. Digital FX is an editing application that works in Digital cameras (such as the Kodak DC265 and Minolta Dimâge EX cameras) and lets you apply effects and adjust the brightness and contrast of an image (see Chapter 20). You can now transfer images from CompactFlash and SmartMedia cards to Iomega's Clik! drive, though once you get them onto the drive you'll have to transfer them to your computer to do anything with them.
>
> A really interesting innovation is the development of printers that will print directly from the memory card. Lexmark's Photo JetPrinter 5770 and Epson's PrintOn PT-100 inkjet printers both accept memory cards from most digital cameras and will print images directly from the card. The Photo JetPrinter also accepts Zip drives connected directly to it so that you can transfer the images from the memory card to a Zip cartridge without using a computer. Both these printers are covered in Chapter 9. They can also be connected to a computer and used for regular printing, so you get the best of both worlds.

One thing you need to think about when looking at connection options is how you plan to use the camera. How frequently will you use it? How many images will you take? Do you need to control the camera remotely?

Often I take a series of images and simply want to transfer them to the computer so that I can delete them from the camera's memory. Other times, I take a specific picture and want to open that image immediately in an editing application and work with the image.

Some applications require controlling the camera from a software application. As an example, a digital camera being used to take photographs for an employee database would be most useful if it were set up connected to the computer. Pressing a button on the computer could trigger the camera to take a picture, download the picture immediately, and maybe even paste it into a database. This may or may not be possible depending upon the software you have and how your camera is connected to the computer. But if your camera can be connected and controlled by software from your computer, then you can probably use a macro scripting application (or for Macintosh users AppleScript) to write a macro that does just that (see Chapter 17 for more information about this).

## CONNECTION OPTIONS

Options for connecting your camera to your computer depend upon the type of camera you have, the type of computer you have, and how much you want to spend to improve transfer speeds. Note that some of these options are limited by the make of camera.

### SERIAL CONNECTION

Nearly all cameras come with serial connections. There should be a serial plug (often an 8-pin DIN plug), and a cable provided to connect the camera to the computer. Macintosh and PC serial cables are different because the serial port on these computers differ. Some early cameras are limited to serial port connections only. Kodak's new digital cameras (DC220, 260, 265) have a single plug that supports both serial connections and USB connections (which are much faster than serial).

While the serial connection has been the defacto method of connecting cameras to computers, it suffers from one major defect; the speed of the connection. With many camera images being 400K or larger, it can take more than a minute to download a single image.

**How to use:** Simply plug the serial cable into the camera and into the computer. It's recommended that you turn off the camera before doing this, though most cameras will happily work if you insert the cable while they are on. Be careful when inserting these cables; if you're careless it is possible to damage the plug by carelessly inserting the cables and pushing too hard to seat the plug (I've bent pins on plugs doing this). You can often "fix" the problem with a pair of needle-nose pliers.

If you already have a printer and a modem connected to your computer then you may be short in the serial port department. There are four solutions to this problem: 1) Find another connection method for the camera, or get rid of one of the other devices; 2) manually disconnect and reconnect an unused device as needed; 3) use a switch box to connect two devices that are only occasionally used to the same port, then switch between devices as you need them; or 4) buy a card with additional serial ports and install it in your computer. Note that unless you need additional ports for other equipment I'd recommend spending the money on a card reader rather than extra serial ports.

Once you connect the serial cable you don't have to unplug it every time you are finished using it. I've left a serial cable connected and hanging down the front of my desk so that I can simply grab the cable and connect it to the camera rather than having to mess around at the back of the computer when I need to connect. Some new computers now feature serial ports on the front of the computer; which makes a lot of sense!

Most computers have two serial ports; on Windows PCs they are often labeled COM1 and COM2 and are 9-pin plugs. Providing they are enabled (the operating system recognizes them—this should be the case unless some other software has disabled a port), all you need to do is connect the camera and run the camera software (see below). You may need to check that the speed of the COM port matches the speed setting in the camera software. The port setting can be checked by going into the Device Manager of the system control panel. Choose the COM port and click *Properties* to get to the Port Settings.

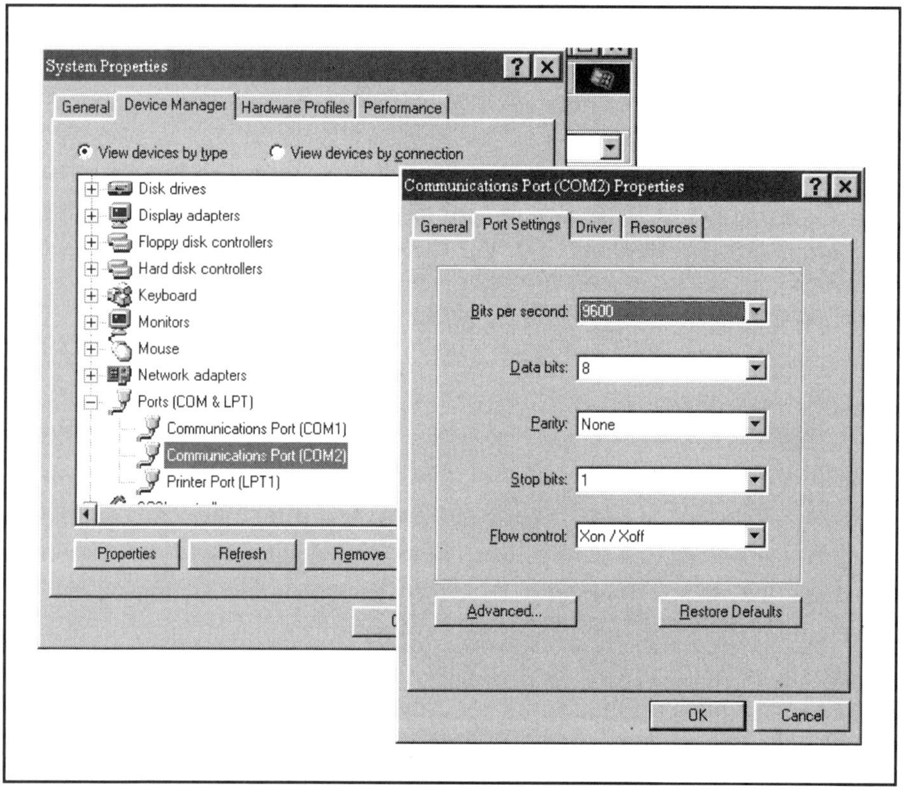


*Figure 5-1. Window port setting control panel.*

Most Macintosh computers also have two serial ports (labeled with a printer and a telephone symbol), though a few models have only one port. These are 8-pin DIN plugs. While you can use either port with a camera, the printer port is reserved for AppleTalk printers. If AppleTalk is turned on (in the Chooser control panel) the port labeled with a printer icon will not be available for another device even though there may be nothing else connected to it. Simply turn off AppleTalk if you want to use that port. There is one reason for using the printer port; it can transmit at slightly higher speeds than the other serial port.

> NOTE: New iMacs and G3 computers do not include serial plugs. Instead, they have USB and FireWire connectors (see below). USB is desirable because it is considerably faster than serial, and you can connect multiple devices. However, not all cameras support USB. You can, however, get USB to serial converters if your camera only has a serial connection. If the camera uses CompactFlash or SmartMedia cards, spend the money on a USB card reader rather than a USB-to-serial connection. It will be faster and easier to use.

**Upgrade options:** If the camera uses CompactFlash or SmartMedia flash memory cards, purchase a card reader.

## SCSI

Used on some of Kodak's high-end cameras, SCSI (Small Computer Systems Interface) provides a high-speed connection between the camera and computer. If your computer lacks an SCSI connector you can buy an SCSI card to plug into a slot in the computer. These cost about $100. PCMCIA cards that support SCSI are also available for notebook computers, though they are much more expensive. Macintosh computers used to include built-in SCSI, though Apple recently stopped including it in their newest models in favor of other connection standards; USB and FireWire. SCSI uses a very large connector and fairly thick and bulky cables. Up to seven devices (including the computer) can be connected in a chain; each device must have a unique number. Most SCSI devices include a selector that lets you choose a number for the device, though some devices come with a set number that you can't change (Zip drives only let you select between SCSI 2 and SCSI 3). Note that SCSI connectors come in two varieties; a 50-pin plug normally used at the computer, and a much larger connection used on SCSI devices. Cables are available in two varieties; computer-to-device and a device-to-device version.

**How to use:** As with serial, you simply plug the SCSI cable into the computer and camera. You MUST turn off both devices before connecting them. Though I know many people who plug and unplug SCSI devices while they are turned on, doing so risks damaging the electronics in both devices.

You must make sure that each device in the chain has a unique number, and that only the last device in the chain is terminated. Termination should be described in the documentation with the SCSI card, the computer documentation, and/or the computer manual.

SCSI setup and termination can be a bit of a "black art" when you have multiple devices; if you are having trouble with the camera connected to other devices, try connecting the camera directly to the computer and see if that eliminates your problems.

**Upgrade options:** If the camera uses a PCMCIA or other memory card you might be able to use a card reader. Though it isn't as quick and easy to use as plug-and-play technologies such as USB, SCSI is very fast. If you have it and it works for you, then you might as well continue using it.

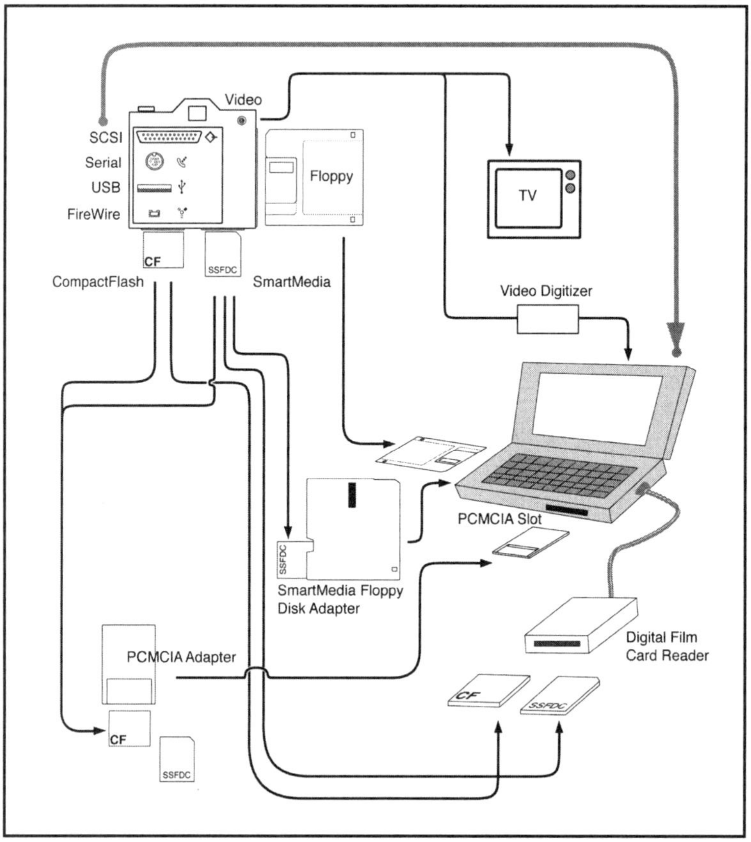


*Figure 5-2. Connection options.*

## USB

USB, like SCSI, is a bus; a communications protocol between computers and computer devices. It was developed by a number of companies including Intel, Microsoft, Compaq and IBM. The objective of this standard was to provide a plug-and-play interface to external peripherals. Other methods for connecting external peripherals (notably SCSI) are not always easy to configure or use. Ideally, USB will allow new devices to be recognized without having to update system software or perform any diagnostics. USB also allows many devices—up to 127—to be connected to the same device, though you will have to buy USB hubs to allow you to do this.

USB can be used for many different kinds of devices, from cameras and scanners to keyboards, hard drives and video products.

As with all products in the computer field, the reality isn't always the same as the ideal. While it is possible that your USB-equipped computer will automatically recognize a new device connected to the computer, it's also just as likely that you will first have to install special drivers or software to work with your device. So you still want to read that "User Guide" with the camera before you connect your new device for the first time!

USB can even be used to connect two computers, though you'll need a special cable that includes an adapter to make this happen.

Even if your camera doesn't support USB, if your computer does, consider buying a USB card reader if your camera uses CompactFlash or SmartMedia cards. A USB card reader should be easier to set up and use than an SCSI or parallel port-based card reader.

USB can transfer at up to 12Mbits/s, though you won't necessarily see this when connecting your camera to the computer. As an example, transferring an image through USB from a Kodak DC260 takes 5 seconds for each image to download, or a rate of 4.7MB per minute. This is considerably faster than using the serial connection.

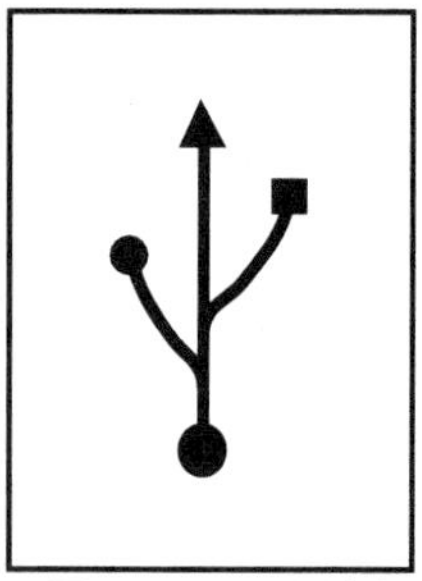

*Figure 5-3. USB connection symbol.*

**How to use:** The Kodak Cameras that support USB (DC260, 265, 220, 240) have a single plug that is used for both serial and USB connections. Different cables are used to connect the camera to either the serial port or the USB port. The USB port is an especially odd looking plug; it looks like a metal rectangle. Without the delicate pins found in serial plugs it should be much better suited to repeated plugging and unplugging.

Kodak's software must be told whether you are connecting using the serial or USB connection, and special drivers are installed for the USB connection.

The real joy of USB is the speed; it's a lot faster than serial, transferring multiple images in a minute rather than many minutes.

USB is already being used for hard drives and CDRs drives, though the hard drives are very slow because of the limited transfer rate of USB when compared to typical hard drive. If you want to be able to easily move devices between computers, buying a USB hard drive or CD-R might make sense.

**Upgrade options:** Only SCSI and FireWire are faster than USB; and SCSI is much more trouble. The USB 2.0 specification has been announced which is considerably faster than USB 1. Unless you're working with video, USB should be all you need.

## FIREWIRE/IEEE-1394

FireWire is a communication protocol that shares many of the attributes of USB; plug-and-play, and multiple devices connected together easily. FireWire, however, has a much faster transfer rate of up to 400 Mbits/s (though Intel has announced USB 2.0, which improves USB's transfer rate to 120 to 240 Mbits/s). At the moment USB is positioned for general-purpose use while FireWire is primarily being used in DV (Digital Video) camcorders and for high-speed hard drives. Because of its more limited adoption rate FireWire is more expensive and so may continue to be used only for high-speed peripherals, whereas USB will remain popular for consumer devices.

FireWire external devices, including hard drives and CDR drives, are starting to appear. They are slightly more expensive than USB drives, but should offer faster transfer rates. You should check the specifications first; some models have been released that don't take advantage of the high speeds FireWire offers. FireWire is also covered in Chapter 19.

Kodak has released four high-end digital cameras with FireWire: DCS 315, DCS 520, DCS 560, and DCS 620.

**How to use:** Much like USB, FireWire promises a simple cable connection, though FireWire adds one wrinkle—it supports both a 4-pin and 6-pin connector. FireWire can be used to connect a video camera to a second camera for copying video. So far, FireWire has only been used for video cameras; there are no still digital cameras.

**Upgrade options:** None.

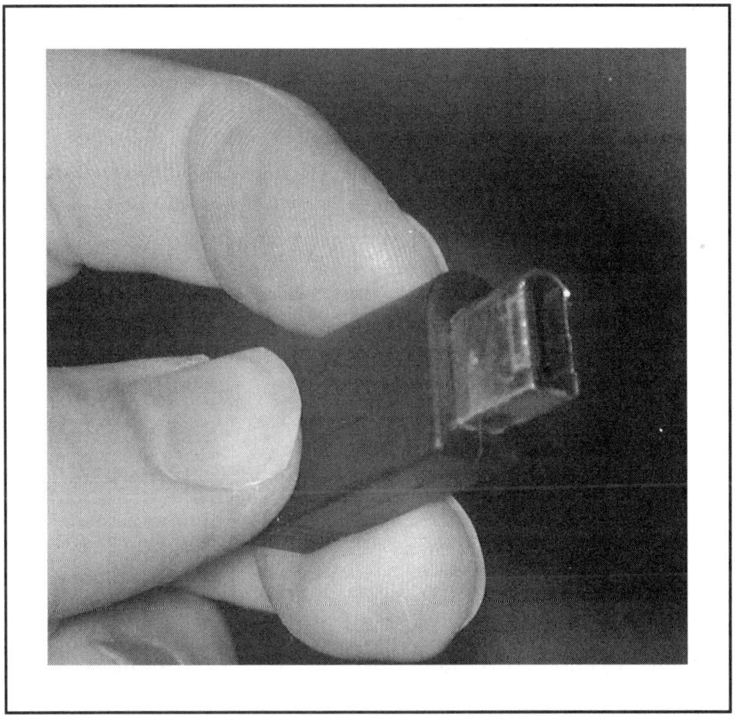

*Figure 5-4. FireWire connection.*

# MEMORY CARD READERS AND ADAPTERS

With memory cards replacing built-in memory, digital cameras become a lot more flexible and a lot easier to work with. Now you can keep adding memory and taking more pictures until your money or your batteries run out. Another advantage of memory cards is that they make it possible to transfer images to the computer a lot faster.

## PCMCIA ADAPTER

If your notebook has a PCMCIA slot then you can insert CompactFlash and SmartMedia cards into the slot, providing you buy the appropriate adapter. The CompactFlash adapter is very inexpensive (about $15) because CompactFlash includes a controller and uses the ATA interface supported by PCMCIA. An adapter for SmartMedia cards costs about $50 and turns the SmartMedia into a Type II-ATA card.

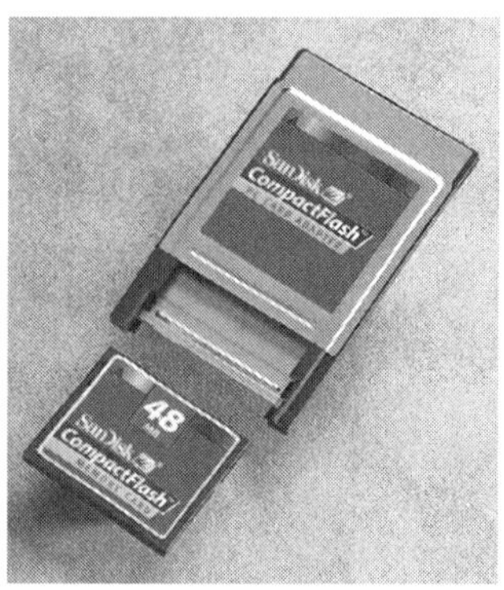

*Figure 5-5. CompactFlash adapter (photo courtesy SanDisk Corporation).*

**How to use:** Insert the card into the adapter and insert the adapter into the PCMCIA slot.

**Upgrade options:** You could switch to a separate reader that attaches to USB or parallel port, but unless you have to use the PCMCIA slot for another device at the same time, this option is the best.

## FLASHPATH FLOPPY DISK ADAPTER

SmartMedia users can buy a FlashPath adapter for their cards. The adapter looks like a floppy disk drive. Several companies sell the same device. The software is better on the Windows side than for the Macintosh. While it's reasonably easy to use, it's not as fast as other options.

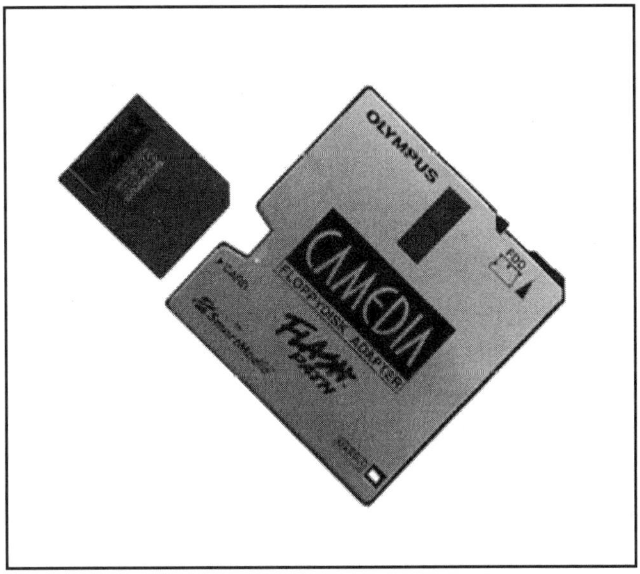


*Figure 5-6. FlashPath adapter.*

**How to use:** You slip the card into the adapter and put the adapter into the floppy disk drive of your computer. You must install special drivers and software before using the device. On the PC the cartridge appears like a regular disk drive and you can copy the files from the card to your hard disk. For Macintosh users, a separate application must be used which features a very sparse interface for selecting the files and copying them to your hard drive.

The cards use a small watch battery which may be used up very quickly if you use the adapter for extended periods of time (don't leave it in the computer unless you are using it). Also, SmartMedia comes in 3.3V and 5V configurations. While it appears that the adapter can be used to erase 3.3V cards, there may be problems with 5V cards. If the adapter starts acting erratically it is probably due to a low battery.

**Upgrade options:** The adapter works well, but there are faster options; the card readers have a much faster transfer speed. Since the cost of a card reader is about the same as a FlashPath adapter, consider a card reader first unless you don't want to mess with connecting the card reader to your computer.

## MEMORY CARD READERS

There are several varieties of card readers available for PCMCIA CompactFlash and SmartMedia cards. The PCMCIA readers tend to be more expensive, and if you actually are using a CompactFlash or SmartMedia card then you'll need to buy an additional adapter.

The CompactFlash and SmartMedia readers generally cost less than $100; some read only one kind of card, though there are readers that support both. The card readers are available with parallel port connections, SCSI connections and USB. In terms of speed, SCSI should be the fastest, followed by USB, with parallel last. All are faster than using a serial port. In terms of ease of use the USB option is probably the best, provided your computer has a USB port.

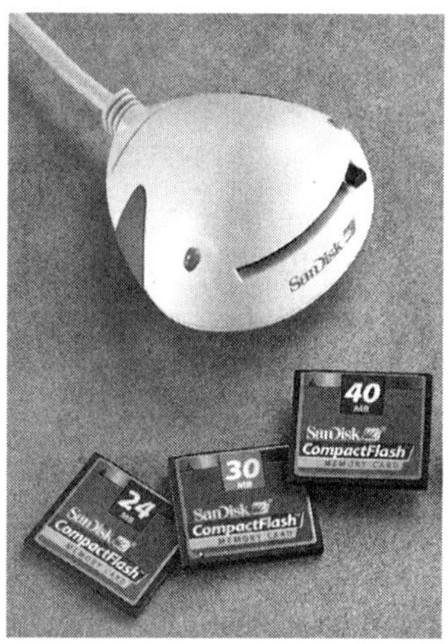


*Figure 5-7. CompactFlash card reader (photo courtesy SanDisk Corporation).*

**How to use:** Connect to the computer following the instructions included with the reader. If you have other parallel devices (Zip drive or printer) you should chain the devices together. Similarly, if it's an SCSI device and you have other SCSI devices, you'll have to pay attention to the vagaries of SCSI connections (see the SCSI section above). USB should be the easiest provided you have a spare plug; most computers come with two. If you have more USB devices than USB ports then you'll need to buy a USB hub to add extra devices. The reader comes with driver software that will need to be installed before use.

**Upgrade options:** About as good as it gets, though parallel isn't nearly as fast as USB or SCSI.

### COMPACTFLASH DIRECT TO USB

Lexar Media has announced a line of USB-enabled CompactFlash cards. These special cards work like a regular CompactFlash card, but using a special JumpSHOT USB cable (list price of $19.95) the card can be connected directly to a USB computer. Lexar provides USB drivers for Windows and Macintosh platforms. The JumpSHOT USB cable will only work with the special USB-enabled CompactFlash cards, which should retail at the same price as a regular CompactFlash card. Find out more about Lexar Media products at: *http://www.digitalfilm.com*

### CLIK! DRIVE

The Iomega Clik! drive provides an indirect method of connecting your camera to your computer. This portable drive can be used to transfer images from a SmartMedia or CompactFlash card to a Clik! cartridge in the field. Then once you get back to your computer you can connect the drive to the parallel port of the computer and transfer the files from the disk. The Clik! drive is described in detail in Chapter 10.

**How to use:** Connects to the parallel port much like a card reader. Requires software to be installed, but drives appear like regular hard drives on your system. No Macintosh version available.

**Upgrade options:** A second flash card (and a card reader) might make more sense than a Clik! drive.

## SOFTWARE

Your camera should come with the software necessary to connect your camera to the computer. This software may not be the best way to connect your computer, see Chapter 17 for alternatives). When you get the camera you should always first use the software that came with the camera just to make sure everything works before you start experimenting with other software.

All camera applications assume that you are connecting the camera directly to the computer and then transferring the images from the camera to the computer. If you want to insert a memory card into a card reader then you will use the software that comes with the card reader (which may simply "mount" the card on the desktop as though it is a regular hard disk).

Using the direct connection between camera and computer has certain advantages; the software may even be used to control taking pictures. The downside is that this connection is usually slower than transferring images from a memory card.

The software that manufacturers provide for transferring images from the camera usually fall into one of three categories; dedicated application, camera mounter, or TWAIN driver. The following sections describe these options. Note that which option(s) is provided depends upon the manufacturer and model. The same manufacturer may provide very different software among their cameras.

## DEDICATED APPLICATIONS

Dedicated applications let you control the camera, preview the images, and download the files. There seems to be a drift away from this kind of software. I think manufacturers are realizing that it doesn't make sense for them to spend a lot of time and resources writing software.

## CAMERA MOUNTERS

These system-level utilities cause the camera to appear on the computer desktop as if it was a hard drive attached to the computer. To transfer images you simply click and drag from the imaginary disk to your hard disk. The advantage of this method is that it is simple and easy to use.

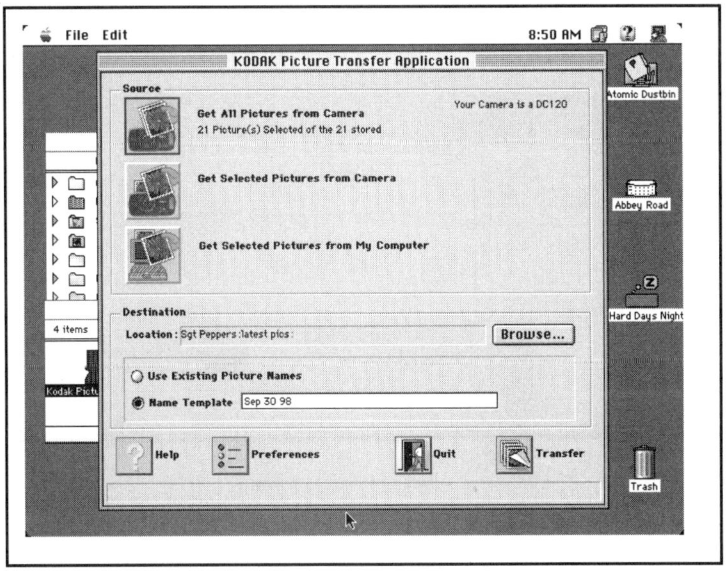


*Figure 5-8. Kodak control software for DC120.*

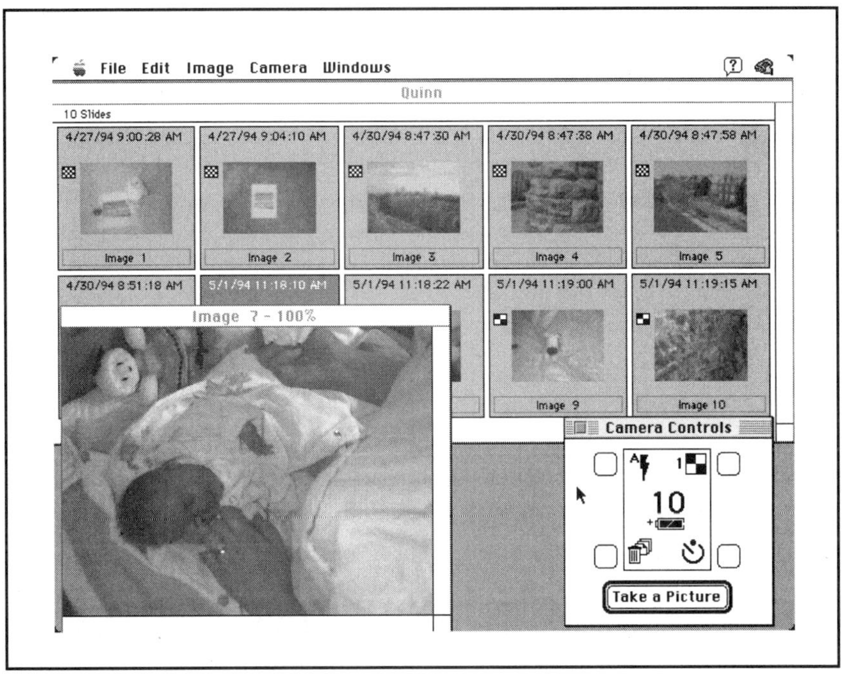


*Figure 5-9. QuickTake 100 control application. It could be used to view images as well as take pictures.*

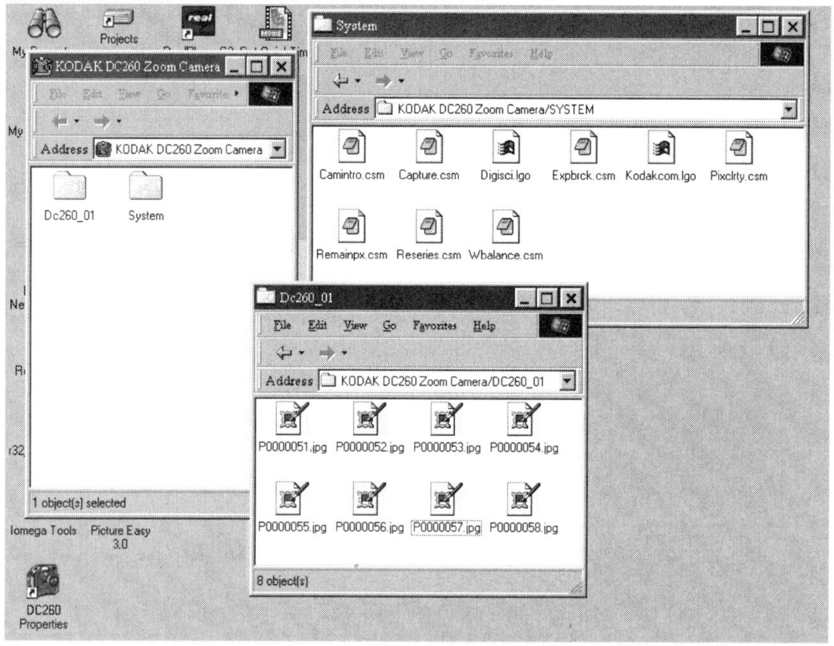


*Figure 5-10. Kodak DC260 mounted on the desktop (icon at lower left) using the USB connection. The camera CompactFlash card appears as a disk drive that contains two subdirectories. One contains the images, the second, called System, contains Digita scripts.*

# TWAIN DRIVERS

TWAIN is a communications protocol for communication between software applications and imaging devices (scanners and cameras). TWAIN evidently doesn't stand for anything, though it has been suggested it stands for "Tool Without An Interesting Name."

Most software editing applications will support TWAIN plug-ins, so scanner and camera developers can develop a single plug-in that is TWAIN compliant, and that plug-in should work with any editing application. Usually, you launch the TWAIN plug-in from within the editing application. The plug-in provides an

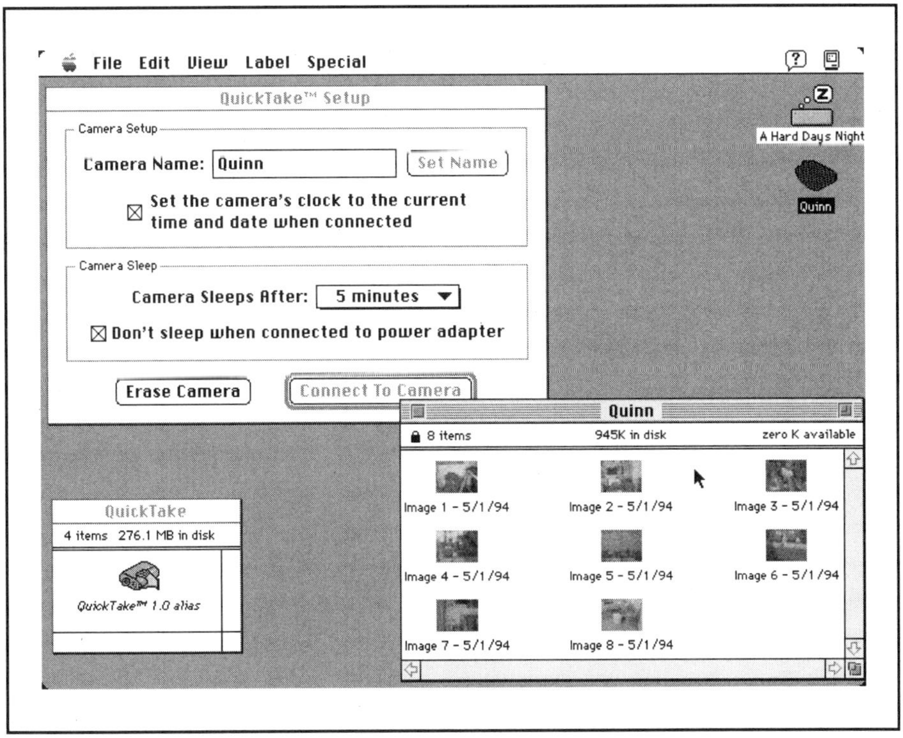


*Figure 5-11. QuickTake mounter control panel. Mounts the QuickTake as a camera on the desktop using the serial connection.*

interface that lets you choose the images that you want to download, and when you select the image it is downloaded and transferred to a new document in the editor. It's then up to you to save the new file.

On the face of it, TWAIN provides a number of advantages; the plug-in will work with a variety of applications and it takes less time to write a plug-in than an application because all the plug-in has to do is handle the selection and downloading of images. It doesn't have to handle display or editing. However, TWAIN drivers can be annoying to use if they don't include a feature to automatically download and save images. Also, if your machine has only a limited amount of memory, the combination of the editing application and TWAIN plug-in running at the same time can tax your computer.

If I can, I prefer to download the files and avoid TWAIN plug-ins.

# CONCLUSION

There are many different ways to connect your camera and computer. There's no "right way" to do it, but some options are better than others.

Once you've downloaded your images you'll probably want to edit them. But before we do that, it's important to make sure you understand concepts of resolution and color management. The next chapters cover these aspects of digital imaging.

# CHAPTER 6
# ADJUSTING RESOLUTION ON YOUR COMPUTER

Perhaps no other feature is more examined when choosing a digital camera than resolution. Resolution suggests the quality of the camera; the larger the resolution, the better the images. But while everyone seems to understand that higher resolution means better quality images, many people get lost when it comes to working with resolution once they get the pictures onto their computer. How does the resolution of the picture fit with the resolution of the monitor? Why do printers have a higher resolution than computer monitors, yet the quality of the images produced isn't nearly as good?

This chapter will attempt to answer these questions. Effectively managing resolution is important both for printing and when working on images for use in web pages or computer-based presentations. Using the correct resolution is important. But while everyone knows that images that lack resolution will look blurry, using files that are larger than you need is inefficient.

As soon as you save images to your computer you will have to deal with resolution. And whatever your final intent; whether it's to use the image on a website, or to print it in a brochure, you'll probably spend a fair amount of time adjusting and manipulating the resolution of a file. That's why it's important to understand the basics and what resolution really means.

When looking at the specifications for a digital camera, you'll find several pieces of information which seem to say the same thing, but don't always match up. Take *Figure 6-1*, which shows some of the specifications for Nikon's CoolPix 950. The first number is the resolution of the CCD. This is an important number as it indicates the resolution of the device used to capture the image. This number may be displayed in terms of the horizontal and vertical resolution of the chip (i.e., 640 x 480) or as a total number of pixels (i.e., 300K). The latter number can be calculated by multiplying the two numbers together. So a CCD with a resolution of 640 x 480 probably has about 300,000 pixels (written 300K, where K=1000).

**CCD Resolution** `1/2 inch high density CCD, Total number of pixels: 2.11 million`

**Image Resolution** `1,600 x 1,200 pixels (UXGA`
`1,024 x 768 pixels (XGA)`
`640 x 480 (VGA)`

**Image Quality/Compression**
Hi = Uncompressed RGB TIFF
Fine = 1:4 JPEG
Normal = 1:8 JPEG
Basic = 1:16 JPEG (Approx.)

**Image Storage**
(on 8MB card) Hi = 1 image
Fine = 8 images
Normal = 16 images
Basic = 32 images
(Number of images will vary depending on the image captured and the resolution selected.)

*Figure 6-1. Nikon CoolPix 950 specifications.*

If you have an early digital camera that has a maximum resolution of 640 x 480 it's very probable that the CCD is little larger than 300K. Many of these cameras used CCDs based on chips built for video cameras. Video camera CCDs usually have a slightly higher resolution, often about 380K. Interestingly, if you multiply the maximum out of the image dimensions in *Figure 6-1* you get a result of 1,920,000, which is less than the 2.1 megapixel resolution claimed in the marketing. Where are those extra pixels? There can be several reasons that pixels are ignored, including camera design and fabrication issues. In the case of the Nikon, the 1,600 x 1,200 size is used because it is a standard computer display size, and is 2.5 times the standard 640 x 480 VGA screen size.

Quoting the CCD (or image) size in terms of total pixels became popular among manufacturers when the megapixel cameras appeared. These cameras were the first to have more than a million pixels. Megapixel cameras are now common for consumer units, with 2-megapixel cameras becoming standard. This number will probably keep increasing over the next few years.

Some cameras include software that interpolates extra pixels; that is, it takes the original image captured by the CCD and actually inserts extra pixel values, creating an image that is larger than the CCD captured. This process is also called *resampling* and is explained below. Kodak's DC120 does this. Canon includes software called PhotoGenie which several people (including the manufacturer!) say produces incredible results when enlarging an image. But whatever algorithm it is using to do this, it is still adding pixels by expanding the original image data, and it's arguable whether any program can accurately add information when it doesn't exist in the first place. While the size of the file may be increasing, the actual information in the image is effectively remaining the same. Carried to extremes, if you took a file and "blew it up" four times its original size, while you'd end up with a much larger image, you'd see no increase in image detail. This process works great for posters, but not for photographs or computer presentations.

Since most graphics editing applications include the capability of enlarging a graphic, this process should be left to the user, rather than performed in the camera.

Before we continue, it's vital to remember that the CCD resolution is based on a raw pixel count, and that in most digital cameras each of these pixels can be used to measure only one color (see Chapter 3). This fact impacts the accuracy of the image, for while you might get a 1,600 x 1,200 pixel image from the camera, it may not be as color accurate as a 1,600 x 1,200 image from a Photo CD or a flatbed scanner. Noise and other problems can also affect the apparent resolution of an image.

## IMAGE RESOLUTION

The next number is the Image Resolution. This indicates the size of the files that the camera produces. The Nikon produces three possible sizes (which you can select before taking the photograph). How does it do this? The camera captures the image, then uses software in the camera to downsample the image and create the lower-resolution image. Capturing at a lower resolution will let you store more images in the camera's memory, and will also reduce capture time because it doesn't take as long to save the smaller file to the comparatively slow memory. However, you can't "get back" the information later (despite "upsampling"), so when you take a picture you need to be mindful of how you plan to use it, and what resolution you will need.

### SHOULD YOU CAPTURE AT A LOWER SIZE?

This depends, primarily upon what you are using the image for and how many images you need to capture. If you're only going to use the images for rough placement, or as tiny thumbnails of a web page, yet you need to capture lots of them, then using the smaller size makes sense. At the same time, it would be frustrating to take the photograph and then realize later that you really need the picture at the larger size (and have to go out and take the picture again).

## IMAGE QUALITY/COMPRESSION

The last parameter that should be mentioned is the image quality/compression setting. The number of pixels in the image remains the same, but the amount of information being saved is changing (for more information about JPEG compression, read Chapter 13). In a sense, resolution of the image is changing, because the amount of detail being saved in the picture is being lowered. An image at the highest compression setting may have detectably lower apparent resolution than the image saved at the lowest setting; though this will depend upon a number of factors, including the actual compression rates, and the subject of the photograph.

## RESAMPLING

We've mentioned resampling (changing the number of pixels that make up the image and redistributing the image information from the original pixels across the new size) in the context of both enlarging and reducing the size of the image.

It's important you understand how best to use resampling algorithms. Most graphics programs include resizing functions you can use to resize an image. The results that you achieve with resampling depend upon three things;

1. The bit depth of the image
2. The type of algorithm used
3. The percentage change

As with all editing, you'll get the best results if you use a high bit-depth image. Even if you have an 8-bit image, if you are going to resize it you should first convert the file to a 24-bit image. This allows the resampling algorithm to create new pixels that have colors based on averaging the surrounding pixel colors. If you resize with an 8-bit palette you greatly restrict the accuracy of the routine.

Even if you want to use an 8-bit palette for the final graphic, switching to 24-bit, resizing, then converting back to 8-bit produces better results than simply resizing in 8-bit. This is particularly true if you can remap to a different palette when you convert back to 8-bit. If you are reducing the pixel size by a large enough amount (50% or more), the resulting image may almost appear as though it was originally captured in 24 bits.

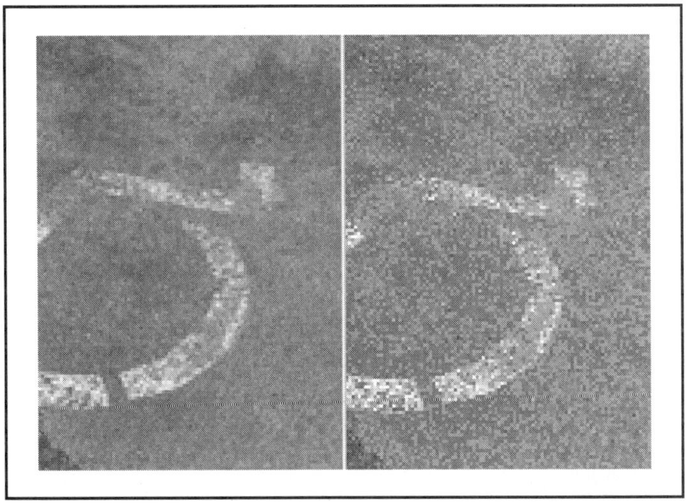

*Figure 6-2. Resizing in 24-bit vs 8 bit. The same 8-bit graphic was resized. The left image was resized in 24-bit color and has lost much of the 8-bit dithering patterns, while the right image was resized in 8-bit color and retains the dithering patterns.*

There are several algorithms used when resizing a graphic. Adobe PhotoShop provides a choice of three: nearest neighbor, bilinear, and bicubic, which are variations on either smart resampling or dumb resampling. With *smart resampling*, the software takes the value of pixels around the new pixel it is creating and creates an average value. With *dumb resampling*, the software just chooses one of the original pixels and uses that color to represent the new colors.

To see how these algorithms work, take a graphic made up of four pixels; two are white and two are black. If it's resized using a "smart" algorithm, the result will be a single pixel that's 50% gray. If you resize using a dumb algorithm the result is either a white or black pixel. Bilinear and bicubic are smart algorithms, while nearest neighbor is "dumb."

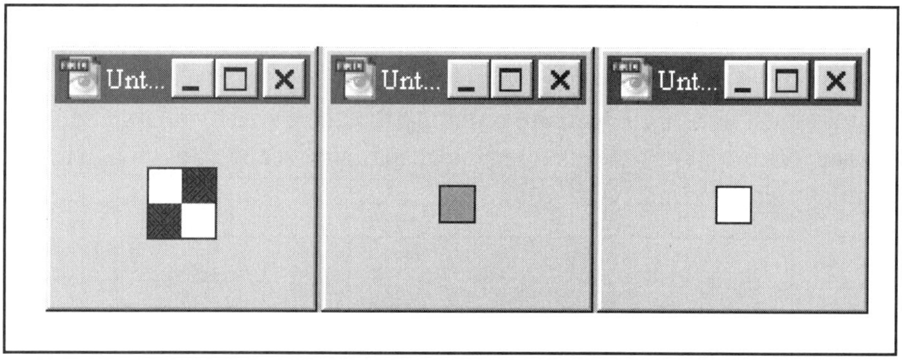


*Figure 6-3. Resizing using Bicubic and Nearest neighbor. The original graphic (left) was resized by 50% using Bicubic (center) and Nearest Neighbor interpolation (right).*

PhotoShop defaults to bicubic, which provides the best results, though it takes the longest time to resize a graphic. Bilinear produces results that are almost as good, and does it in a little less time—a consideration if you have really large files. Nearest neighbor essentially throws away pixels, or doubles existing ones. The examples below show the effect of resizing a graphic using different routines.

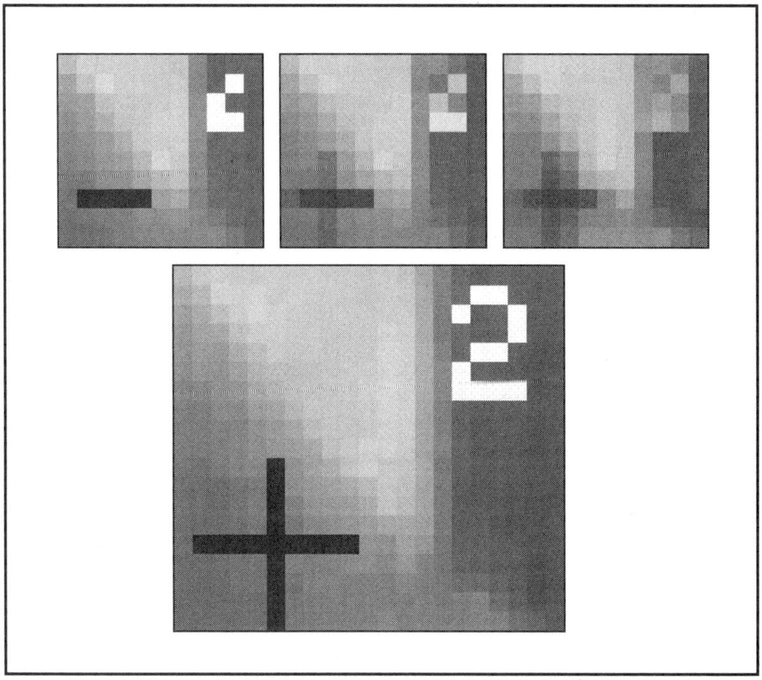


*Figure 6-4. Resizing to 50% using Bicubic, Bilinear and Nearest neighbor. The original graphic (bottom) was scaled to half size using Bicubic (top left), Bilinear (top center) and Nearest neighbor (top right).*

## HOW GOOD ARE THE RESULTS?

The smart resampling algorithms can perform pretty amazing jobs when working with most photographic images. It's usually images that contain fine detail or patterns of pixels that these algorithms will have problems. A pattern like the one used in *Figure 6-3* when resized by only a few percent becomes fuzzy; all the program can do is use off-color pixels to create an illusion of the original image. The results, when enlarged, don't look very good (see *Figure 6-5*). Fortunately, most photographs don't have such sharp, tiny details.

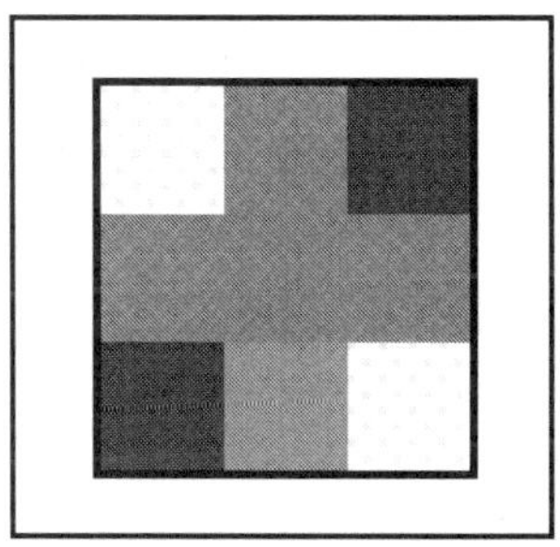

*Figure 6-5. Resizing a 2 by 2 graphic to 3 by 3 pixels.*

As you can see in the example of *Figure 6-6*, enlarging the image to 115% of its original file doesn't dramatically alter the photographic portion of the image, even though the computer graphics (the black cross and the white numeral) suffer from moderate distortion.

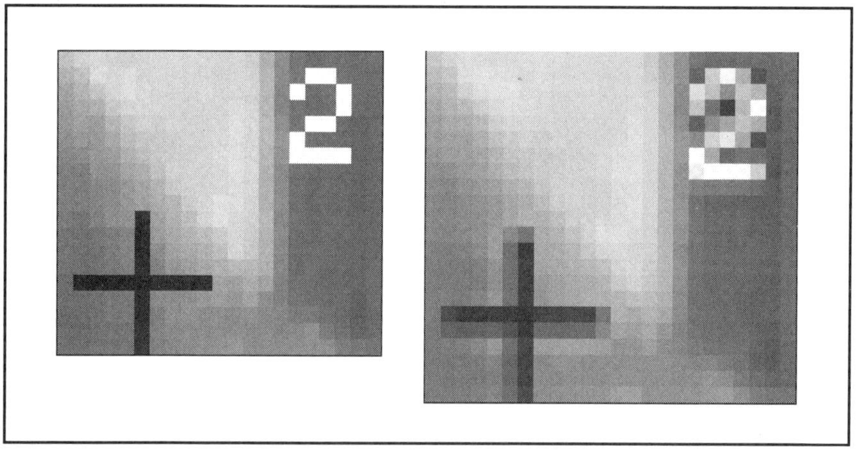


*Figure 6-6. Resizing to 115% using Bicubic.*

**Tip:** After resizing—whether increasing or decreasing the image size—the resulting image often looks a little soft and blurry. You can fix this problem by applying a sharpening filter to the image (be careful not to over-sharpen the image). Sharpening filters are covered in the next chapter.

**Conclusion:** Resizing algorithms perform well when scaling an image downwards as long as the image is photographic in nature. They also do a good job when enlarging an image by a small amount, though the results won't have as much detail as original images viewed at the same size.

## DISPLAYING AND PRINTING PIXELS

While you'll often hear the resolution of an image described in terms of being 1248 x 986 or 640 x 480, that number in many respects is only half the story. It's only when it is time to display the image onscreen or print it that you will know how many pixels you need in your image to produce acceptable results.

At that time you will know two things; the physical size you want the image to fill and the resolution of the display device. Monitors and printers both reproduce images as a grid of color points. The number of these points per inch is the *resolution* of the device.

For monitors it's usually described as *pixels per inch* while printers use the term *dots per inch.*

When displaying images on a monitor, images should ideally be displayed at a 1-to-1 correspondence. One pixel in the image should be displayed as 1 pixel on the computer monitor. When displayed in this way, you will be able to see as much detail as exists in the image. If the image is displayed larger, then the image will appear to lack detail, while if you display it smaller there is detail in the image that you won't be able to see.

Unfortunately, when it comes to printing an image, things become more complicated because printer technology is very different from video display technology. Color monitors mix red, green and blue light in different intensities to create a single pixel of almost any color. But printers use cyan, magenta, yellow and black ink to create their output. Most printers can only place one dot at a single location, and each dot can only be one color.

To create the illusion of many colors, printers use dithering (patterns of many different colored pixels) to create the illusion of many different colors. Effectively, the color information is not accurate at the dot level. So a group of 4 or 16 pixels might more accurately reflect the actual color in that area of the image. For this reason you don't have to send a one-pixel-to-one-dot file to your printer. If you have a 600 dpi printer, using a 150 or 300 dpi resolution image will look produce output that looks as good as using a 600 dpi file. For more information about how different printer technologies work, check Chapter 9.

## RESOLUTION OF YOUR MONITOR

If your display card and monitor support it, you can often switch the resolution displayed on your computer screen. A multi-sync monitor, which is common today, is needed to be able to switch resolutions. The operating system and software should handle switching between resolution relatively seamlessly. Unfortunately, resolution is not usually defined as pixels per inch, instead it's designated by the number of pixels displayed on the entire monitor. Which resolution you work at is primarily a matter of personal preference. I have a 17-inch monitor which can be set to a range of values from 640 x 480 up to 1600 x 1200. I can see the whole image of a 1600 x 1200 pixel file when it's set to 1600 x 1200. To see the whole image when the screen is set to 640 x 480 requires zooming out to 35%. While the size of the image on the screen is the same, more detail is actually visible when the screen is set to 1600 x 1200, because the monitor is displaying smaller pixels—and more of them. However, I find it difficult to read text and icons when the screen is set to that size, and since I zoom in most of the time to work on images, a happy medium is often somewhere in between these two values. This depends upon personal preference, and the size of the monitor you have connected to your computer.

Be aware also, that older computers are often limited in the amount of video RAM that is included. Video RAM is used to store the images being displayed by the monitor. When you set the monitor do display more pixels, more video RAM is needed to store that image. For this reason, some computers, when you switch to higher resolutions, will limit the number of colors they will display, perhaps limiting you to thousands of colors (16-bit color) or even 8-bit/256 colors. Adding more video RAM, if your computer allows it, can solve this limitation.

## CHANGING MONITOR RESOLUTION UNDER WINDOWS

Go to the Start Menu and choose *Settings > Control Panel*. When the Control Panel window opens, double-click the icon *Display*. When the Display control panel opens, click on the *Settings* tab. Resolution is adjusted by sliding the *Screen Area* slider left or right. Note that the Colors choice might change depending upon the amount of video RAM you have available. Once you have chosen a setting, click the *Apply* button. Depending upon the version of the operating system, the type of graphics card, and what you change, the change will either happen immediately, or a warning message will appear. The warning

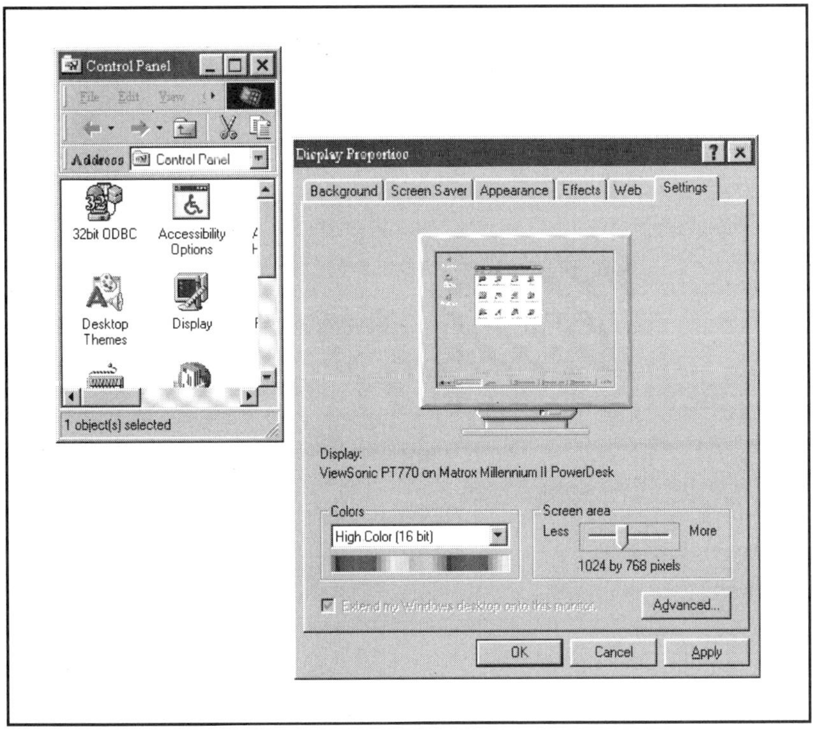


*Figure 6-7. Changing monitor resolution under Windows 95/98.*

message tells you that you are about to change the windows settings, and that the screen will go black. It should then reappear and you should click *okay* on the dialog box to save the settings. If the screen does not reappear, just wait. After 15 seconds, the computer will reset itself to the original settings.

Click *Okay* to exit.

## CHANGING MONITOR RESOLUTION UNDER MACINTOSH OS

Go to the Apple Menu and choose *Control Panels > Monitors & Sound*. (If you have a pre-System 8 version of the Mac OS it will just say *Monitors*). A dialog window appears. To change resolution or color bit depth, simply click on the new setting. The computer will update the display. Click the *Close* box to exit.

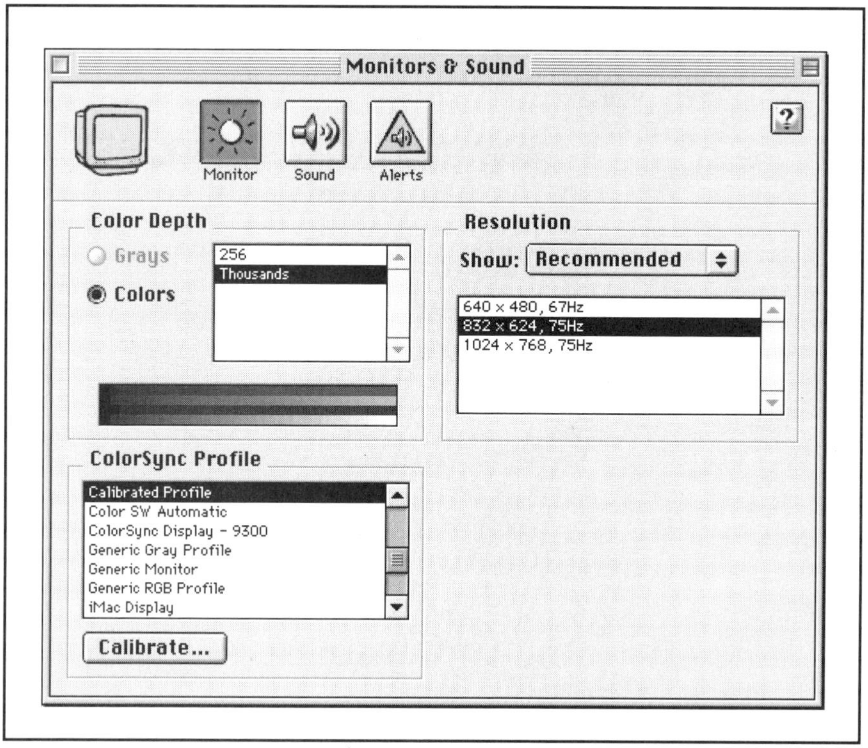


*Figure 6-8. Changing monitor resolution under Macintosh OS.*

# THE INS AND OUTS OF MESSING WITH RESOLUTION

Once you have the image in your computer, it's highly likely that you will have to change the size of that image at some point. If you're going to send the picture to someone or use it on a web page then you'll probably need to use an editing application to change the resolution of the image. Later in this chapter, working with web graphics is specifically covered in depth.

## CHANGING DPI DOESN'T DO ANYTHING

Most graphics file formats include a dpi (dots per inch) value for the image. Both JPEG and TIFF support this; but GIF files do not. The dpi is most commonly used when printing the image. It indicates how many pixels of the data file should be used to print an inch of the image on the printer. Though the printer prints dots, there is not always a one-to-one relationship between pixels and printer dots.

Note that changing the dpi of an image does not change the physical size, or the image data in the file.

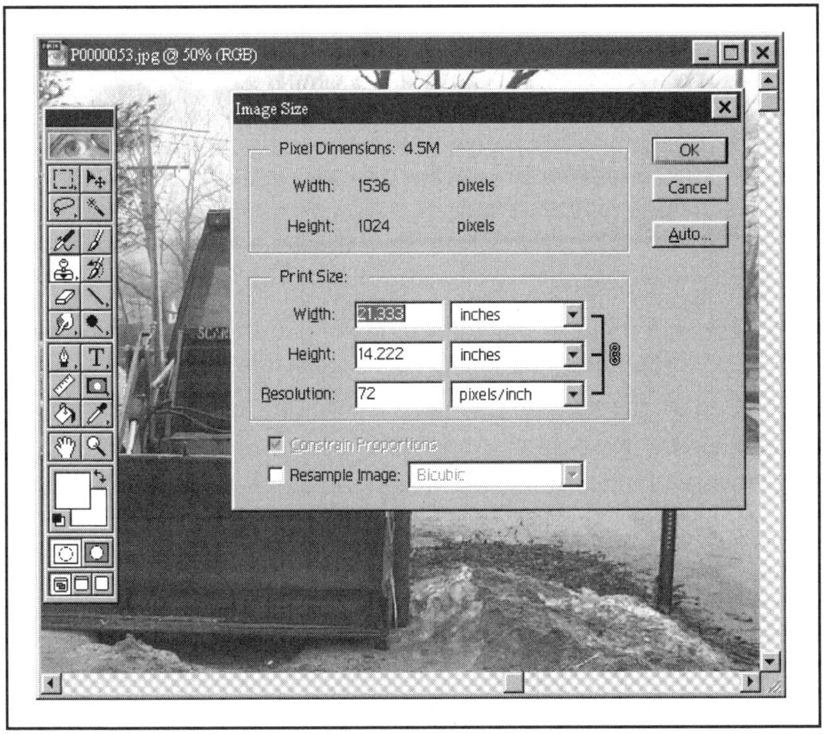


*Figure 6-9. Adobe PhotoShop image size dialog.*

*Figure 6-9* shows the image size dialog from Adobe PhotoShop editing an image captured using a Kodak DC260. We're going to focus in this section on PhotoShop because it's one of the most popular editing applications. However, nearly every graphics editing program has a function that will let you edit your images in this way. You should refer to your user manual to find the equivalent feature.

The four important pieces of information are:

**Size of the file:** This is displayed at the top of the dialog next to the text "Pixel Dimensions."

**Image resolution:** Displayed in the section *Pixel Dimensions* next to *Width and Height.*

**Resolution:** In the section *Print Size*, labeled *Resolution.*

**Print size:** In the section *Print Size* next to the fields *Width* and *Height.*

This dialog lets you alter both the dimensions and resolution of the image. Notice that the "Resample Image" checkbox is off. With Resample Image off, only the dpi of the image will be changed. With it on, the number of pixels in the image will be changed. We're going to be messing with just the dpi first, so make sure that the checkbox is off to start with; it's easier to get a grasp of what is going on.

Before we begin, remember that the number of pixels in the image is directly related to file size. If we increase or decrease the height and width in pixels of the image, then the file size will change. But changing the Print size—either the width/height or the "Resolution" (dpi)—may or may not change the number of pixels in the image, and therefore the image size. This is because the pixel width of an image, divided by the dpi, gives the print size width.

$$\frac{\text{Pixel width}}{\text{dpi}} = \text{print width}$$

If the image has 1536 pixels and the resolution (dpi) is 72 pixels per inch, then the print size width is 21.333. But if you change the resolution to 144, then the print size drops (see column 1 in *Figure 6-11*) in half to 10.6 inches, but the file size remains the same.

So changing the print resolution without changing the number of pixels in the file only changes the size of the printed file.

Using this dialog, it's also possible to change the print resolution, but leave the print size the same. To do that, we have to check the "Resample Image" option. If we then type a new number into the resolution field, the print width will remain the same, but the number of pixels that make up the image will change. If we choose to enter 144 dpi, the file size increases to 18M and the width and height of the file in pixels doubles (see column 4). Note that if we then drop the resolution back to 72, while leaving the Resample Image checkbox unchecked, then the print resolution changes to match the new number of pixels in the file and the dimension is now 42.667 inches.

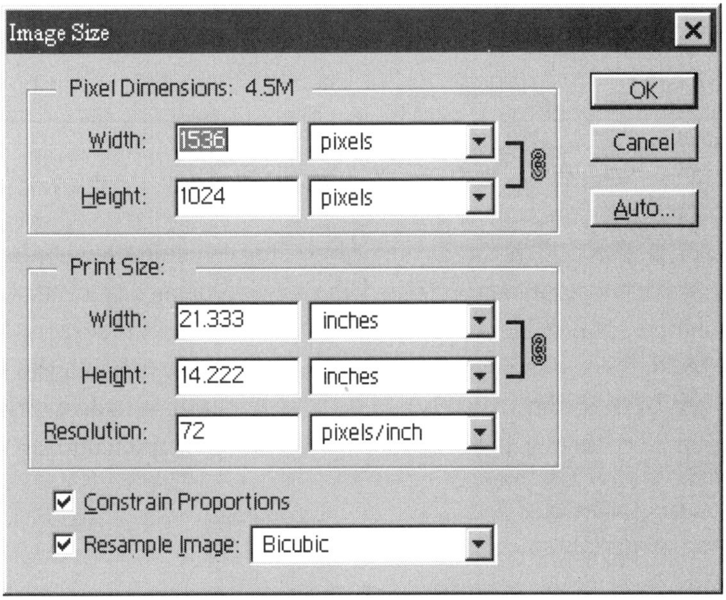


*Figure 6-10. Adobe PhotoShop Image Size dialog. With the Resample Image dialog box checked the pixel dimensions of the graphic can be changed.*

It's important to note that I don't recommend going up (increasing resolution) if all you want to do is print the file. Increasing the file size just makes the file larger and increases the time it takes to print, save and edit the file. In most cases, it's better just to let the printer driver handle scaling the image.

## WHAT'S THE @50% ALL ABOUT?

Astute readers, or those with a passing familiarity with PhotoShop, will have noticed that in the top of the graphic window in *Figure 6-9*, next to the file name, is the text "@50%" (if you didn't notice, take a quick look at the graphic now). What does this mean? Does this mean the graphic is going to be dis-

| File Size | 4.5M | 4.5M | 18M | 18M |
|---|---|---|---|---|
| Width | 1536 | 1536 | 3072 | 3072 |
| Height | 1024 | 1024 | 2048 | 2048 |
| Print Resolution | 72 | 144 | 72 | 144 |
| Print Width | 21.333 | 10.667 | 42.667 | 21.333 |
| Print Height | 14.222 | 7.111 | 28.444 | 14.222 |

*Figure 6-11. Image resolution chart.*

played or printed at 50% its actual size? No. This is just PhotoShop's way of telling the user the magnification at which the image is being displayed. At 100% means that one pixel in the image matches one pixel on the screen. At 50% each pixel on the screen represents four pixels in the image (the image is half the size both vertically and horizontally). This figure has nothing to do with how the image will print or the size it will be displayed in another applications.

## WHAT ABOUT SCREEN RESOLUTION?

One thing the image dialog doesn't show is screen resolution. PhotoShop assumes a 1-pixel-to-1-pixel match, and generally you should do the same. But what resolution is your screen? This gets complicated, because while most printers are sold with a fixed resolution (i.e., 600 dpi), screens are rarely measured in that way. Usually they are sold as total inch size and an aperture screen measurement in fractions of a millimeter (.28mm is common). Even this becomes meaningless with multi-sync monitors that let you choose different display resolutions. You can probably work out the dpi for your screen by using a ruler and dividing the width by the display setting you are using. A 12-inch display (that's the horizontal measurement of the image, not the quoted size which is measured diagonally) at 800 pixels = 66 dpi.

But this number isn't very useful, because the measurement changes from monitor to monitor, machine to machine.

## WHEN IS RESOLUTION PASSED FROM APPLICATION TO APPLICATION?

As you can see, when PhotoShop opens a document it displays the resolution and print size of the image. However, that information is not always used by other applications. Some file formats don't let a program store a resolution, so when you save the graphic and open it in another application then you're back to square one (though obviously the actual number of pixels in the file does not change).

There are some strange limitations. For example, when Macromedia Director, a multimedia authoring tool, imports a Macintosh PICT file it does take the resolution value and scale the graphic accordingly. The program assumes that your monitor is set to 72 dpi (even though it probably isn't), so if you import a graphic that's 300 dpi the graphic will be shrunk significantly. The worst part of this "feature" is that the scaling algorithm Director uses is not very good.

Most page layout programs, such as Quark Xpress and Adobe PageMaker, observe the resolution when importing a file, and will scale the image accordingly in the layout. You can still change this setting once the graphic is imported. Interestingly, while Adobe PageMaker will scale TIFF files, it won't do it for PICT files.

## WHAT ABOUT FILE SIZE?

Many first-time users don't understand why the size of files varies so much. This primarily depends upon compression and the subject matter of the image. If the image is saved using an uncompressed format, then the size of the file should be almost the same as the resolution of the file. So a 640 x 480 image file should be about 900K. Why?

640 X 480 = 307200 (the number of pixels in the image)

If it's an RGB image, each pixel requires three bytes (one byte each for red, green and blue).

307200 x 3 = 921600 bytes

But if compression is used (whether lossy or lossless) then the size of the file will vary depending upon the subject matter and compression used. An image containing a lot of one color will compress much better than an image with a lot of detail. You can read more about compression in Chapter 13.

One thing to remember: no matter what the size of the file on the disk, the computer has to uncompress the file to display it on screen. Let's say you have a file that's 3000 x 2000 pixels. That's equal to 18MB (3000 x 2000 x 3 = 18MB) of data. If you use heavy JPEG compression you might be able to get that file down to less than a megabyte in size. But when an application opens that file it has to unpack the data and reconstruct the 18MB of information that it will then display to you. So be aware that even though JPEG and other compression formats save disk space and Internet transmission times, opening and editing large files remains a problem because of the processing that needs to be done. If you don't have enough free memory, some editing applications won't even let you open a very large file.

Some file formats (notably FlashPix) try to solve the problem of displaying large files efficiently by storing smaller versions of the image within the file. An application, if it only needs a 640 x 480 version of the file, can read in just that data, and not have to read the entire 18MB.

## WHAT SIZE SHOULD MY GRAPHIC BE?

The next section covers working with images for screen display. For those wanting to print files, more discussion of how to work with files is covered in Chapter 9.

Images are often only displayed on screen; whether for a multimedia presentation or a web page. Unfortunately, as with color, knowing at what resolution a given monitor will display an image is virtually impossible.

However, 72 ppi (pixels per inch) is often used by default. In reality, a screen could be set to almost anything, though most commonly it will be between 72 and 86.

Some applications, particularly on the Macintosh, assume a screen is 72 ppi because that was the original resolution of the Macintosh screen. Having 72 pixels in an inch matched the number of points in an inch, which is a unit of measurement used in printing. Points are used to measure the size of type: a 72-point typeface is an inch high.

## GRAPHICS IN WEB PAGES

Because no number is correct, and you have to start somewhere, when I create graphics for screen work I still use the 72 dpi number. When creating web graphics, you should always resize the graphic to the size you want and then save the new copy using JPEG compression or in GIF format.

### DOESN'T HTML LET YOU RESIZE A GRAPHIC ANYWAY?

Yes, it does. By including the width and height parameters in the image tag you can scale the graphic to any size that you want. But there are two reasons why you should avoid this. Firstly, if you are sending a file at a higher resolution than you are displaying, then you are wasting download time. Secondly, the scaling in the web browser may not look as good as can be done using a graphics editor. Web browsers tend to be more concerned with speed, so they don't necessarily perform as good a rescale as can be done if you use your editing application. In fact, even editing applications, when you zoom in and out, don't produce as good results when redrawing the graphic on screen at different sizes as they do when resampling the image. The example in *Figure 6-12* shows the same graphic viewed zoomed out at 50% and scaled (resampled) to 50%. The resampled image is clearly better.

Chapter 16 provides an introduction to graphics and the Web.

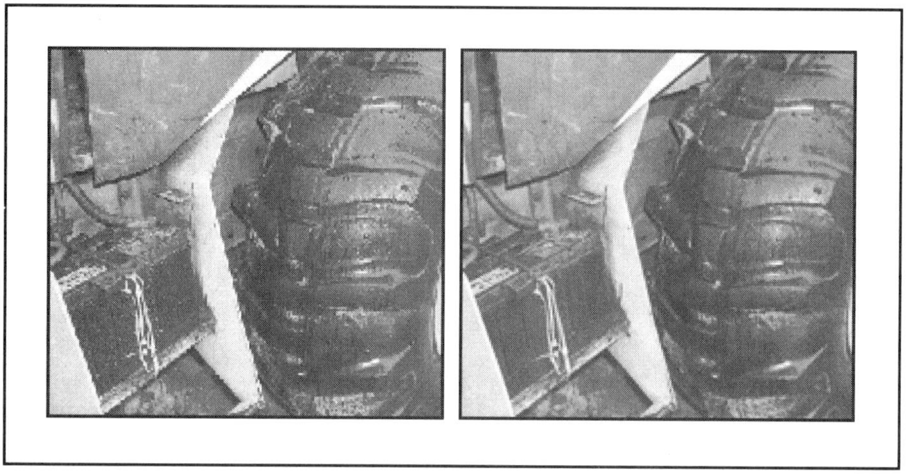

*Figure 6-12. Fast resizing (left) and bicubic resampling (right). Notice that the fast resizing produces jagged edges.*

## MOVING ON

Now that we've mastered resolution, we can move on to image editing.

CHAPTER 7

# THE DIGITAL DARKROOM: IMAGE EDITING

In traditional photography, taking the photograph is only half the process. While most casual users take their pictures and leave the developing to the photo lab, the serious photographer knows that in the darkroom contrast and brightness can bc adjusted, negatives can be sandwiched together, and other tricks can be used to create better photographs. Often times, having access to a darkroom ensures that the photographs are reproduced the way they were intended to be, and not butchered by the automatic processing of the photo lab machine.

Digital cameras and image editing software bring the darkroom to you, and in the process provide you with control and options yesterday's photographer could only dream about. You can fix the color balance and adjust the contrast. Or sharpen the image and remove unwanted objects or people from your pictures. Image editing can make a good photograph great. Unfortunately, it can rarely take a bad photograph and make it good. If an image is severely underexposed or badly out of focus there's often very little you can do except apply an effect filter and claim it was what you intended all along! This chapter is intended as an introduction to the topic of image editing. Hopefully it will give you an idea of what you can do (and how) and get you started on the road to the digital darkroom.

The first section of this chapter introduces editing applications, and describes how graphics file formats store their information. Color bit-depths, specifically 8-bit and 24-bit color, are introduced, as well as CMYK file formats and Alpha channels. Suggestions for file formats to use while editing, as well as the resolution you should be editing at are provided. The remainder of the chapter covers actual editing; removing red-eye, removing objects from an image, adjusting color balance, brightness and contrast adjustment and sharpening images.

There are several image editing applications available which offer all kinds of features, from simple editing to special effects to power tools. Which one is right for you? That depends upon a number of things. What do you need to do? How often will you be using the tool? What do you already have, and what did the camera come with? Frequently, cameras and scanners come with applications which can do most of the things that will be covered in this chapter. Before you rush out and buy "the tool," check to make sure you don't already have something that will do the job.

If you want to buy something, there's lots to choose from. While a power tool such as Adobe PhotoShop (Macintosh/Windows) is perhaps the easiest recommendation to make simply because it can be used to do just about anything, it's an expensive choice. Much of its power is in prepress, which is something many people don't need—or want—to get involved with. It doesn't make much sense for someone who might use the program only a couple of times a month to buy such a tool. Programs like Adobe Photo Deluxe (Mac/Win), MetaCreations Kai Photo Soap 2 (Mac/Win) and Microsoft PhotoDraw (Win), offer many of the functions found in PhotoShop, in a friendlier package at a much lower price.

Adobe PhotoShop (Mac/Win)

Adobe Photo Deluxe (Mac/Win)

Corel Photo-Paint 8 (Mac/Win)

MetaCreations Kai Photo Soap 2 (Mac/Win)

Microsoft PhotoDraw (Win)

ULead PhotoImpact (Win)

Jasc Software Paint Shop Pro (Win)

For those who aren't so interested in image retouching there are also some fun tools for manipulating images in weird and wonderful ways. MetaCreations Kai Super Goo is a great example of such a tool; it lets you wildly distort your images.

# IMAGE EDITING: WHAT'S AN IMAGE?

Digital editing software provides tools that let you manipulate the digital information that represent your images. It's important to understand some basics about the files that you are working with so that you can understand how these tools work.

Computer graphics files can be separated into two primary kinds of file; bitmapped and vector format. In bitmap graphics, a picture is made up of a grid of pixels (or dots) and the color of each pixel is stored in the data file. Each dot has a single color value. As long as the dots that represent the scene are small enough, this makes for a very accurate, but large file. Bitmap graphics are very good for storing photographic images, but they are not very efficient and it's difficult to go back and make even minor changes to bitmap data. Another limitation of bitmap files is that they are limited by the resolution (number of

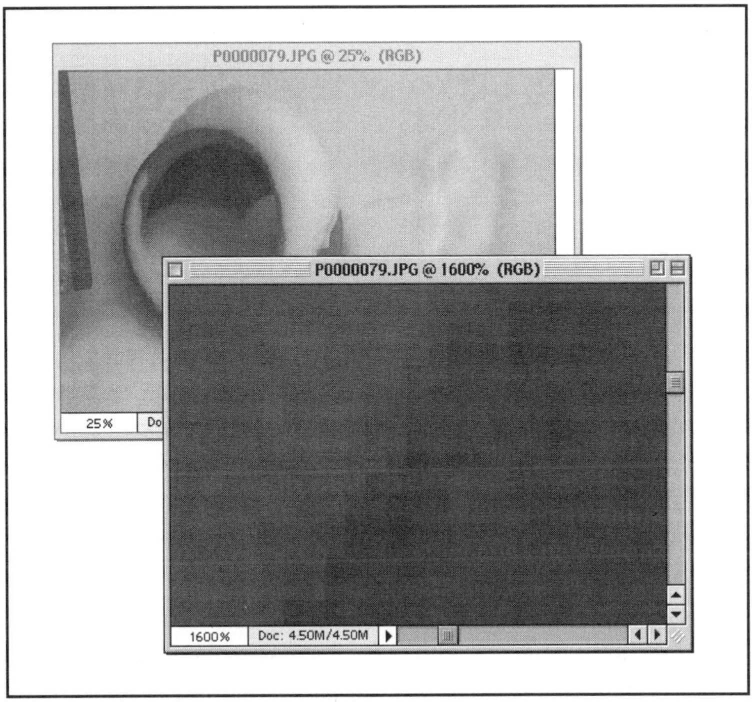


*Figure 7-1. Bitmap graphics. PhotoShop lets you view you image at different enlargements. As you zoom in you can see the dots that make up the file.*

pixels) that they were created with. If the file you've been working on is 640 x 480, you can't press a button and enlarge it to 1600 x 1200 and have the image look sharp.

Vector graphic formats use mathematical equations to store line segments. This is a very efficient way to store graphics created from scratch using line drawing tools (circles, squares, text, etc.) and it's very easy to go back and edit these graphics. But this format is not very efficient at storing photographic images because photographs usually contain very small details and subtle changes in color which are very hard to reproduce other than at the pixel level. For this reason, both graphic formats continue to be widely used for different purposes.

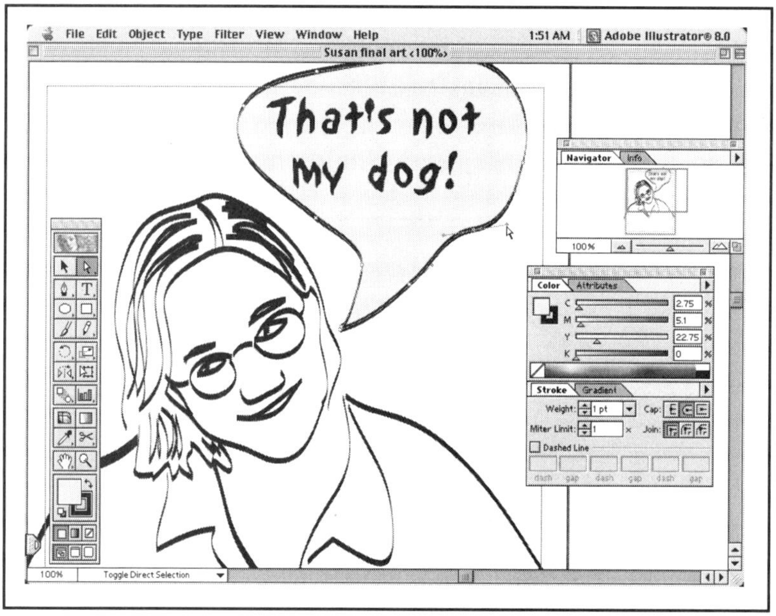


*Figure 7-2. Vector graphic. Adobe Illustrator (pictured) and Macromedia FreeHand are two programs that let you manipulate vector graphics.*

Since this book is about digital cameras, most of the rest of the discussion will center on bitmap graphics.

While some programs can work with only bitmap graphics, and others can only work with vector-graphics, a few programs (for example Macromedia Fireworks) contain elements of both graphic formats and can work in vector or bitmapped modes. One reason such tools are rare is that most file formats only store one graphic type. These file formats have appeared over the course of several years, and are often developed for a specific purpose; GIF was developed to make small 8-bit images, TIFF was developed for scanned images, and JPEG was developed to save compressed photographs. More detailed information about specific file formats can be found in Chapter 13.

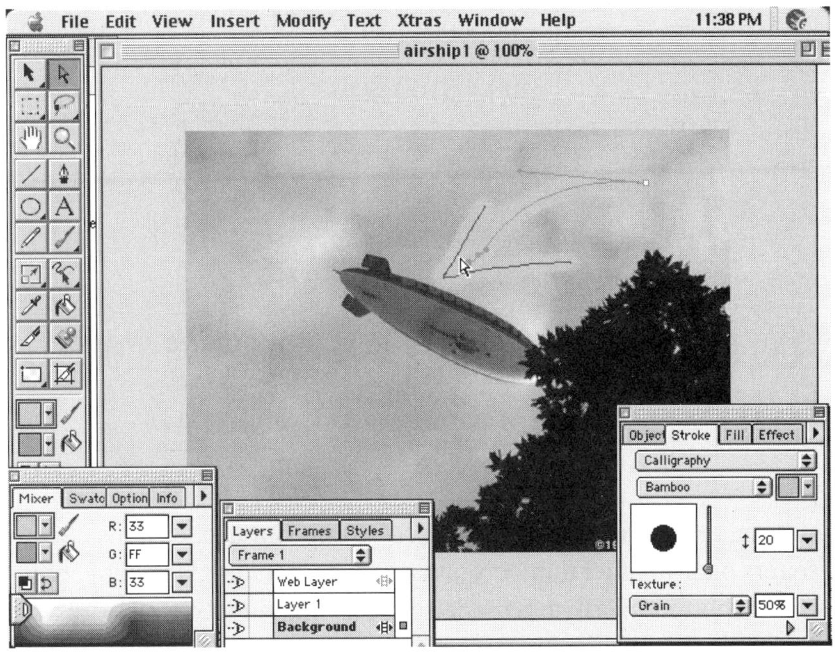


*Figure 7-3. Macromedia Fireworks. Macromedia Fireworks supports both bitmapped images and vector graphics in the same file. The arrow has been drawn using a vector-based paintbrush. Fireworks file format, PNG, supports both types of information. If you save the graphic as a JPEG or GIF file the program converts the vector graphics into bitmaps.*

Most computers store photographic images as 24-bit images. For those not familiar with computers, computer memory is made up of bits and bytes. A bit is the smallest piece of memory and can store two values; 0 and 1. To store text several bits must be used; a byte consists of eight bits. Because the 8 bits in a byte can each be 0 or 1, a byte—using base-2 math—can store a value between 0 and 255 (256 values).

By using bits in groups of 8, each 8 bit chunk of memory can represent one of 256 values:

| | |
|---|---|
| 00000000 | = 0 |
| 00000001 | = 1 |
| 00000010 | = 2 |
| 00000011 | = 3 |
| 00000100 | = 4 |
| etc. | |

*Figure 7-4. Bits and bytes.*

## BITS AND BYTES; HOW A COMPUTER USES MEMORY TO STORE CHARACTERS.

To represent the roman alphabet you need 26 uppercase letters plus 26 lowercase letters plus 40 other numbers and symbols, equaling a total of about 96 values. For this reason a byte is used to store an individual character in a piece of text. (Note: Since only about 96 characters are really needed for English text, it's possible to make a text file smaller by saving it in 7 bits. This is sometimes done by text compression routines.)

So a byte doesn't store characters, it just stores a value that represents the character, and the computer program(s) know that a value of 36 = "$" (the dollar sign).

You're probably wondering what this has to do with computer graphics. Since bytes are used to represent text, computer processors have been optimized over the years to deal with memory in byte-sized chunks. Therefore, when it came time to store graphic information, it made sense to use the 8 bits/1 byte scheme to store color information.

At first, color was often handled using palettes. A palette is a table that contains a set of colors (i.e., it actually contains the RGB values for each color stored in the table). The value stored for a pixel is actually a pointer to the color table. It points to the entry in the table that contains the color value for the pixel. Since a byte contains 8 bits, the color table contains 256 colors. And so 8-bit color was born. In fact, palette color can also be used with less than 8 bits. Doing so not only reduces the number of colors that the graphic can contain, it also reduces the size of the file. The GIF file format provides options for saving a graphic with different palette/bit sizes.

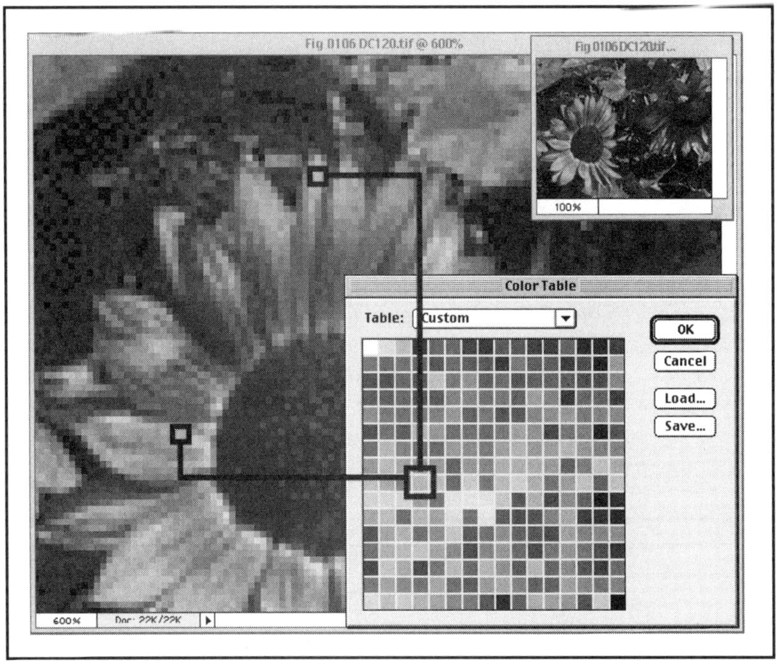


*Figure 7-5. Palette Graphic. The top right graphic has been mapped to an 8-bit palette (bottom). The file is stored as a grid of bytes, each byte holding a number which points to one of the colors in the palette. The palette is usually also stored in the file.*

There are some standard color palettes available. Both the Mac and Windows operating system have a standard 8-bit palette (which are similar, yet different) and there is also a "browser safe" palette (see Chapter 16). Most graphic programs also provide the option of creating an adaptive palette—a palette that contains the closest set of colors that will best represent the colors in the original image.

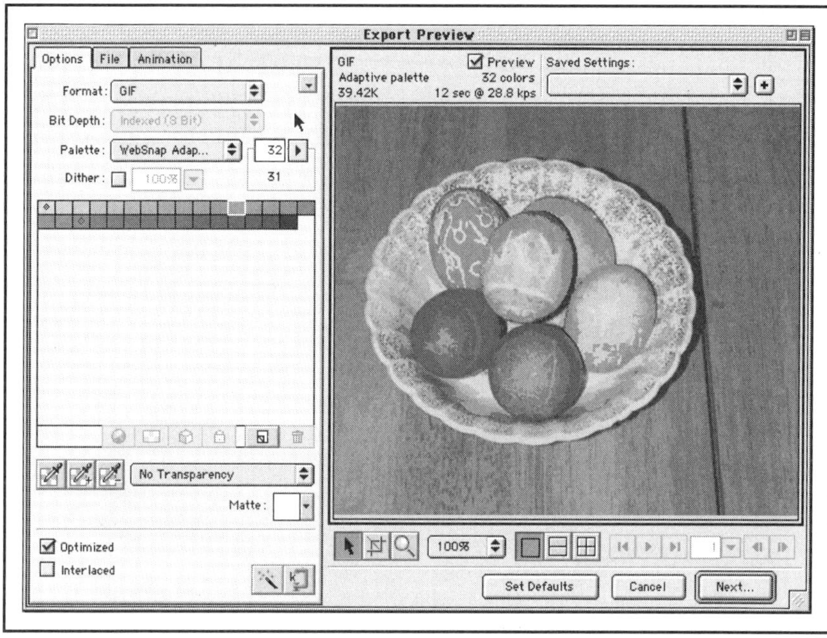


*Figure 7-6. Different sized palettes. A graphic that has been converted to a 32-color palette using Macromedia's Fireworks program. Notice that the image suffers from posterizing (bands of pixels that are all the same color) because dithering was not used. If dithering was used the colored banding would be reduced.*

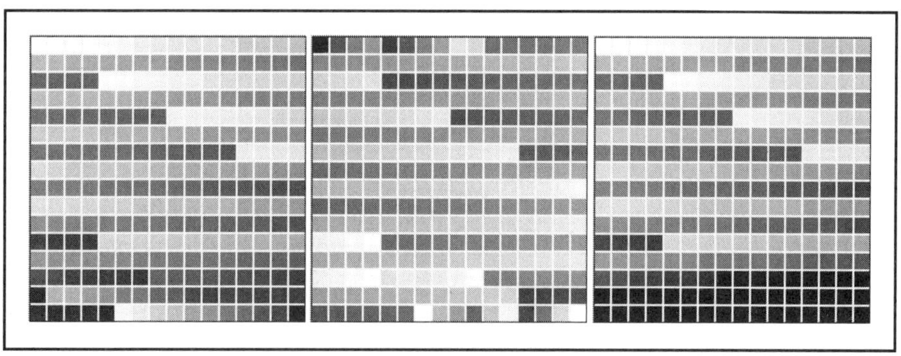

*Figure 7-7. Macintosh, Windows and Web palettes. These are the most common 8-bit palettes (along with grayscale). The Web palette shares many colors with the Mac System palette. It contains fewer colors because some colors have been reserved for the system colors used by the Macintosh and Windows systems. If these system colors weren't reserved, the toolbars and palettes of the web browser would display incorrectly.*

## WHAT'S AN ADAPTIVE PALETTE?

An adaptive palette is a color palette that contains colors that closely match the colors found in the photographic image you have created. You first start with a 24-bit image that you want to convert to an 8-bit palette. Programs that will create an adaptive palette first create a frequency chart for the colors in the image, and then create a palette based on those most frequently used colors. This isn't done simply by using all the most common colors that are in the image. If a palette were created that way then colors that are small but important parts of the image (for example, the red stop sign) may not be saved and the image will be drastically different from the original photograph. So the software must analyze the variation between the colors as well to determine which colors it can discard.

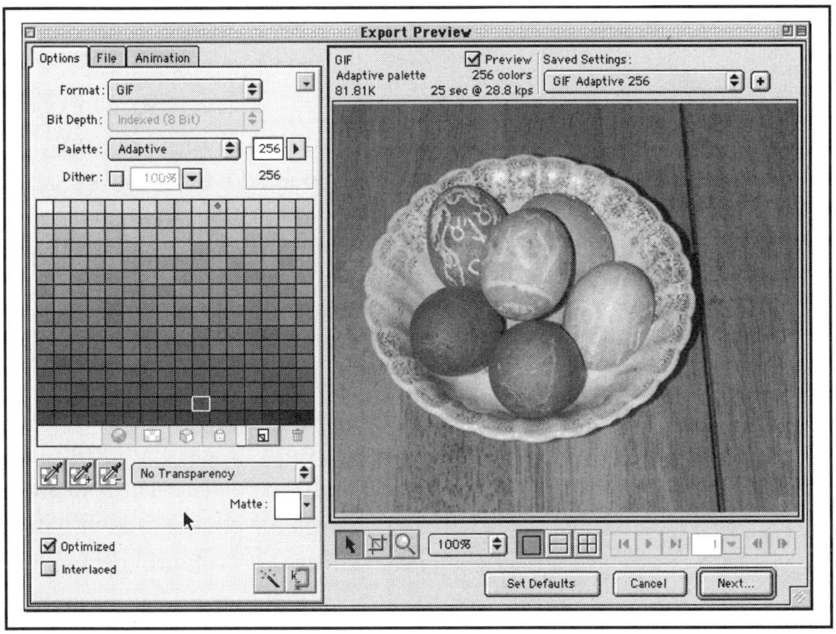


*Figure 7-8. An adaptive palette created using Macromedia Fireworks. The palette contains only colors found in the image*

Using an adaptive palette can create an image that is very close to the original image at a third the size of the RGB image; however there are some drawbacks.

When viewing an adaptive palette image on an 8-bit display, the palette for the video display must be changed to view the image correctly. This can result in strangely colored windows, tool bars and other applications windows that are visible while the graphic is open. Similarly, if you have two 8-bit graphics and want to display them at the same time in an 8-bit display, then either some general-purpose palette must be used, or one image will look correct and the other will be badly distorted. For this reason, web browsers usually use a single palette and display the graphics in that palette.

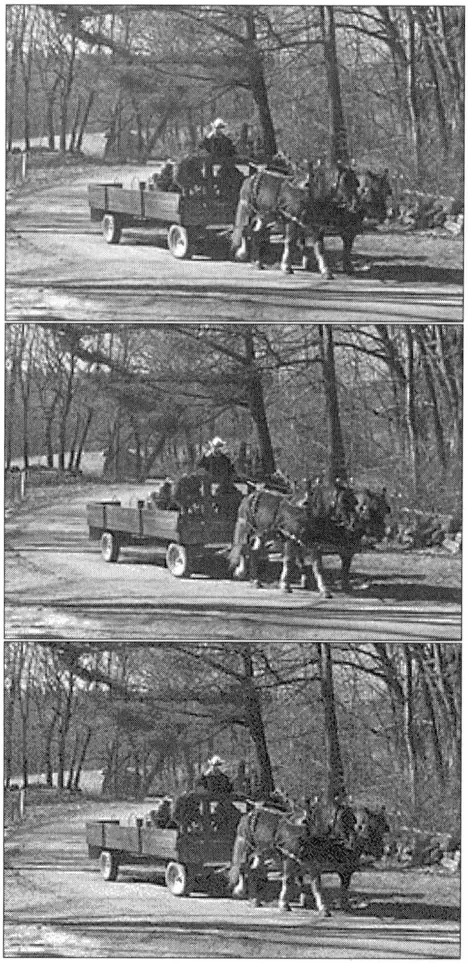

*Figure 7-9. Top to bottom: 50% enlargement of a portion of an image saved in 24-bit, 8-bit adaptive, and 8-bit Web palette.*

With an 8-bit image, a 640 x 480 graphic is 307,200 bytes. As a 24-bit graphic it would be 921,600 bytes because each pixel requires 3 bytes to store the RGB color, so storing a graphic as an 8-bit graphic makes the file a third of the size. Also, a lot of early color computers only displayed 8-bit color because the video RAM was expensive. The cost of equipment has fallen so much that the vast majority of new computers support millions of colors. The most likely reason that you will want to use 8-bit graphics is for special applications such as on a website, but even in that case you should probably use JPEG for photographs if you can, and let the browser use dithering if the user isn't looking at the site in true (24-bit) or high (16-bit) color.

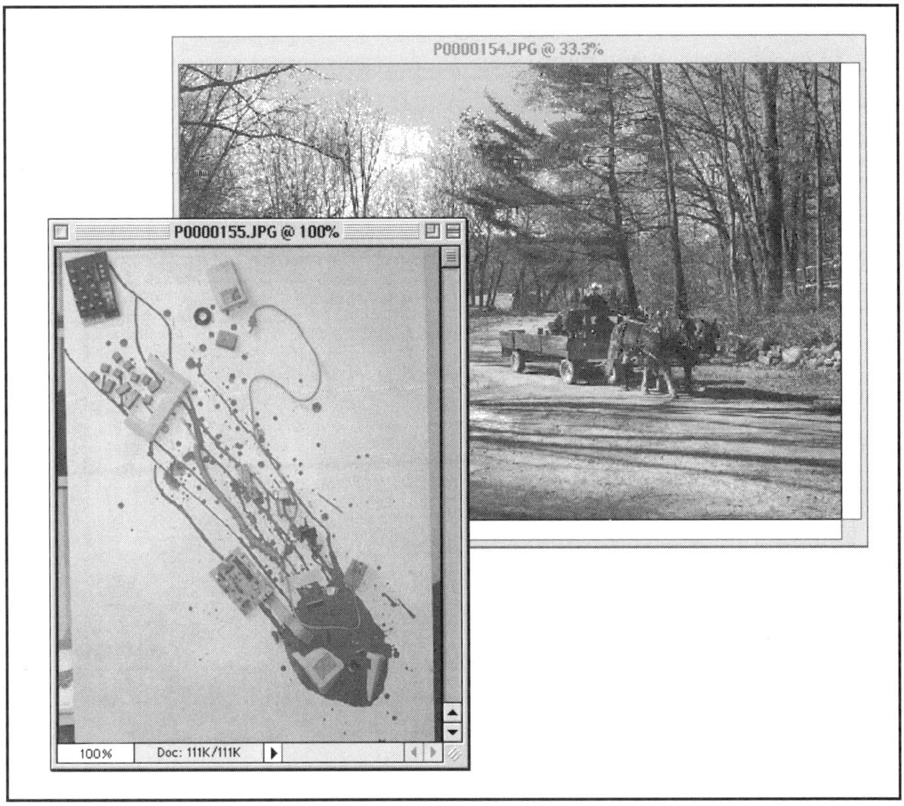


*Figure 7-10. Distorted colors in toolbar and second image when viewed in 8-bit color. Notice the posterized effect in the sky of the second image, which is not visible when the images palette is being used.*

Note: even if the display is only set to 8-bit color, you can still display and edit a 24-bit image; it will just be dithered to display on the 8-bit display.

## 24-BIT COLOR

As machine power increased, and memory prices went down, it became possible to work with 24-bit files. With 24-bit color, each pixel is represented by three bytes, each byte storing the value for red, green and blue of the color of that pixel. The red, green and blue colors are often referred to as *channels*. In theory, a 24-bit RGB system can represent 16 million different color values.

This results in more subtle changes in color than can be perceived by the human eye (in most cases; some people say there are colors where it is possible to detect limitations in a 24-bit storage). Even if your final graphic is to be in 8-bit, if you are working with photographs you should perform all your editing with a 24-bit image, then convert to an 8-bit palette at the end. This will result in the best quality image.

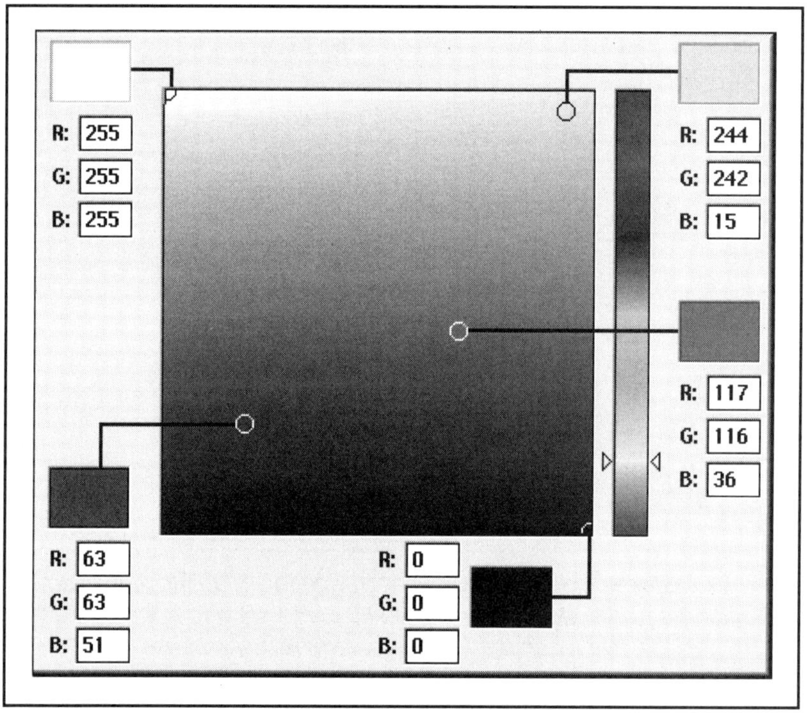


*Figure 7-11. 24-bit color palette. The color swatch represents shades of yellow. The RGB values at different points are shown. White is represented by the value 255, 255, 255 while black is represented by 0,0,0.*

Most computer graphics are stored in either 24-bit or 8-bit graphics, but there are some other methods of storing color.

For a 16-bit image, only two bytes are used, and only 5 bits are used for each color channel. This results in only 32 steps for each color. At that level, particularly with color gradients, it is possible to see banding in the image, but for many color photographs this effect will be almost impossible to detect.

Some graphics programs and digitizing hardware support color at larger bit depths than 8 bits per channel. For digitizing applications, capturing at 10 bits per channel then mapping to 8 bits has some advantages (see Chapter 8). Adobe PhotoShop also supports 16 bits per channel. The theory behind using more than 8 bits per channel is that while 8 bits provides very high quality images, if you start applying filters or contrast adjustments, the number of levels in an image can be reduced dramatically. This can eventually result in banding (sudden changes in colors). By using more than 8 bits per channel while editing, this banding can be eliminated or at least reduced.

## CMYK AND OTHER COLOR FORMATS

Many programs support storing graphics in CMYK format as well as formats such as LAB color. Unless you have a specific need for using those color formats they should be avoided. CMYK is used in production printing environments and there are differences in the way color is represented between RGB and CMYK. Since most low-priced color printers use RGB drivers, switching to CMYK will actually degrade your image. In some applications not all effects work in non-RGB color spaces, and in the case of CMYK the files are a third larger than RGB because there are four channels instead of three.

Also, apart from image editing applications, few programs support these color formats.

## ALPHA CHANNELS

Alpha channels provide a method of masking an image; indicating what portions of an image are visible, and what parts are transparent. While alpha channels are beyond the scope of this book, they can be very useful when merging an image with another, or when merging images with video. Alpha channels are normally stored as an extra 8-bit channel, so the size of the graphic file is increased by a third.

The alpha channel is usually displayed as a gray scale graphic, but there's really no standard way of displaying the alpha channel; both gray scale and red have been used to display alpha channel information. The alpha channel indicates transparency: if the value in the channel is white, then the alpha channel is totally transparent; if it's gray it is partially transparent; and black indicates no transparency. This transparency is then used when either painting onto the graphic or when merging the graphic with another file. In areas where the alpha channel is transparent or partially transparent the image will be changed, but where there is no transparency no change will occur.

## WHAT FILE FORMAT SHOULD I USE WHEN I'M EDITING?

While most digital cameras save graphics in a lossy compressed format such as JPEG, you should not save files in that format while you are editing them. To understand the difference between lossy and non-lossy compression, read Chapter 13. Instead, use the native format of the editing program. For example, PhotoShop uses the .psd file format, while MetaCreations Painter uses .RIFF. Alternatively, use a non-lossy file format such as BMP (Windows), PICT (Macintosh) or TIFF (either platform). This should be done to maintain the quality of the image while editing. Even if your original image has already been compressed, once you start adjusting colors and applying effects you are changing the information in the file. This information should be saved while you are editing because each recompression of the file will create color change. Constantly recompressing, even though you don't change the image much, results in a loss of image quality. Also, JPEG does throw away color information, which is something that the effect filters work with.

Save using a non-lossy file format while working, then save a copy in JPEG format when you have finished editing the file. I usually save the native file in case I want to go back and edit it some more. This is particularly useful for programs like PhotoShop which let you create layers; storing different changes in different layers can make the editing process easier. Layer information cannot be saved in JPEG, GIF or TIFF files.

## WHAT RESOLUTION SHOULD I USE?

Resolution is such a confusing topic that it has its own chapter. Since your digital camera will provide an image in a specific resolution, then you are limited by that resolution to start with. However, there are some other considerations which may affect what you do.

If you have an image that's 1024 x 968, and you are creating a graphic that's going to be 240 x 180, should you resize the image before you do any other retouching? Unfortunately, there's no hard and fast rules and several factors are involved:

**Editing on larger work and then shrinking it will provide the best results.** If you need to touch-up an image, try to work on as large an image as possible. A touch-up job that looks only passable at the original resolution of the file will often look a lot better once the image is scaled down. This is also true for effects such as rotating parts of an image.

**Hardware performance.** If your computer is only a year or so old, and you're using a medium-resolution digital camera, then performance should be acceptable while working on your files. But if you have a high-end digital camera that creates 18-megabyte files, or an older computer, it doesn't make much sense to work with such a file when all you want is a 240 x 180 image for a website. In general I recommend working on a file that's 2 to 4 times the final size at most.

**Shrinking images way down alters the appearance of the image.** Shrinking an image from 640 x 480 down to 240 x 180 shouldn't affect how your image looks, but if you're shrinking it down to 32 x 32 (a thumbnail) then you might find you are wasting your time doing editing on a 1600 x 1200 original. Working on a 240 x 480 file will go a lot faster. At the same time, editing thumbnails is very difficult; if you're creating a thumbnail you should work a few times larger than the final size and let the software do the hard work to create the thumbnail for you. Editing 32 x 32 thumbnails pixel by pixel to create realistic images is very difficult.

**Some effects should be applied at the end.** Sharpening an image is a function that should be left until you scale the image at the end, unless you are creating a collage and need to match one image to another, or your image is seriously out of focus. Applying a slight sharpen to the original image makes little sense as you will probably need to do it again once you rescale the image.

What if you only have the image at the size you need it to be? Will it be easier to edit if you blow it up to 200% its size, then scale it down when you're finished? While a case can be made for doing this, I've found that it often doesn't work. The problem is that you've lowered the effective resolution of the image, but all your painting tools will be working at the current resolution and creating much subtler effects than are in the rest of the image. When you resize downward the additions will look different because they will have a sharper appearance; the scaling algorithm, even though it throws information away, does use algorithms to calculate the new pixels based on surrounding pixels—it doesn't just throw away every second pixel. The result is that the area you have worked on will look as though you have applied a selective sharpening filter to that area of the image. This is particularly true if the quality of your image is poor to begin with. Of course, results differ, depending on what you are doing to the image, so feel free to experiment first. For a description of resizing algorithms, and how to use them, read Chapter 6.

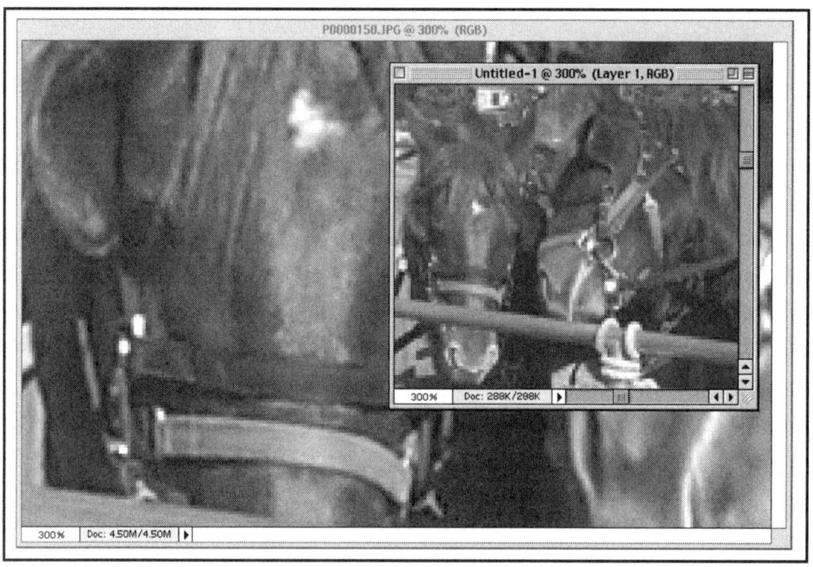


*Figure 7-12. An image before and after resizing to 33% of the original size. Both are being viewed at 300% so that the pixels are visible.*

## RED-EYE REDUCTION

Red-eye is one of the most common problems with pictures taken by the casual user. If you're lucky, your software might include an automatic Red Eye Reduction feature. Unfortunately, these features usually work by dropping the brightness of the color in the eye (since most color is gone from the eye, it's impossible to know what color eye the person is supposed to have). Your subjects will end up with brown eyes—but it attracts less attention than bright red ones.

If you don't have such a feature, then use a selection tool to select the region of the eye that needs to be altered. Then use the color adjustment and brightness/contrast tools to adjust the color of the eye to a more pleasing effect. The real secret is to maintain the highlights and shading in the eye while adjusting the color. What you don't want to do is simply dab a different color on top of the eye using a paint brush. This will remove the subtle light effects, and will make it appear as though the person is blind or has some kind of eye problem!

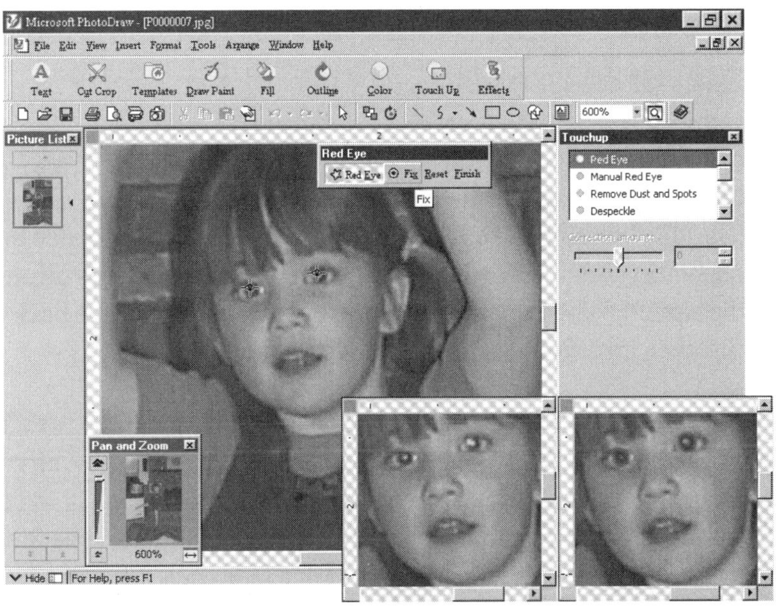


*Figure 7-13. Red-Eye Reduction with Microsoft PhotoDraw. PhotoDraw lets you select the eyes by clicking on their centers, then using a slider to adjust the amount of adjustment. Alternatively, you can use a manual mode and draw the region that you want to adjust, then move the slider to adjust the amount.*

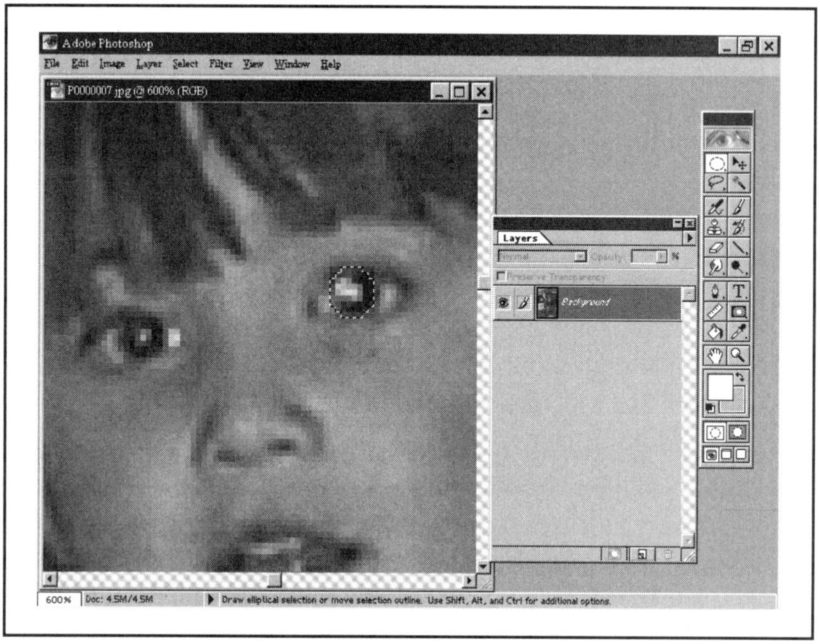


*Figure 7-14. Manual red-eye reduction.*

## WHAT IS RED-EYE AND HOW TO AVOID IT?

Rather than remove red-eye from the image, how can you avoid it altogether?

Red-eye is quite literally the reflection of the red blood vessels inside the eye. This occurs most commonly when the flash of the camera is very close to the lens of the camera. The light travels to the eye, bouncing off the back of the eyeball and then traveling straight back to the camera.

Part of the problem is flash is usually used at night or in dark places, exactly the kind of place that the retina of the eye opens up to allow the most amount of light into (and out of) the eye. Many cameras have a red-eye reduction feature, which is really a retina reduction feature. It flashes the flash multiple times before firing the shutter. The theory is that the bright light causes the retina to close down, thus reducing or eliminating the red-eye effect. Unfortunately, it also increases drain on the batteries, and more severely, eliminates any impromptu flavor you might hope to achieve with your photographs. The mini-strobe effect can be annoying, and it's hard to remain smiling naturally while that's going on.

The other (and better) solution is to move the flash away from the lens. In other words, if your camera will accept an external flash, use that and place an external flash slightly away from the camera. It doesn't have to be far; just hold the flash a foot or so away and you'll be fine. Light that's reflected from the eyeball will no longer be seen by the camera. That, though, requires that you: a) buy an external flash; and b) then carry it with you. Also, your camera must work with an external flash. It's also not as convenient as using the built-in flash, but you'll find that you'll eliminate red-eye, and you'll get better pictures too because you can bounce the flash off walls or ceilings and create more natural lighting.

## REMOVING UNWANTED PEOPLE OR OBJECTS FROM A SCENE

Face it, you didn't buy a digital camera to save money on film or to cut your processing costs. Being able to immediately view your pictures wasn't a big deal either. What you really wanted to do was to edit your pictures once they were taken. Take that almost perfect photograph and turn it into a masterpiece by adding the perfect sky, adjusting the color of the grass, or removing Aunt Emily from the group shots.

Digital editing applications will let you do all that, and a whole lot more. All you have to do is spend a little time learning how to use the tools and away you go!

This section takes you through the basics of editing an image. We're not talking about color or brightness and contrast editing here, which is described later in this chapter; instead, we're going to talk about manipulating pixels—adding or removing things from photographs.

While this example uses Adobe PhotoShop, many other editing applications (Microsoft PhotoDraw, Corel PhotoPaint and others) include the same tools. Only very basic editing programs lack any method to copy and paste parts of an image. That's all you need to get started, though a clone tool (also called Rubber Stamp) and a Smudge (or Finger) tool are very useful additions. Being able to flip and distort selections is also something that you'll find useful.

The great thing about image editing is that it often isn't that difficult. You don't have to create new work or even be able to draw. There are usually other parts of the image you can copy and adjust to fill in the areas you want to change.

Two things before we start:

1. If at all possible, work on an image that's at a higher resolution than you want the final image. For example, if your original image is at 1500 x 1200, but you only need a 640 x 480 final image, work on the larger size, then scale down the image at the end. The scaling process actually obscures some of the mistakes and minor errors that you make.

2. Work in small areas. The smaller the amount you change, the easier it is to blend in with the rest of the work.

The sample photograph (*Figure 7-15*) was taken using a Kodak DC260 camera, and is going to be used to demonstrate some of the techniques used in editing an image. It's a pretty good photograph, but it suffers from one minor defect. The straw in the foreground is bright green and stands out, drawing the eye away from the rest of the picture. This might not be so obvious in black and white when reproduced here, but if you look at the sample on the CD you'll see what I mean.

There are other problems that could be adjusted; the finger is blurry for example, but that's another project.

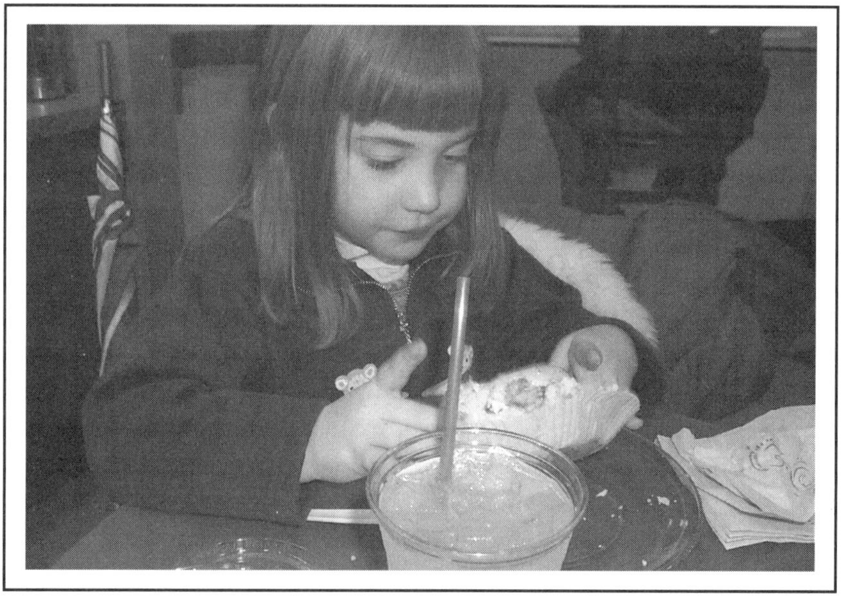

*Figure 7-15. Original photograph.*

To fix the photograph we're going to remove the straw.

First we're going to use the clone brush to remove the top of the straw where it covers the hair.

**How to Use the Clone Brush**

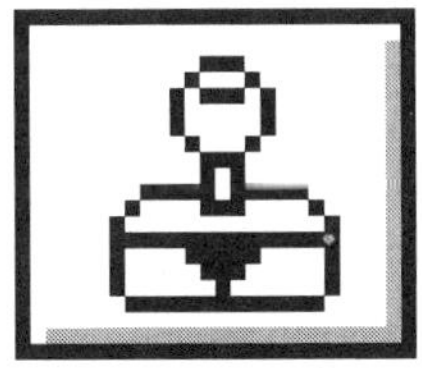

*Figure 7-16. Clone brush.*

The clone brush copies a section of an image and paints it over another area. To use the tool you must first indicate the region you want to copy from. In PhotoShop you do this by selecting the tool, moving the tool to the area you want to copy, then holding down the Alt key (Windows) or Option key (Macintosh) and clicking. The tool actually changes shape slightly to indicate when you are going to set a new copy point. You then move the tool to the place you want to copy too and click and drag. As you drag, the copied section will appear at the new location your are drawing. Notice that as you paint, the copy-from and copy-to point maintain the same relationship; you're not just copying the same point, you're copying the same relative area.

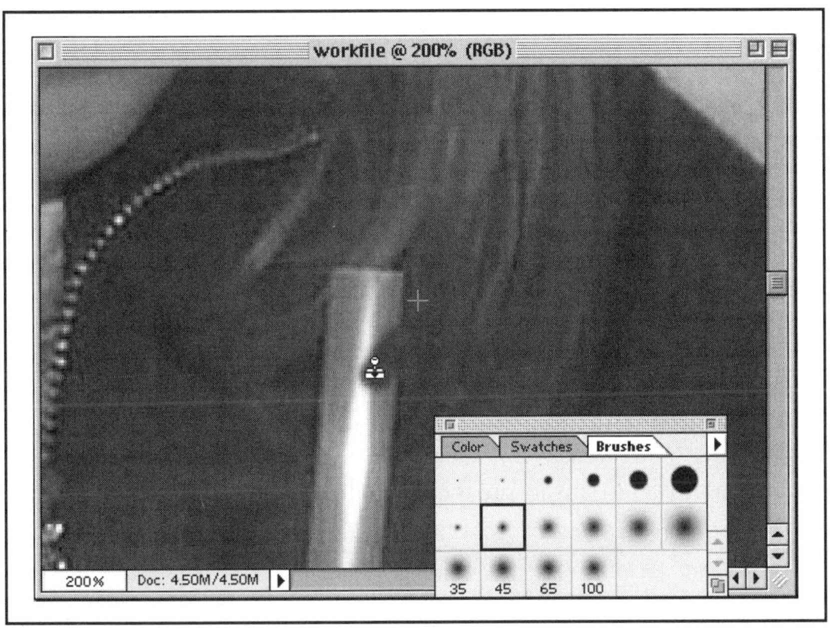


*Figure 7-17. Clone brush. The clone brush paints on what it is copying from another area of the image (indicated by the cross-hair).*

Notes:

With PhotoShop, as you paint the tool copies from the original image, not the adjusted image, so if your clone point reaches an area you have painted over, then it will still copy the original image, not the new image. Not always what you want it to do. Simply stop painting so that the program updates the file, and start painting again.

Use small brush sizes (at least compared to the area you are changing). Also, using a soft brush makes it much easier to create realistic effects except in regions where you have a sharp border. In those cases you have to switch to a different brush shape.

Paint in small regions rather than paint by holding down the mouse and dragging and dragging. It would be a shame to paint for five minutes, then make a wrong turn and have to undo the whole thing.

Some people use the opacity control to adjust the level of cloning. Depending upon the effect you are trying to accomplish this can be useful (though I rarely use it myself).

Rather than painting in long strokes, using a stipple effect by just painting very small sections can create very realistic results.

Watch for subtle changes in color; a large area of one color might look to be the same, but when you copy one region to another you may see a noticeable difference. You often have to move close to the region you are copying.

## Copying a Section

Sometimes it's faster to simply use the selection marquee and copy large chunks. The bottom of the straw, where it enters the ice, is a great place to do this.

Using the freehand selection tool, draw an irregular shape, then copy and paste it to create a copy, and drag the copy over to cover the straw. Creating a irregular selection really helps in this case.

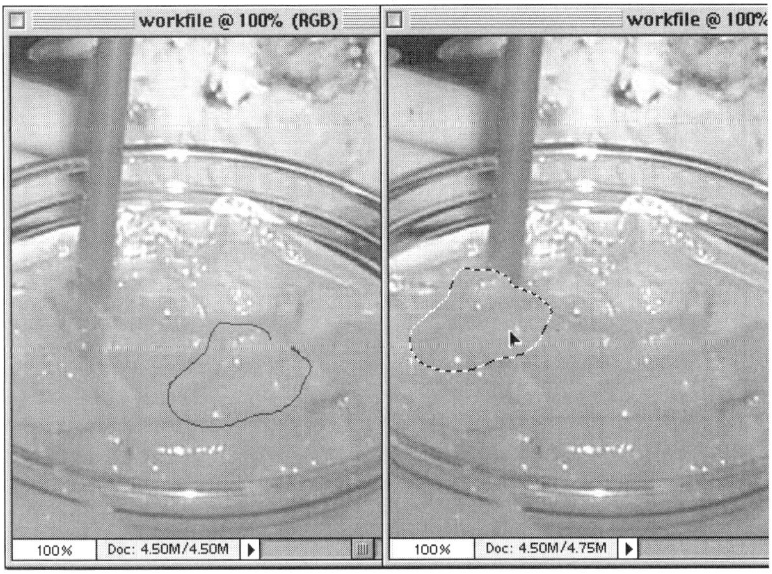


*Figure 7-18. Copying an area of the image.*

> Notes:
>
> This technique works better for some places than others. In this case, both the ice and the pastry work quite well for this technique.
>
> Once again, you want to use the smallest areas you can; though in this case you can get away with fairly large selections.
>
> Experiment with feathering the edge. This is an option of the selection tool in PhotoShop.

### Tidy Up Using Smudge Tool

Using the selection tool doesn't always create a seamless fit with the rest of the image. Using the smudge tool you can obscure the edges and really hide the fact that you edited the photograph.

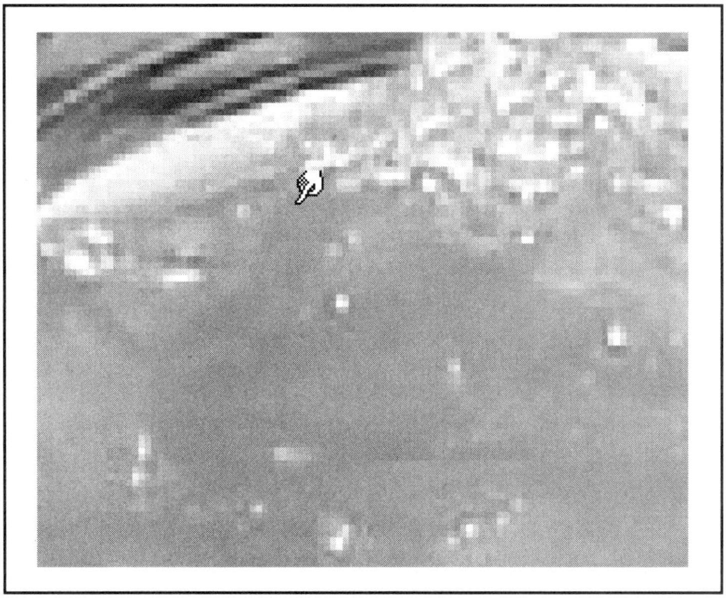

Figure 7-19. Smudging.

## Manipulating a Selection

After removing much of the straw, one of the problem areas in this photograph is the bear embroidery on the shirt. Fortunately, we have half of the face of the bear and can use that to fill in the rest of the bear (this is where we get really lucky; if the straw went down the middle of the face this would have been a lot tougher!).

Use the selection tool to select and copy the half of the face that can be seen.

Copy and Paste the copy into the document. PhotoShop 4 and 5 automatically add this to a new layer.

Select the image in the new layer and use the Flip Horizontal menu to flip the selection. Move the piece into position.

Use the Distort function to stretch the face to better approximate the face of the bear.

Use the Smudge and Clone brush to fill in any missing part of the face and clean up any problems.

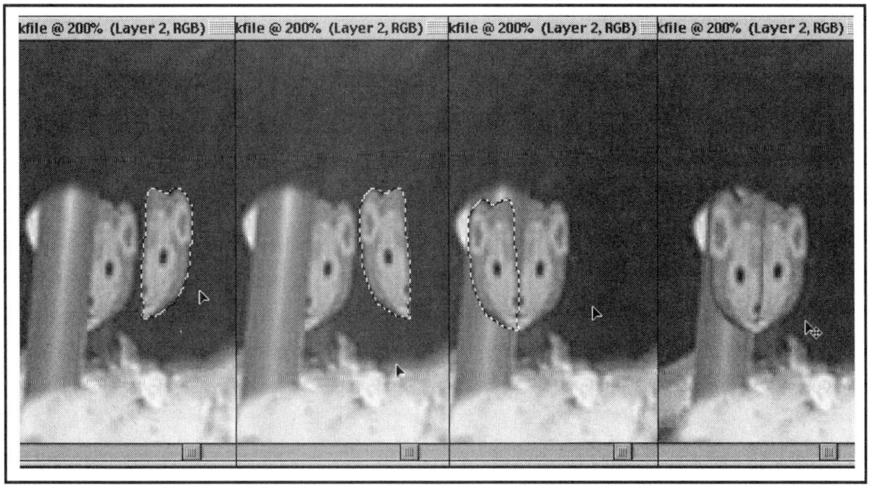


*Figure 7-20. Copying and stretching.*

## Advanced Cloning and Smudging

The toughest part of this image is the lip of the cup, which features two curved edges. That's hard to replace.

Use the clone tool to extend bits of the cup from the left and right. Here you need to increase the curve during the copy. You can do this by offsetting the clone point so that the cloned material curves away from the original. Copying small sections makes this work.

Still don't have it quite looking right? If you have it close to what you want, you can use the smudge tool to slightly alter the way the segment moves by clicking on it and simply pushing a little. If you only move the brush a tiny distance the tool will "move" the pixels underneath the brush without distorting very much.

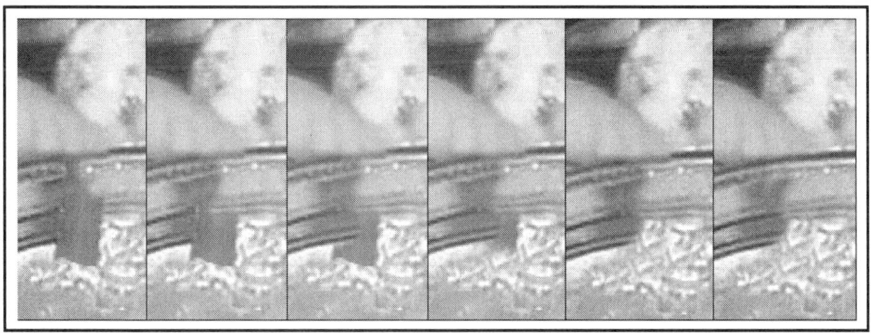

*Figure 7-23. Clone tool sequence.*

### Examine your Work

Most of the time I work on the image zoomed in. Every now and then you should zoom out to see how the image is coming together. Often it looks a lot better zoomed out than it does up close.

Once you have completed the changes, save the file (in case you need to do further work), scale it to the final size and save the new image with a new name.

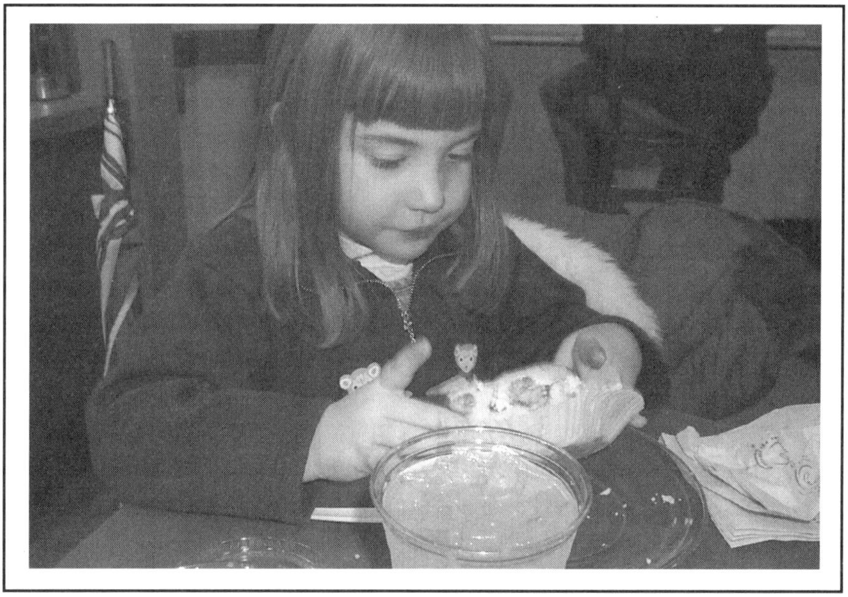

*Figure 7-22. Final image.*

**TIP:** Work Small. Yes, this is repeating what was said above; just checking to see if you are paying attention!

Remember, like any great artist, step back (zoom out) to inspect your work as you go.

**TIP:** If your image editor doesn't keep track of the history of your changes—and even if it does—save your work as you progress so you don't lose everything if the machine crashes. And save versions of your graphic along the way too. Being able to go back six steps can be useful if you suddenly change your mind, or realize that the pimple you were trying to remove has now turned into a molehill and it's time to try another technique.

# ADJUSTING COLOR BALANCE

In image editing, learning how to adjust color balance is one of the hardest things to do. Unlike brightness, where you adjust along a single dimension (light to dark), color correction can be very difficult. This is because while an image may appear slightly "yellow," if you don't adjust the image correctly you can end up not just with an image that has too little yellow, but maybe is starting to look a little blue. It can be very easy to end up with an image that looks worse than the one that you started with.

Software editing applications usually provide a color adjustment tool of some kind that can make the task simpler. Adobe PhotoShop includes a Variations dialog which shows different copies of the same image with slight changes (you can adjust the amount of change). You can interactively keep choosing an adjustment until you have an image you like. I find that it works well for large, course adjustments, but is not as useful for small changes.

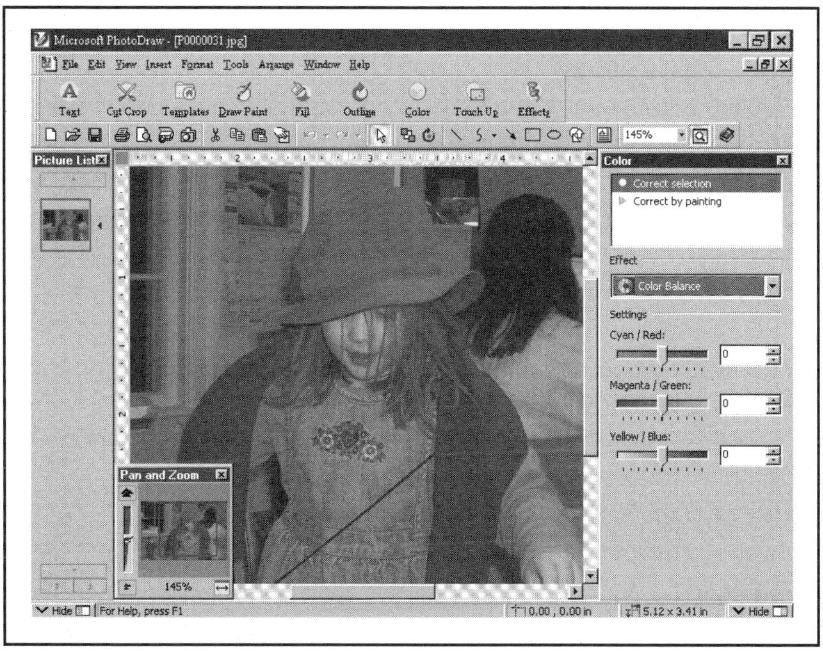


*Figure 7-23. Adjusting color balance in Microsoft PhotoDraw. Most editing tools let you adjust the color using sliders that change the cyan/red, magenta/green and yellow/blue cast of the image.*

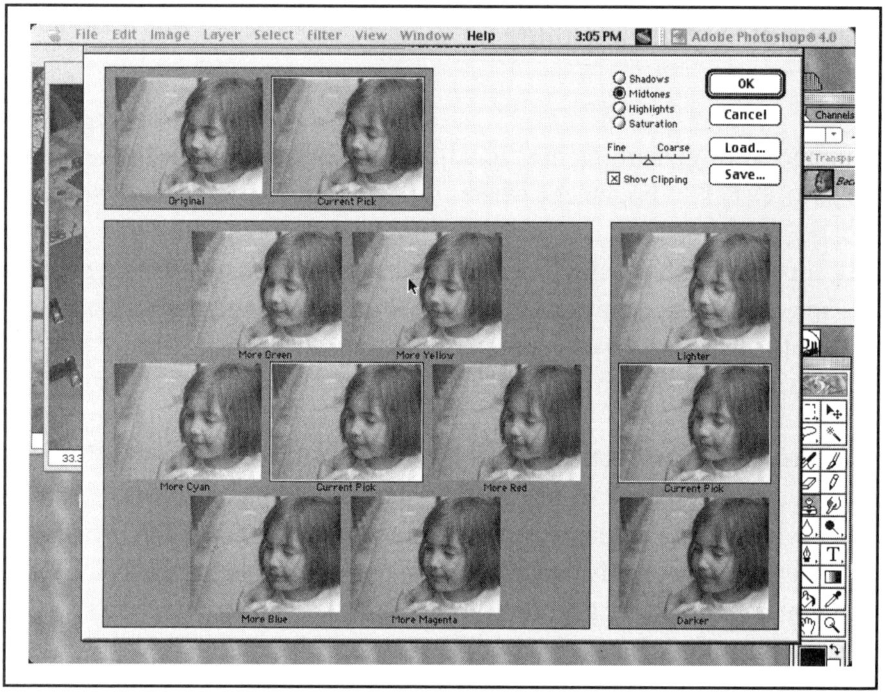


*Figure 7-24. Photoshop Variations tool.*

PhotoGenetics is an interesting tool that is designed to let you interactively adjust the color balance and lightness and contrast of an image. It works by displaying a before and after image and asking you to choose which one is better, and by how much. It then presents another pair, and you continue through cycles of images until you reach a point where you think the image is as good as it can get.

It is certainly an interesting concept, and can very quickly adjust some images. Somewhat like the Variations function however, it works well with large changes, but is less useful when the changes that need to be made are small.

Learning how to color correct an image can be difficult. Practice is the best way to learn; and remember to only make small changes. Also, work on a copy of the original file, and keep the original file open in a second window. That way you can see how you are progressing—or whether it is getting worse!

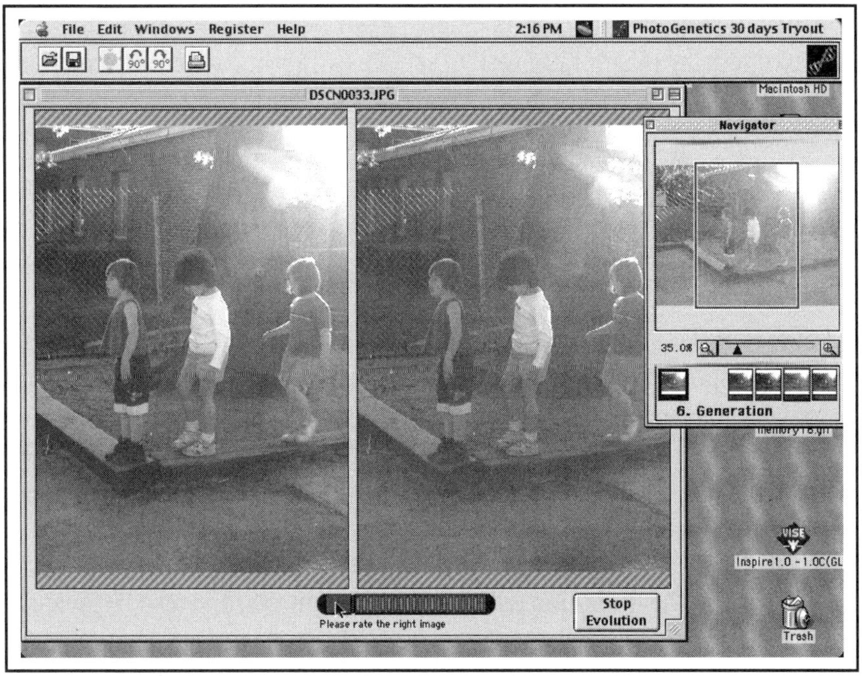


*Figure 7-25. PhotoGenetics image adjustment program*

A set of color gels, available at some photographic stores, can be helpful. You look through them at the image to see how they alter the image. However, since you can do the same thing almost immediately using software, it's questionable whether it is worth the expense.

Most editing applications provide an editing dialog that consists of sliders marked 'Cyan – Red', 'Magenta – Green' and 'Yellow – Blue' Each slider either adds one color component, or appears to subtract the other.

If you find that you are moving all three sliders in the same direction, then you are doing something wrong; moving all three sliders the same distance in the same direction will cause no color change.

One of the hardest skills to learn is determining what is wrong with the color balance. Look for white or neutral areas in the image to see color shifts. Sometimes you can see that there's a color shift, but determining whether it's magenta or red (for example) can be hard. Your best bet is to experiment; if it looks like red, try magenta as well to see the difference when subtracting the two colors.

## BRIGHTNESS, CONTRAST AND LEVELS

One of the most common editing functions performed on digital images is adjusting the brightness and contrast. I think nearly everyone likes to have images that look bright and high-contrast; it makes them look more "real." Some might argue that life is never that bright, but photography is an art form, and you're just trying to convey how you feel about that moment. If you want to make it brighter, then feel free to do so!

Just be aware of the limitations of these image controls, and also be aware that the levels control (if your editing program has one) is often a better tool for manipulating images than using the brightness and contrast tools.

Levels shows the distribution of the values in the pixels in the image. One side represents black and dark gray pixels, while the other side represents white and light gray pixels. For most images the vast majority of values will be in the middle, with some darker and lighter values. If the histogram cuts off at one end or the other, then that means there are no "absolute blacks" or "absolute whites" in the image. That's not necessarily a problem, but you can improve the contrast and brightness of the image by adjusting the distribution so that the darkest pixels are black and the whitest pixels are white. This will "punch up" (make more vivid) the image. To do it, you simply adjust the sliders at either end of the levels diagram.

An interesting side effect of using these tools is that you soon find out that 24-bit color—while it can do a great job of representing an image for display—doesn't hold as much information as you might have thought. It's very easy to overly compress the information in the image and end up with posterizing effects (regions of an image that all have the same color value and look unnatural).

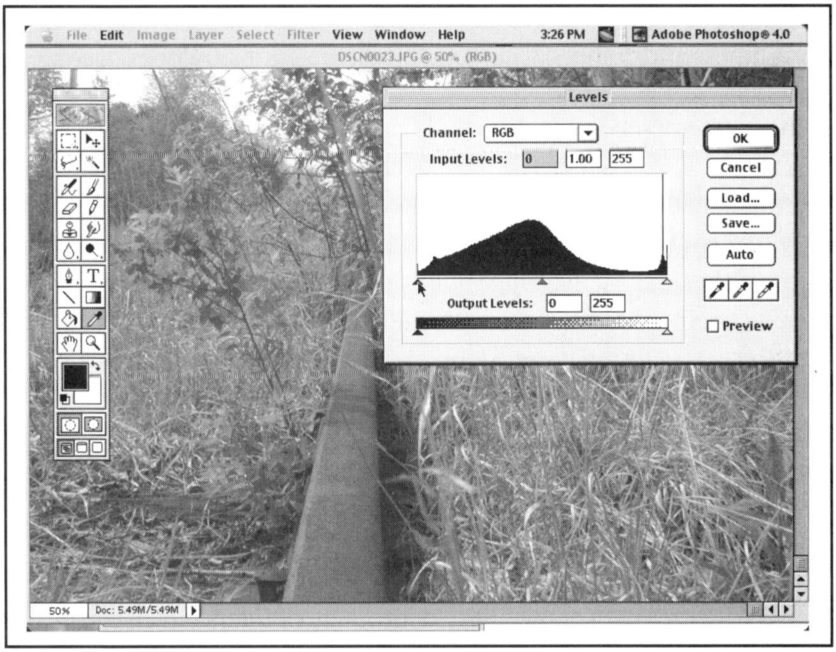


Figure 7-26. Levels tool.

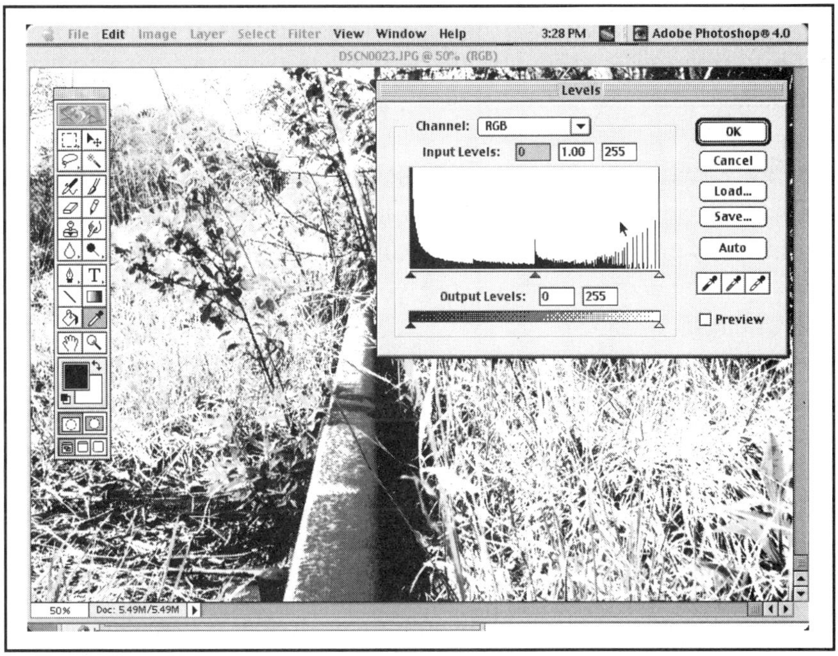


Figure 7-27. Posterizing in an image caused by compression of the brightness range of an image.

I don't know whether there is any good advice you can give someone about the brightness and contrast tools, other than to take it slowly.

One thing to remember, contrast also changes when you sharpen an image (see below), and contrast will also appear to change if you remap the image to an 8-bit system palette. Because of the limited number of colors available in 8-bit palette (except for adaptive palettes), when you switch to an 8-bit palette the contrast of the image will change quite dramatically. Before making your own final adjustments to contrast for an image that will be saved in 8-bit, convert the image to see what the image will look like. Sometimes you'll find that you don't need to adjust the contrast before performing the conversion.

## SHARPEN FILTER WORKS LIKE A CONTRAST ADJUSTMENT

Editing programs often come with a wide range of filters which can be fun for special effects. However the most useful filter is the Sharpen filter— provided you can adjust it to apply just the right amount of sharpening. Sharpening is nearly always needed after resizing or rotating an image; if you don't find that true it's possible that your editing program applies a sharpen filter automatically after resizing

The sharpen filter included in most paint programs acts like a contrast adjustment between certain ranges of pixels. After scaling an image, the sharpen filter can sometimes produce similar results to adjusting contrast. If you have scaled an image and it needs to be sharpened, you should do that first, before adjusting the contrast.

Remember that most filters can be applied to a selective area of an image (by using the selection tool to select the areas and then choosing the sharpen tool). If your program supports it, make sure that the selection edge is "feathered" so that the sharpen effect doesn't suddenly appear to end at the edge of the selection.

**TIP:** If you are going to be scaling your image to make it smaller, adjust the brightness and contrast before you scale the image. You'll end up with an image with more levels and more apparent detail.

## PREPARING AN IMAGE FOR PRINTING

The *Color* chapter (8) and the *Printing* chapter (9) provide more advice that relates to image editing. The *Resolution* chapter (6) also covers adjusting image size, which is usually done with the image editing tool.

### IMAGE EDITING IN THE CAMERA AND PRINTER

Up until recently you had to download your images to a computer to edit them. This is now changing with cameras becoming more sophisticated. The Digita operating system (see Chapter 20) turns cameras and printers into more intelligent devices. Their program, Digita FX, allows you to apply effects in the camera or printer. While it's not as subtle and powerful as using your computer (you're editing using the small LCD panel on the camera), it does allow you to perform simple operations.

## NEXT: MORE COLOR

The next chapter briefly examines the problem of Color; what it is and how to improve its accuracy. It also includes information on color calibration. Or, if that sounds too complicated, skip straight to Chapter 9 to find out how to print your work.

# CHAPTER 8
# COLOR

The goal of anyone working with color images is to reproduce colors as accurately as possible. Unfortunately, most color reproduction systems can reproduce only some of the colors that are detectable by the human eye. Worse, not only are these devices limited in their ability to reproduce color, but they are not consistent within themselves. One color monitor may reproduce colors very differently from another color monitor. Printing devices reproduce the fewest number of colors and use a different color model (CMYK) from the one commonly used by digital cameras and computer monitors (RGB). And to top it all off, different operating systems use different gamma levels, resulting in different results when viewing the same image on different systems.

All of which causes problems for any photographer trying to produce accurate color on their monitor and color printer.

The mantra of color technology is:

***Different devices display color differently,***
***OR, color is device-dependent.***

One thing you have to accept is that the color accuracy of inexpensive inkjet printers is limited. While manufacturers do their best to create a tool that works well for everyone, they are limited both by their audience and the cost of adding features. Finding accurate color profiles for most consumer devices is unlikely, though this is changing as the quality of the low-end printers is improving, and service bureaus are seeing the advantages of using inexpensive printers.

The first half of this chapter describes the principles of color models, calibration, and conversion between models. The last section, Calibrating Your Monitor and Printer, provides advice on getting the most consistent color results from your printer.

# COLOR TERMS

When working with color and color devices, you'll encounter a number of color terms:

**Calibration**. Adjusting a device so that it reproduces colors accurately. Most devices change over time and need to be recalibrated—monitors in particular. Some devices offer very little calibration options, particularly the consumer color printers.

For monitors, calibration involves adjusting the contrast and brightness of the display. It may also involve adjusting the relationship between the red, blue and green intensity.

Calibration of the monitor is designed to remove color casts, and set a known white point for the monitor. To get best results you have to use special hardware, and the monitor needs to be calibrated frequently (typically every two weeks). Room conditions also affect the image on the monitor. A monitor hood can reduce reflections and glare.

**Calibration Device.** Used to calibrate and/or characterize a device. A *Colorimeter* measures light and is used for monitors. A *Densitometer* measures the density of film or printed ink and is used with output devices.

**Color Model.** How colors are stored and represented. There are a number of color models, the most common are RGB (Red Green Blue), CMYK (Cyan Magenta Yellow Black) and CIElab (or Lab).

RGB is common in computers and scanners, CMYK is common for printing, and CIElab is common as an intermediary between different color models.

HSB (Hue Saturation Brightness) is also a color model. Though it is rarely used to store color data, it is often offered as a method for adjusting colors in image editing programs.

**White Point.** A measurement of the value of red, green and blue light at full intensity. This is measured in degrees Kelvin and is based upon the color of light emitted by a blackbody radiator heated to that temperature. There are several different "standard" white points including:

4100K fluorescent light

5000K page white

6500K daylight in overcast sky & Windows default

9300K Macintosh default

**Characterization.** Creating a map of the colors that a device can reproduce. This map is called a *profile*, which contains the information about the device.

**Profile.** The color map for a device created during characterization of a device.

**Gamma.** Relationship between voltage and brightness of a monitor (see below).

**Clipping.** Colors that cannot be reproduced by a device will be clipped; reproduced as another color.

**Color Management Module**. A software system that handles translating colors from one device to another using color profiles for the devices.

**RGB Color.** RGB color is an *additive* model. Red, green and blue light are combined to create different colors. Mixing all three colors at full brightness creates white. An absence of all colors results in black. Most digital cameras use RGB when capturing color, and all monitors use RGB to display color.

**CMYK Color.** Printers use CMYK, which is a *subtractive* model. Cyan, magenta, yellow and black ink are printed on paper. The ink acts as a filter, and reflects only the light with the same color as the ink. All colors combined mean that no light is reflected, giving black. If no ink is printed, the white of the paper is seen. The color of the paper can have a significant effect on the appearance of the final print; if the paper is off-white the image will have a similar cast.

While the three colors (cyan, magenta and yellow) should combine to make black, it's difficult to do due to limitations in the inks and color mixing. The resulting color often is closer to brown than black. For this reason, and to increase contrast, black is added to the process (hence the CMYK, rather than CMY model).

RGB and CMYK have different color gamuts. The *gamut* is the range of colors that can be reproduced. While RGB generally can display many more colors that CMYK, there are some colors that CMYK can print that most RGB monitors cannot reproduce accurately.

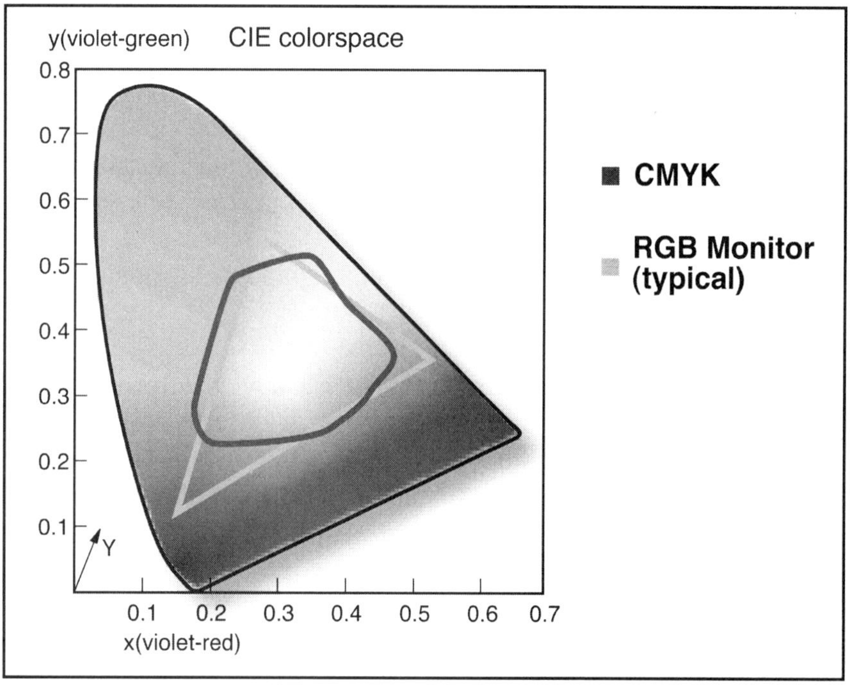


*Figure 8-1. Color gamut shows the colors visible to the human eye. Lightness, the third dimension, is not shown.*

## WHERE COLOR DIFFERENCES OCCUR

There are several points at which the color in an image can change.

**Capture Device**. All digital cameras see color differently from each other. How the color is captured (single chip, multi-chip or multi-shot camera) has some relation to the color accuracy produced by the camera. Also, the quality of the CCD and the algorithms used to convert the analog information captured by the chip into a digital color image are central to the accuracy of the color images. The correct color balance and exposure are contributing factors.

In most cases there is little you can do about the color accuracy of the image that is created—short of buying a new camera. If the camera has color balance and exposure controls, adjusting these can improve the quality of the image produced. Similarly, the type of lighting in the scene and the use of a flash will alter color; generally the flash will produce more accurate color than interior lighting.

**Computer Monitor**. If the monitor is not correctly adjusted the images seen on the monitor will not match the images as they will appear when printed. Make sure that brightness and contrast are correctly adjusted; using a grayscale band as a test image can be the best way to evaluate brightness and contrast. Display the image on the monitor and then see the effect of adjusting the monitor.

**Printer.** How accurately does the printer reproduce the colors in the images? This is often the greatest limitation of a system, and the most difficult to correct for.

## COLOR MANAGEMENT AND CALIBRATION

Because color representation is very different between different color models, a color model for independent color was developed by the Commission Internationale de l'Eclairage (International Committee on Illumination, or CIE.) The system, CIElab (or Lab for short) is based upon how the human eye perceives color, and is supposed to be color neutral.

Color management systems were developed in an attempt to handle moving images from one device to another. Such a system requires:

**1.** A device-independent color space.

**2.** Color profiles for the devices being used.

3. Software (called a Color Management Module) that converts the original image to the device-independent color space, then handles all conversions for other devices using profiles created for the devices.

Each device should have its own profile. There are three types of profile:

Input Devices (i.e., cameras and scanners)

Display Devices (i.e., monitors)

Output Devices (i.e., printers)

The Color Management Module handles conversion of out-of-gamut colors to colors supported by the output device.

Apple's ColorSync is an example of a Color Management System, which is based on a Color Management Module developed by Heidelberg Color Publishing Solutions (formerly Linotype-Hell). ColorSync also supports others CMMs from Kodak and Agfa.

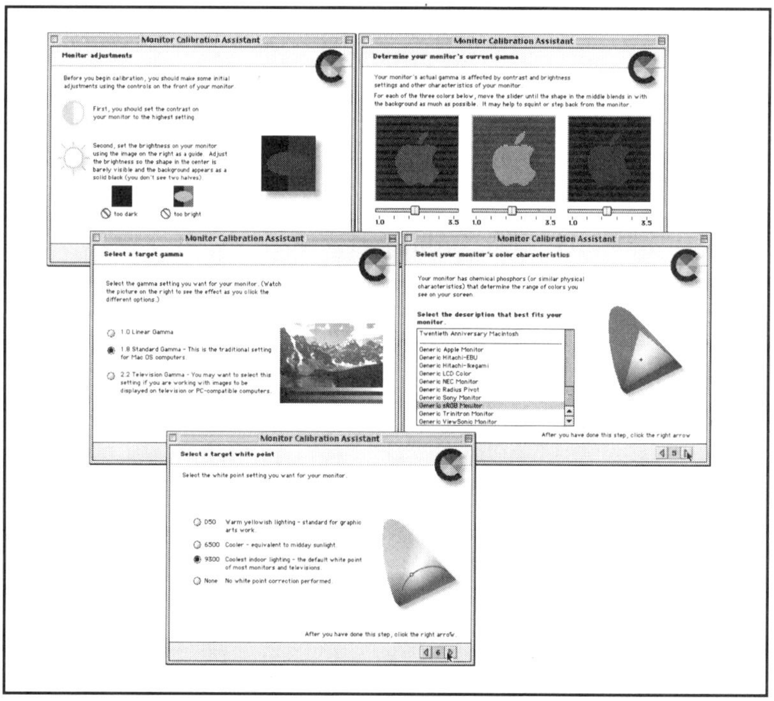


*Figure 8-2. Setting up options for ColorSync.*

Windows 95 and NT 4 do not support color management, though some applications use Kodak's KCMS and Agfa's FotoTune. Microsoft has developed its own CMS for Windows 98 called ICM 2.0. It is also based upon Heidelberg's CMM.

Color Management systems are vital if you are producing a lot of color work, demand absolute color accuracy, or are working with printers or service bureaus to produce your final copy. They are the best way to guarantee that you get the results you expect. However, they will only work if you can characterize your equipment. Unfortunately, this can be difficult, as well as expensive to do.

Both Adobe PhotoShop and Apple ColorSync include basic tools for calibrating your monitor.

Adobe Gamma for Windows may or may not be able to calibrate your monitor, depending upon your system and hardware. Windows NT 4.0 does not work with this software, and some video cards do not work with it. While this means no system-wide calibration is possible, it can still be used within Photoshop 5.

Pictographics ColorSynergy is a tool for calibrating color. The tool can be used to create profiles for a digital camera. To create a color profile you first photograph a MacBeth ColorChecker Color Rendition Chart. This color chart consists of 24 differently colored squares. ColorSynergy then analyzes the photograph and matches it to the known values for the colors. It can then produce a profile for the camera. There are no special camera-specific requirements needed to use ColorSynergy 4.0—it will work for any camera. However, a new profile will need to be created for different light setups; tungsten, fluorescent, flash and daylight. Find out more at: *http://www.picto.com*

## CONVERTING BETWEEN COLOR SPACES

Switching between color spaces isn't necessarily an easy process. If the image contains colors outside the gamut of the display or print device, a decision has to be made about how to handle the out-of-gamut colors. No solution is necessarily perfect; if you're losing colors then your image won't look the same as the original. There are several ways to do this:

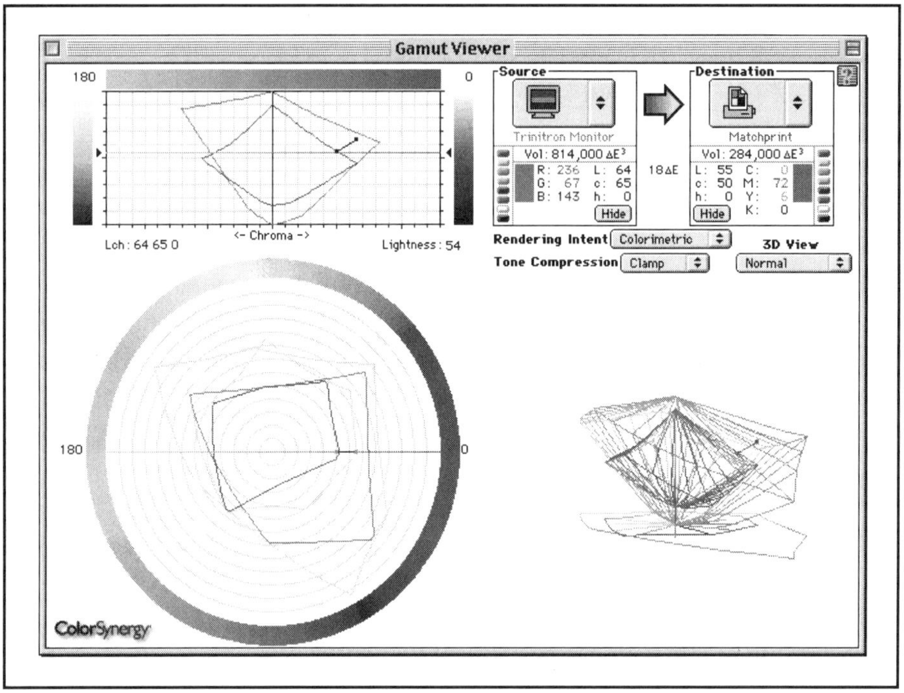


*Figure 8-3. ColorSynergy Gamut Viewer window allows you to visualize and compare the gamuts of any devices. Two cross-sectional views are provided as well as a 3D model that can be rotated in real time and even viewed in stereo (3D glasses are included!).*

**Compress the entire gamut.** This takes the entire gamut of the image and redistributes it over the gamut of the output device. This produces a good overall match to the original image, though many colors may change slightly.

**Map to closest color**. Only colors that are out-of-gamut are changed. They are mapped to the closest color supported by the printer. This will result in some colors that should be slightly different being rendered as the same color (an error called "clipping"). This method should be used when you want to maintain the accuracy of the majority of colors within the gamut of the image.

**Maintain saturation or brightness.** Adjusts the color but maintains either the saturation or brightness of the colors within the image. This method retains the relative values of the colors within the image. *Saturation* is a measure of the amount of gray in a color in proportion to the pure color (hue). A vivid color is highly saturated and has little or no gray, while a pastel color with the same brightness and hue has a large amount of gray. *Brightness* is a measure of the lightness of a color. To adjust a color, hue plus either saturation or brightness must be changed. This method of adjustment is used primarily for graphics or when a specific color must be maintained as close to the original as possible.

Conversion between color models of colors within both gamuts is not always lossless, due to roundoff errors. With roundoff errors, values stored as integers in one color model may not convert to an integer in another color model, resulting in slight differences when the color is converted back to the original color model. For this reason, you should avoid performing unnecessary conversions from one model to another.

## GAMMA: THE BIGGEST DIFFERENCE

One area that causes a lot of confusion—and also creates big differences in images transferred from Macintosh to Windows computers—is Gamma. *Gamma* is the relationship between voltage and brightness of a monitor.

A monitor is an analog device, but it is not linear. While a voltage of 0 = black, and a voltage of 1 = white, a voltage of 0.5 does not produce a 50% gray. Instead, it is closer to 80% gray. For this reason, when displaying an image, the digital values (0 to 255) must be adjusted so that the image displayed on the monitor will provide an accurate reproduction of the image.

**Brightness = voltage\gamma (where gamma is a power function)**

Gamma makes the image darker; the higher the value of gamma, the darker the image.

Unfortunately, gamma is factored in when storing data in a computer file (see below). While this doesn't cause problems when moving within systems with the same gamma, it does cause a problem when moving files from one system to another. This is because different gamma values are used by different systems. The following are standard values for gamma:

| | |
|---|---|
| Windows: | 2.2 |
| Mac: | 1.8 |
| Television: | 2.5 |

Using gamma encoding means that the brightness values aren't equally distributed across the digital values. In other words, there's not the same amount of difference in brightness between a value of 0 and 1 as there is between the values at 120 and 121. Interestingly, if the changes between each step were the same, then our eyes would be able to detect the changes in brightness. By a seemingly fortunate coincidence, gamma compensates for the characteristics of monitors, as well as providing a more efficient model for storing visual information in 8 bits for our eyes. The differences between brightness levels because of the gamma encoding gets smaller in the region that our eyes is most sensitive. JPEG uses a gamma of 1.8 and 2.5, depending upon the system being used.

This characteristic also explains why cameras with a 12-bit A/D converter can claim to produce better results than an 8-bit device, even though they both produce 24-bit (8 bits in each channel) files. This is because the CCD captures linear data, which is then converted to RGB using gamma encoding. Since an 8-bit Gamma encoded file can store more information than an 8-bit linear system, the extra information from the 12-bit camera will result in more detail in the final images.

Fortunately, in practice, you need to know little about gamma unless you are moving files from one system to another.

The most common place that users run into problems is with websites and multimedia projects where the color of images are different depending upon the system you use to view the pages. While it's theoretically possible to create two different copies of graphics and load them depending upon the operating system of the user, the easiest solution is to adjust your images slightly to compensate for both systems. If you have a Color Management System, adjusting the gamma to 2.0 might be the best solution. Otherwise, Macintosh users should slightly lighten their images, and PC users should slightly darken their images.

Finally, this problem might be solved in the near future. Internet Explorer 4.0 supports ColorSync on the Mac.

## CALIBRATING YOUR MONITOR AND PRINTER

Setting up your equipment so that it produces consistent results can be very difficult. It's also a little counter-intuitive. When first working with color images, most people start out very happy with how the image looks on their monitor, but are disappointed by how it looks when printed. They want to make the prints look like the image on the screen. But it's very hard to do that, particularly with low-end printers. Most advice therefore, centers around adjusting how the image is displayed on the monitor so that it closely matches what the printer will produce. This is done either by adjusting the monitor display settings and/or using a color management tool which will adjust the image displayed on the monitor. Color management software is making its way to all computers in the form of Apple's ColorSync and Windows Color Management tools. Adobe PhotoShop includes its own color management technology. Unfortunately, much of this technology is of limited use unless you can get color profiles for your input (camera) and output (printer) devices. Software that creates profiles is generally expensive and/or not easy to use.

So what to do? There are three options:

**Calibration software.** If you're creating lots of images (and making money doing it!), then you should consider using color management software and getting color profiles for your devices. This is vital if you are producing work that will be printed on color presses.

**Adjust your monitor so it mimics the printer.** In theory this is a good idea, but it means that you will have to adjust your monitor when working on other work (web graphics). Besides, the monitor will never exactly match the printer.

**Set up you monitor and learn the differences between it and the printer**. For most people, as long as they make sure the monitor is set up correctly (brightness and contrast are set correctly), it's reasonably easy to learn what you need to do to produce the kind of image you want using your printer. This takes time, and you have to be a little rigorous about this process. First, start with everything set to basic "factory" settings. In particular, use the ink and paper recommended by the printer manufacturer. In almost all cases, the manufacturer's products will provide the best quality from their printer—or at least as good as the manufacturer believes it can produce. Before you start exploring other options, start with the manufacturer's consumables. Consult the printer manual and use the correct settings for the paper you are using.

Use a good photograph with a wide range of colors and that is correctly balanced to test the printer. Ideally, this should not have been manipulated on your computer (otherwise, your monitor could have already entered the calculation and added some effect to the image). Print it.

How does the photograph look? If it looks really bad, then you need to check your printer settings first. If it's "a little" off, then you can start experimenting in your image editor to adjust contrast, brightness and color.

Also, how does it look on the monitor? If it looks okay printed, but looks bad on the screen, then perhaps your monitor is out of adjustment.

The next step is taking a picture from the camera and seeing how that works (a digital camera does not necessarily create an accurate color image). Using a color chart such as a MacBeth ColorChecker, take a photograph, then print it (again without manipulating it). How does it compare? Now trying adjusting the image using your editing software to get it as close as possible to the original.

## CONCLUSION

The next chapter, *Printing*, provides additional information and advice for producing the best quality output.

# CHAPTER 9
# PRINTING

Digital cameras invite you to take pictures and save them on your computer. The computer makes for a great way to file and view images. If you use a good naming scheme and file your images by date then you'll be able to find pictures much faster and easier than if you keep all your pictures in a shoebox. And you can email the pictures to friends without having to worry about going to the drugstore to get another copy made.

But at some point you are bound to want to get a hard copy; a print that you can hang on the wall or send to someone. You might even want to put them in a frame and hang them on the wall. So what's the best way to do this? This chapter covers the various options, as well as providing advice for how to choose an output device, and how to improve the quality of your output. Different printers are also examined.

The chapter is organized into the following sections:

**Output Options**. The different technologies available for turning a digital image into hard copy.

**Before You Print**. How to prepare a file for printing.

**Evaluating Printing Options**. Things to consider.

**Print Longevity**.

**Printers**. In-depth coverage of printer choices.

**After It's Printed**. Things to do after the print is made. There's also a short diagnosis section.

**Computerless Printing**. You don't need a computer to print your pictures anymore.

Before we begin down the road of color printing it's vital to remember the following:

***Different color devices render color differently.***

Chapter 8 explored this in a little more depth, but don't forget that color monitors and color printers use very different technologies and color models to produce color. These differences, as well as the limitations of most consumer hardware and software, result in differences between the image you view on the screen and the one you print. Calibration and experience should enable you to get consistent results, but it can be difficult to get an exact match.

## OUTPUT OPTIONS

There are a number of choices for turning a digital file into hard copy. The following list outlines the broad categories. More information about individual printer types is provided in the "Printers" section:

**Purchase a color printer.** Color printers come in a range of types: inkjet, thermal, laser and dye sublimation. Each technology offers different quality and price per image.

**Use a service bureau.** Local service bureaus offer access to different kinds of color printers, including high-end IRIS inkjet printers, dye sublimation and laser printers that you might not be able to justify purchasing. These are good solutions for occasional use or for special projects.

**Digital image kiosks.** Now appearing at a number of drugstores and department stores, digital imaging kiosks consist of a flatbed scanner, computer (with floppy, CD-ROM and maybe a flash memory card reader), and a dye sublimation printer. You can use the kiosk to print images using the high-quality dye sublimation printer (Note: that's what the Kodak stations use; you should always confirm the quality and type of printer before you use it). Just put the images on a floppy and bring it down to the kiosk. If you have a Macintosh, make sure that you use a PC-formatted floppy, not a Macintosh disk, and name the file with ".jpg" extensions. Some kiosks are based on Windows computers which cannot read a Macintosh disk.

A kiosk is quick and convenient, but you may get very little support; it's possible that the store personnel are less familiar with the device than you are! Also, the printing options provided by the software in the kiosk are often limited. A kiosk I tried allows you to print multiple copies on the same 8.5 x 11 page, but you can't print different images on the same sheet. The image quality of the dye-sublimation printer at the kiosk I tried was very good—so good that printing a 1500 x 1000 resolution image to 8 x 10 proportions resulted in a print in which you could see resolution artifacts (i.e., limits in the quality of the image caused by the low resolution of the source file).

**Photographic print**. A number of service bureaus, including services on the Internet, will produce a photographic print from a digital file. The digital image is exposed to photographic paper using an LED display and then developed like a regular photographic print. Using this method offers the high quality and long print life of a "regular" photographic print. If you're getting many copies it can be reasonably cost effective too. Several of the Internet services offer special software that you can download that lets you choose the print options, choose your file, and then the software uploads the file to the service for printing.

Companies that offer this service include:

Signature Color: *www.signaturecolor.com*

Mystic Color Labs: *www.mysticcolorlab.com*

Sedona Digital Print Service: *www.sedonadps.com*

Print My Picture: *www.printmypicture.com*

**Photographic negative.** A digital image is converted to a negative in a similar method as the photographic print but with a device called a Film Recorder. The negatives can then be used to produce multiple traditional prints. Some of the companies that offer photographic prints also offer this service.

### WORKING WITH A SERVICE BUREAU

If you are going to use a service bureau, whether it's a local service bureau that offers dye sublimation and color laser printing, or one of the on-line print services, it's important to find out as much as possible before you send them a file. What resolution do they recommend for the printer and print size that you want to use? Obviously, you will be limited by the resolution of your original file and the size that you want the final print, but it can be helpful to know what resolution is recommended for a particular printer. Also, the bureau may be able offer other tips for getting the best quality results from the printers that they have. This is particularly true of a local printer that you can stop by and visit. Some of the on-line services have FAQs which answer a lot of questions.

## BEFORE YOU PRINT

Before you send the file to the printer or service bureau, there are things you may need to do to prepare the file. You also need to understand how the printing process works.

### CMYK

Most printing processes are based on the subtractive model; light strikes the paper and is absorbed by the inks on the paper. Three colored inks; cyan, magenta and yellow, are used. Mixing the colors together produces black. If no ink is used, the white light reflected by the paper is seen. Due to limitations of the inks, and to improve contrast, black ink is usually used as well, resulting in the designation CMY**K** (blac**K**).

Computer monitors mix red, green and blue light to produce colors, and is called an additive process. The three mixed together produce white light. No colors produce black. The differences between these processes is the greatest reason why your output will never exactly match the image you see on the screen. However, there are other limitations of the printer technology which contribute to the differences between looking at an image on the screen and an image that's been printed.

## PIXELS-TO-PIXELS

Both monitors and printers use colored pixels to recreate the digital file. Every color pixel in a monitor can be any one of millions of colors because each pixel is a mix of red, green and blue light. With printers, each dot on the page can only be one color; cyan, magenta, yellow or black. To create the illusion of other colors, groups of colors are dithered. So a 4 by 4 dot square may contain a mix of cyan and yellow pixels to create a shade of green.

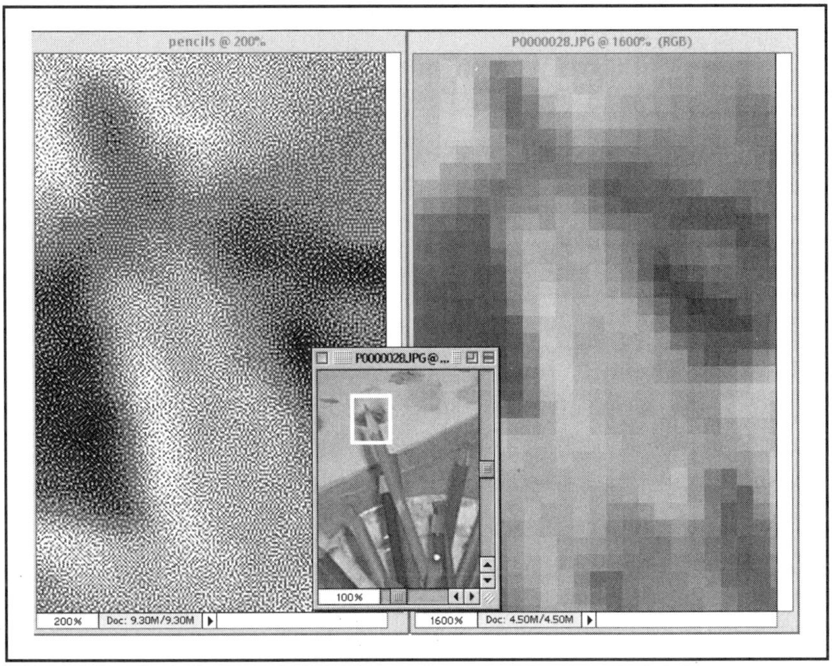


*Figure 9-1. Monitor and printer dots.*

Because of this, a 600 dpi color printer is not capable of reproducing all of the detail in a 600 dpi file. You may see no difference between printing a 600, 300 or 150 dpi file to a 600 dpi printer. While a 150 dpi image should be fine for most photos, you will see a difference in the image where there are high-contrast changes (for example, where one pixel is black and its neighbor is white). This most often occurs with black text. So if your image has black text on a white background in part of the image, you might want to use a higher resolution than you would with a photographic image. Because of variations in how printer drivers work, and how the algorithms convert the original data to the dots the printer places on the page, it's difficult to know exactly how any given printer will handle a particular image without printing it.

There are some printers which don't lay down individual dots on the paper. Dye sublimation printers (described below) merge colors together, so an individual dot can contain a combination of cyan, magenta and yellow ink. This technique effectively improves the apparent resolution of the printer; each pixel in the original file can be represented by a single pixel on the page that is the same color (or close to it).

Hewlett-Packard's PhotoSmart printer, while an inkjet, slightly overlaps the pixels that it lays down on the paper. This helps improve the apparent resolution of the resulting print, though it doesn't quite match a dye sublimation printer.

## DIFFERENCES BETWEEN PRINTERS

Comparing the printing resolution of different printers is a complicated issue primarily because of the differences in technologies. But even two inkjet printers can produce very different results even though they have the same resolution. While many of the common color printers use four inks (CMYK), there are several inkjet printers developed for printing photos that use six inks. These printers add a light cyan and a light magenta color, which reduces the size of the dither pattern needed to produce many colors. A six-ink printer, which has the same resolution as a four-color printer, will produce a print that has much less dithering, and therefore appears to have more detail.

**WARNING: DON'T USE CMYK FOR YOUR FILES.** One of the greatest areas of confusion is whether to send RGB or CMYK files to consumer inkjet printers. Since these printers are CMYK printers, many people think they should edit their files in CMYK and send CMYK files to the printer. While that sounds logical, nearly all inkjet printer manufacturers assume that users will be working with RGB files and will be sending this format to the printer. They therefore write their printer drivers to work with RGB files. If you use a CMYK file, the information is first translated back to RGB, and then translated back to CMYK for the printer. The resulting images won't look very good. Nearly every user who tries sending CMYK files to their low-end color printers reports problems with their printer. Sticking with RGB solves the problem.

## THE IMPORTANCE OF PAPER

Different paper produces different results. Not the least of which, remember, is that the paper provides the "white" in the picture. The whiter the paper, the brighter the picture. While inkjets print on a variety of paper, they produce their best results when using heavy paper that has a glossy surface. Always start with the paper produced by the printer manufacturer when first experimenting with a printer. Using other manufacturers' paper can produce very different results with the same printer, and may require experimenting with the printer settings. Epson's Stylus Photo printer includes a number of different paper choices, the highest quality being Glossy and Film. These work just fine with Epson's glossy paper and transparency film, but other manufacturers' glossy papers sometimes work better with the film setting than the paper setting. So be prepared to experiment when you try a new paper.

This doesn't mean that you should only use the manufacturer's paper with a printer—just that you may need to experiment when you use other papers. In some cases it is true that another paper does not produce acceptable results, but sometimes you'll find that other papers are better than the manufacturer's. Unfortunately, this depends on the printer, paper, and sometimes on personal preference and how you will be using the output.

## BIG DIFFERENCES IN COLOR OR CONTRAST

If you have a large difference in color or contrast, then there are several possible causes:

**Sending CMYK instead of RGB file**. Change the file to RGB and try again.

**Incorrect printer driver settings**. Check the manual for the printer.

**Maladjusted monitor**. An incorrectly adjusted monitor can cause you to over-correct an image. Try an uncorrected image first.

**Gamma problem.** Images created on a computer with a different gamma will appear brighter or darker (depending upon the direction you are moving). This should be apparent on the screen too. Adjust the image (see *Color* chapter).

## CAN YOU ENLARGE A FILE WHEN PRINTING?

When printing to most printers you can use a file with a lower resolution than the printer resolution. Ideally, the file resolution should divide evenly into the dpi of the printer. For most image/printer combinations a 4 to 1 resolution is as low as you want to go, so for an 8 x 10 print on a 720 resolution printer, an image file with a resolution of about 1800 x 1440 is required.

> 720 / 4 = 180 (printer resolution divided by 4 = target resolution)
>
> 180 x 8 = 1440 dpi: width of image
>
> 180 x 10 = 1800 dpi: length of image

If a lower resolution file is used (say 600 x 480), then the pixels in the image will be apparent when you are close to the image.

## ENLARGING A FILE USING EDITING SOFTWARE, OR USING THE PRINTER DRIVER

If you want to print the image at a larger size, should you use the printer software to enlarge the image, or should you first use an editing application to enlarge the image, and then print it? While it's possible that the printing software may have a better resampling algorithm, it's doubtful that the differences will be major. Any advantage is probably offset by the amount of time it takes to adjust the image, and the extra print time.

## PRINTING POSTERS

Many inkjets printers, though they aren't very wide, will accept long pieces of paper. Nearly all Epson inkjets support this, and Epson sells panoramic paper which measures 8.5 x 24 inches long. This makes it possible to print very large work, but you can also create posters (an image printed over multiple pages) using an image editing program.

A few programs will print an image over multiple pages. If you want to do a lot of poster printing, you might want to get a program like:

**PosterShop from Onyx Graphics**

*www.onyxgfx.com*

Available in Pro and Lite versions for PC and Mac. Check that it works with your printer before buying.

**PosterWorks from S.H.Pierce & Co**

*www.posterworks.com*

Macintosh only, requires a PostScript printer.

These programs are worth the money if you're printing lots of posters, but if you only want to make a single one, it might be better to use another technique.

To manually create a poster, use your image editing software to cut the image into pieces, then print the pieces individually, cut them out and tape together.

First, work out the size you want the final image to be. Divide that by the maximum dimension you can print on your paper (probably about 8 x 10.5"), then work out the size of the rectangles you'll need to print. To do that, divide the final length by the actual number of pixels to get the pixels per inch:

Final image: 26 x 19.5

Image resolution: 640 x 480

640/26 = 24.6 pixels per inch

Maximum print size: 8 x 10.5 = 196 x 258

Cut the image into squares that are 196 x 258. PhotoShop lets you crop the canvas size by a specified amount, so you can crop the file into strips by cropping out the rest of the image.

Take the segments and print them at 24.6 pixels per inch. For simplicity you might want to recalculate your poster size at 25 pixels per inch.

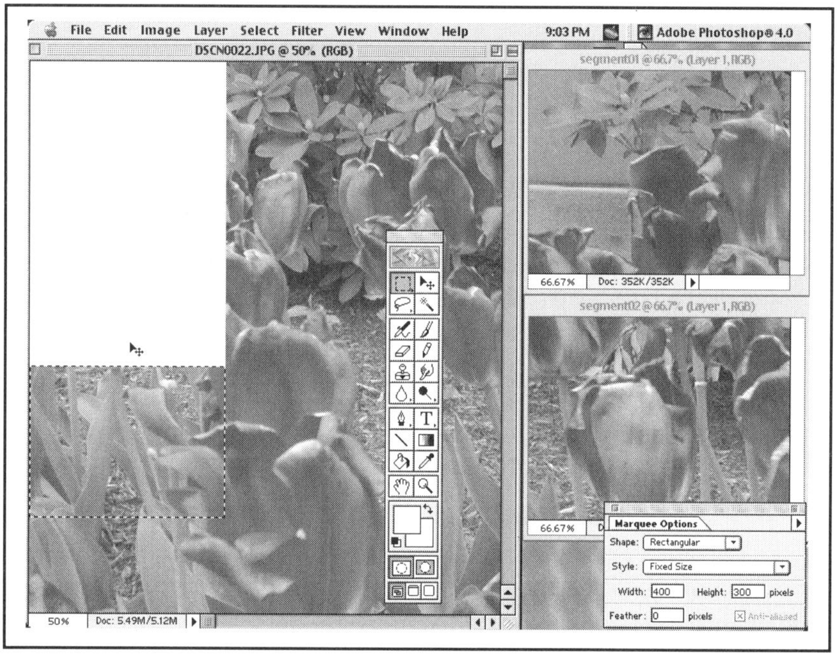


*Figure 9-2. Trimming file into segments.*

## PRINTING TO THE EDGE

Want to print to the edge of the page? No printer available offers that option. The printer needs the edges to grab hold of the paper, as well as to guide the paper in the printer. Consumer printers also tend to be less accurate in feeding paper than might be desired. The alternative is to print on a larger piece of paper and then trim the paper to the desired size. In printing, this is often referred to as printing a bleed (printing outside the area you want and then trimming so that it appears the item was printed "to the edge").

Some software provides printing options that will print multiple copies of one or more images on the same sheet of paper, making it possible to print three 4 x 6 prints on an 8.5 x 11" sheet of paper (see Chapter 17).

Also, some paper comes pre-perforated; you print on the page and then remove the perforated edges to produce an edgeless print.

### CAN'T GET A GOOD BLACK & WHITE PRINT?

Trying to print a black and white print on some color printers can result in something more brown than black because the printer uses the CMY colors rather than the black ink. In theory, mixing cyan, magenta and yellow should produce black, but in reality it creates a brown-black color. Make sure that your file is black and white (i.e., grayscale) and not RGB. That makes sure that there is no color in the image file that will cause the printer to use the colors rather than black ink. Also, check the printer driver to see if there's an option for printing a black and white image.

## EVALUATING PRINTERS

If you're considering buying a printer you'll find a lot of different models and types to choose from. Different printing techniques offer different features and performance. Choosing between these printers can be difficult. While the cost of the unit is often the most important criteria, other things to consider include:

**Cost per Print.** What is the cost of consumables? Include both the cost of the ink cartridge or ribbons, as well as the cost of the special paper that is required for glossy prints. If you use the printer frequently the cost of consumables will soon pass the cost of the unit.

**Print Quality.** What's the resolution of the printer? Higher resolution does not always mean higher quality prints; dye sublimation printers can produce prints that look better than inkjet printers with double the resolution. The number of color inks used can also impact apparent image resolution.

**Print Longevity.** How long do prints last? There are several factors that affect print life, most critically the ink and paper. Inkjet prints tend to deteriorate quickly (see below).

**Paper Options.** Some printers are limited by the paper they can work with. For nearly all printers the highest quality output is achieved only on special paper. But some printers produce acceptable output on only a limited range of paper.

## PRINT LONGEVITY

How long your print will last depends upon a number of factors, not the least of which is, which process was used to produce the output. Inkjets are inexpensive, print on a wide range of materials, and produce great looking prints, but the lifetime of these prints is much shorter than dye sublimation prints. The new printers, particularly those designed for producing photographs, produce amazing prints, and with a digital camera the digital darkroom is a reality. But just as the "traditional" photographer must be concerned with the archival quality of their output, so too the digital artist should be concerned about the longevity of his/her artwork.

The three things that most contribute to the stability of a print are:

**Paper.** How the surface stands up to long-term light exposure, as well as the acidity of the paper. Archival paper has low acid content that resists fading and yellowing of the paper.

**Ink.** Inks are of two types; dye-based and pigment-based. Dye-based inks produce bright colors, but dyes will fade. Pigment-based inks aren't as bright but last longer and are more resistant to water. Most inkjets use dyes. Some dyes include a UV blocker which extends the lifetime of the ink when exposed to sunlight.

At the moment, the ink is probably the weakest link: protecting prints from light and using the correct paper may double the estimated lifespan of a print from 2 to 4 years. But at 4 years the lifespan on an inkjet print is a fraction of traditional photographs, which can last 25 years or more.

**Environment/Exposure to Light.** This is one of the biggest determinants of the lifetime of any print, traditional or inkjet. Exposure to bright sunlight will drastically shorten the lifetime of any print.

Early Epson inks and papers had very short lifetimes; when exposed to bright sunlight they can start fading after only a couple of months. In their early printers, Epson was focusing on accuracy of color reproduction and quick drying inks rather than print longevity. In 1998 Epson released new papers which they claim improves the print lifespan. Epson rates the new papers at 3 years. Unfortunately, paper testing does not seem to be very "standard;" figures ranging from less than a year to more than double Epson's claimed life have been given by different organizations. Hewlett-Packard, with the help of some testing organizations, claims that there Photosmart prints last much longer than Epson's older inks and papers—as long as five years.

As with any photographic print, minimizing exposure to sunlight and fluorescent light, as well as heat, will extend print life.

Of the many printers available, dye sublimation prints have the longest quoted life. Kodak offers a printer that applies a UV clear overcoat layer to protect the final print. The print lifetime is thought to be as long as 25 years.

Things you can do to extend the life of a print:

Keep it under wraps (exposure to light shortens the lifespan).

Cover with a protective film (protect from UV and environmental factors).

Coat with lacquer (protect from UV and environmental factors).

## PRINTERS

This section looks at the types of printers currently available, and looks at some of the most popular ones.

## INKJET PRINTERS

Inkjet printers are the most popular and inexpensive choices for printing color images from a computer. The image quality possible with these devices has improved dramatically over the last few years, though for all inkjet printers the best quality output is possible only with special paper.

Inkjet printers squirt drops of ink from tiny nozzles onto the paper. The ink is wet, though it usually quickly dries. Most of the inkjet inks don't resist water well.

There are a large number of companies selling color inkjets; these are general-purpose printers that use CMYK, or four colors, to produce their output. Printers with a resolution of greater than 600 dpi can produce images that are almost photographic in quality. But even at 720 dpi they may still exhibit noticeable banding or dithering in areas of the photograph. Several companies have produced special "photo" printers which use six inks (CMYK plus a light magenta and a light cyan) that produce even better prints. Dithering is negligible; particularly in the latest printers released. Printers to consider include the Hewlett-Packard PhotoSmart, Epson's Stylus Photo 750, Lexmark's 7200 and Canon's 7000.

Why not use other colors rather than extra cyan and magenta inks? One reason for this is that it makes writing the new printer driver much easier. The developer simply uses the lighter cyan and magenta where there are light cyan and magenta regions. If another color were added (say green), then the whole image would have to be dithered in a very different way.

In terms of performance cost and image, inkjet printers are cheaper and produce better results than most color laser printers. They're cheaper to run than dye sublimation printers, but the prints don't last as long. Expect to pay less than $1 per 8 x 10.

Inkjets are often limited in the types of paper they work with. Limits for paper thickness are usually the result of the feeding mechanism, though there is also the possibility of damage to the print head or nozzles.

## REFILLING INKS

To refill, or not to refill? For inkjet printers one of the greatest expenses is the ink cartridge, which can cost as must as $30 each (depending upon the printer). To save money, should you refill your ink cartridges? Some people swear by it, saying that it saves money and results in prints that are just as good as the manufacturer's inks. Others report problems with print quality and cartridge failures. Unfortunately, I fall into the latter camp, having had very mixed results when refilling black ink cartridges for a Hewlett Packard inkjet printer several years ago. I found that every second or third refill I performed didn't work right for one reason or another, and with that rate of success I wasn't saving any money by buying the refill kits.

A clear advantage of refilling your cartridges is that some ink makers offer archival quality inks.

Some cartridges are reasonably easy to refill, while others seem to be intentionally designed so that it is almost impossible to successfully do it. Yet, there's no reason why manufacturers should be the only ones who can produce the inks for their printers.

The refill manufacturer will provide the ink in some kind of container and instructions on how to perform the refill. Refilling the cartridge should not void your warranty; though damaging the cartridge in the process and then using it in the printer would. The number of times that you can refill a cartridge depends upon the type of cartridge, and varies from 3 to 10 times. Some cartridges include a print head, while others act only as a reservoir; this accounts for some of the differences in cartridge life.

There's one problem you can have if you switch inks; if they aren't the same color as the original inks, then your output will be different. While not a problem for general color work, it could be more noticeable for photographs. To correct for this problem, you might have to use a color management tool and characterize the inks. This won't be easy because you're sending an RGB file to a CMYK printer. You might want to try a profile editing package (ColorBlind Edit, ColorFlow), but these are expensive. Or, find a service bureau which will sell you a test file to print, then they will create a profile for you. You can find a partial list of suppliers that offer such services at this page:

*www.apple.com/colorsync/services*

Some refill kit makers include:

**Definitive Data Solutions**

*http://member.aol.com/ddsmain/ddsmain.html*

(888) 465-5487 or (614) 840-9900

**Repeatotype**

*www.repeatotype.com*

(973) 696-3330

Offers a line of glossy photo inkjet papers.

**Luminos Photo Corporation**

*www.luminos.com*

Offers a line of inkjet photographic paper and inks that they claim will produce prints that will resist fading caused by indoor lighting. Lumijet Fine Art Preservation Inks are available in two sets. The Preservation Platinum series has a fade resistance of 65 years, with their Preservation paper, while the Silver has a life of about 25 years.

## USER REPORT: EPSON STYLUS PHOTO

I have an Epson Photo printer and have been very happy with it. However, many people have had problems with ink head clogging. This could be based on use; our printer is used reasonably regularly; but may go for a few days without use. If the printer is unused for several weeks, then that might cause clogging. The printer has one black cartridge and one color cartridge. The color cartridges last for about twenty 8 x 10s.

Epson papers aren't known for longevity, but they produces the best results on the printer. Fuji has a paper that is supposed to have a lifetime of 5 times that of Epson, while Mitsubishi Artist Glossy is also supposed to be very good.

When using other papers you need to experiment a bit with the settings to make sure you are getting the best results. The printer isn't exactly fast, but then it is on a serial connection.

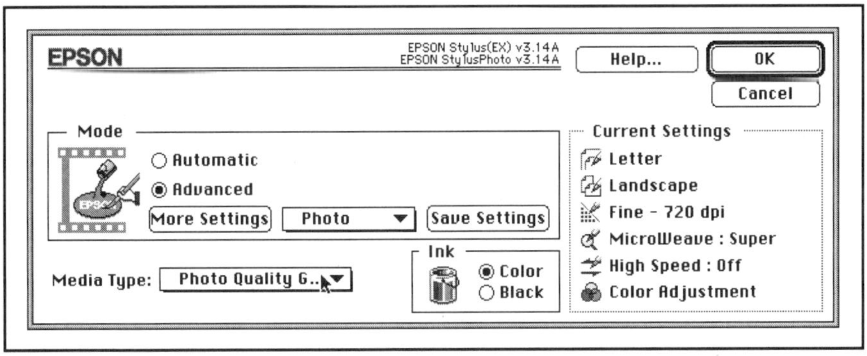


*Figure 9-3. Epson Printer driver.*

## HEWLETT-PACKARD PHOTOSMART PRINTER

Another very popular photo printer, the PhotoSmart, actually has a lower dpi (300), but it overlaps the dots slightly, creating the illusion of continuous tone prints. Ink cartridges last for about seventy-five 8 x 10s, depending on a number of variables, but the cartridges cost more than the Epson cartridges so per-print cost is about the same for both printers. Paper and ink combination has good longevity; reportedly much better than Epson. Unlike the Epson, the nozzles are included in the cartridge, so you replace them each time you replace the cartridge

Because the nozzles are in the cartridge, there is a utility that you run to align the nozzles when a new cartridge in added.

## IRIS

The Iris printer is a large-format, high-quality, color inkjet printer that can print on a wide variety of surfaces. It is available as a service at many print bureaus.

Unfortunately, it is expensive, and as with all color devices it may take some experimenting to get exact color matches, but if you want to print your image onto a thick textured paper the Iris printer may be your only choice. Depending upon the paper used, the IRIS prints have a much longer lifespan than most of the inks used in consumer inkjets.

Surprisingly, the resolution of the Iris printer is only 300 dpi, but because of the way the ink is applied to the paper in a continuous flow it appears to have a much higher resolution.

## COLOR THERMAL PRINTERS

Thermal printers are more expensive to run than inkjet printers, and the printers themselves can be more expensive too. The printer uses a special ribbon that is coated in the separate inks: cyan, magenta, yellow and black. The paper runs through the printer and is pushed against the cyan ribbon while a heating element melts the appropriate places to apply the ink to the page. The process is then repeated for the other colors.

The ink does not soak into the paper, and does not bleed, so depending upon the resolution the output can be of high quality. ALPS makes inexpensive printers which use this technology (described in the following section).

## DYE SUBLIMATION

Dye Sublimation takes a solid ink, turns it to a gas (sublimate), and applies it to the paper. A thermal print head heats a dye ribbon to transfer the dye. The print head can control the size and amount of dye that is transferred, making dots of different sizes, making it possible to create a continuous tone image. This method creates soft edges, so text is not as sharp. The dye-sub process is similar to color thermal, and ALPS makes printers which can work in both modes.

## ALPS

ALPS offers a range of low-cost printers that are color thermal printers, but also offer a dye sublimation option. This is all offered in a printer that costs about the same as an inkjet. Dye sublimation produces much better prints (for photographs) but as a higher per-page cost, about $3 per 8 x 10. The quality of the prints seems to be very good; the color thermal prints exhibit dithering, but this is not seen in the dye-sub prints. Print life should be longer than for inkjet prints.

Some users have reported that banding can start to appear in the prints, particularly as the printer ages. This may be caused by maladjustments in the feeding mechanism. It's recommended that you always use the right paper, feed it in the right direction, keep the printer clean and let it cool down after 8 pages. Also, check that the hookup is not longer than 6 feet. Dye-sub requires a smooth paper surface.

## OTHERS

Several manufacturers have released dye-sub printers that print images up to 4 x 6 inches.

Canon's CD-200 Digital Printer lists at $490 and has a CompactFlash slot for printing directly from a memory card. Parallel and serial interfaces for both Macintosh and Windows is also included. Resolution is 288 x 144 dpi, and it takes about 90 seconds to print a 4 x 6" (3.2 x 4.2" print area). Cost is about $1 per page. The printer can also be used as a video printer; connecting it to a video source and then printing a still image. Canon only guarantees CompactFlash cards used with Canon cameras, though they say it should work with other cameras.

Sony's FVP-1 Photo Printer offers features very similar to the Canon printer, but it includes a floppy disk drive rather than a CompactFlash card reader. It features a 4.4 x 3.3" print area, has a 300 dpi resolution and prints in about 100 seconds. It includes printer software for Windows operating system. Price is about $495.

The Fuji NX-5 Digital Photo Printer prints at 150 dpi on 4.1 x 5.8" in two minutes. It features a SmartMedia card reader and costs $499. It works with both Windows and Macintosh computers.

The Olympus P-330 Digital Home Photo Printer also uses SmartMedia cards as well as working connected to a computer. It takes about two minutes to print a page. Price is $449. Like the other printers, it can be used to print stills from a video source.

## LASER PRINTERS

Laser printers use toner (powdered ink) which is attached by electrostatic charge to a drum. The toner is heated and fused to the paper. Color laser printers use colored toner to produce final output. Laser printers are the fastest printing devices and print on a range of papers, but they are expensive, costing at least $2,000.

### POLAROID PRINTER

Polaroid offers the Colorshot Photo printer. This inexpensive device "prints" on Polaroid instant film, producing a "photographic quality" print. Obviously, it's limited by the size of the Polaroid film, but the quality will match original Polaroids.

The Colorshot is available for parallel port and USB, but only works with Windows 95 and 98. It's fast too, printing up to four prints per minute.

Polaroid also sells the Direct Connect adapter which acts as a bridge between a digital camera and the printer. It lets you print without attaching the camera to a computer, and includes options to "print all" as well as "print index," which prints several thumbnails on one sheet. The adapter can add text or a logo to each picture too. Not all cameras are supported, you should check the Polaroid site to see if the adapter supports your camera.

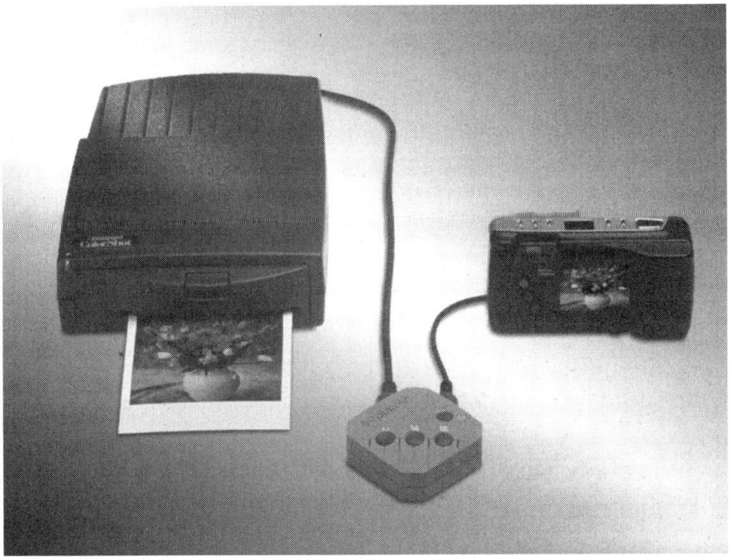

*Figure 9-4. Polaroid Colorshot and Direct Connect Adapter.*

## AFTER IT'S PRINTED

### LAMINATION

Laminating, or coating your prints, may improve their longevity. Just as it's difficult to predict how long a print will last based on the materials used, it's even harder to determine how laminating will affect the lifetime of your print, though informal tests suggest that laminating could double the life of the print.

Lamination should protect your print from environmental effects, and may cut the light rays striking the ink and paper, thus extending the "life" of the print. Unfortunately, lamination changes the look and feel of the print.

There are four "lamination" options. The first is to use a liquid spray, such as Optima's Maxim XE (http://www.optima-int.com). These are relatively inexpensive and easy to apply to a small number of prints. Applying it evenly can be difficult, and you have to be careful that the liquid doesn't interact with the ink or paper you are using.

True lamination processes use a film which has a glue layer applied to the plastic. The plastic is available in a cold lamination format, which requires no special equipment to apply, and in a hot laminate form. For cold mount, the glue is tacky, and covered with paper. To apply, you peel off the paper and put the film onto the work. This doesn't require extra equipment to apply, but it's difficult to apply smoothly to thin paper. Thicker paper, particularly if it has a hard surface, is much easier to work with. You should make sure that the ink has dried before applying. Cold laminates are more likely to separate from the artwork over time, or if the print is rolled or manipulated.

For better protection, a hot laminator should be used. Heat melts the glue and fuses the laminate to the artwork. For best results, this requires a laminator, which can cost several hundred dollars, depending upon the size of the materials you want to laminate. The plastic lamination comes in a number of thicknesses and bases, the most common being polyester and vinyl. Polyester is more rigid and withstands higher temperatures, but vinyl is easier to work with.

Printing prints on a glossy-surface paper makes little sense if you are going to laminate it (apart from the extra thickness of the paper).

## PRINT TRIMMING

While special paper sizes and perforated paper can minimize the need for trimming your prints, inevitably you'll need to trim some photographs to the size you want. A pair of scissors will do the job, but it can be difficult to make long straight cuts with scissors, and tedious if you are cutting lots of photographs.

The adventurous might use a razor or paper knife, a metal straight edge (ruler) and a cutting mat (piece of cardboard) to cut paper. It's the only solution for cutting irregular shapes and is less likely to warp the paper edge than using scissors. But it can be very hard to make a straight edge. If you don't concentrate the blade tends to wander off all over the place. For this reason, a rotary paper cutter is recommended. They're quick, easy to use, and can even be used to trim thin strips of paper.

I have a fairly old one that has a plastic pressure plate that you press down to keep the work in place while using the trimmer. The trimmer runs back and forth on a metal bar. Best results are achieved if you place the paper at the opposite end to the trimmer so that you can get up a head of steam before the trimmer hits the paper. Also, you don't want to cut too many pieces of paper at the same time. I often only cut one page at a time.

Aligning the paper correctly so that the trimmer takes off exactly the right amount of paper can take a little practice. This is partly because of the plastic pressure plate, which makes it difficult to see exactly where the cutter is going to cut!

Guillotine paper cutters look impressive (and can also be used for cutting cabbage!). I've never been a big fan of these devices; the paper always seems to shift during cutting because of the shearing force of the blade. A guillotine trimmer is not something you want to use with children around. While it's possible to hurt yourself with a rotary trimmer, a well designed trimmer covers most of the blade, protecting both small and large fingers.

If you buy a paper cutter, make sure it's fairly sturdy and is large enough to cut the paper that you will be typically using (figure that the most common size is 8 x 11, so get something large enough to fit an 11-inch side). It's worth spending money to get a good one; it should last you many years.

## FRAMING

While you can purchase inexpensive frames at many locations, a really good framing job can make a tremendous difference to a piece of art. If you're framing yourself, using a matte is vital; it creates a space between the glass and the art work. This will prevent the art from sticking to the glass, which can cause all kinds of undesireable effects. Also, using some kind of spray-on adhesive to attach the work to a mounting board will prevent bowing of the paper.

# PRINTING PROBLEMS

The following are just some of the problems you might encounter with your printer. As with other computer hardware, sometimes problems are caused by changes in software or connections. If you are having a problem, check the following:

1. Did anything change on the computer since the last time the printer worked?
2. Is the printer plugged in?
3. Are you using a different printer driver?
4. Have the printer settings been changed? Does changing the printer settings fix the problem?
5. Are you using different paper or ink?
6. Does the printer misbehave with all programs, or just with one?
7. Have you run out of memory or hard disk space? Many printer drivers spool to the hard disk. If you run out of (or are low on) disk space the file may not print.
8. Have your cleaned the printer recently? Is it time you did?
9. Does it need a new cartridge?
10. Does it work with another computer?

## THIN WHITE LINES

There are at least two reasons why you might get thin white lines in your print-out. The first is a clogged print head, the second is a software problem with Adobe PhotoDeluxe, which can print thin white lines after rescaling a file. Check to see if the problem occurs for another unaltered image; if it doesn't, Adobe's recommended solution is not to scale the image!

### EVERYTHING HAS A STRANGE TINT TO IT

Have you tried printing a standard image as a test? Have you checked all of your printer settings? Have you tried printing from another application? Cleaned the print head (check the manual for instructions)? Replaced the ink cartridge?

If you're still having problems, has the printer ever worked? This could be a hardware problem. You may need to clear the clogged head. Most printers have a test page you can run to see if a head is clogged or not printing (is the cartridge low on ink?).

### PRINT SKEWING

Consumer printers don't always feed the paper very accurately, which can result in the image being slightly skewed; not square with the paper.

Does this problem occur with regular printer paper? Are the print guides aligned correctly for the paper? Check the rollers for dirt, lint or hair. Try cleaning the print rollers.

## COMPUTERLESS PRINTING

Two companies have already announced, or are shipping, printers that don't need a computer to print your digital images. Lexmark's Photo Jetprinter 5770 is a 1220 x 1220 resolution inkjet printer which can be connected to a Windows PC using a serial connection. However, it also reads SmartMedia and CompactFlash cards and will print multiple picture formats, frames and text messages. A parallel port Iomega Zip drive can be connected to the printer and you can transfer images from the memory card to a Zip disk.

Epson's PrintOn ET-100 works in a similar way to the Lexmark. It also has a reader for CompactFlash and SmartMedia, but it includes the Digita operating system (see Chapter 20) which makes it possible to perform simple editing in the printer as well.

Both of these printers make it possible to work with a digital camera, and print photographs, without the aid of a computer. You might not even need to buy a computer.

For quick printing of images, these printers are attractive; however, it might be too soon to do away with the computer. Managing your image collection will be more difficult without a computer. Also, your image editing options are much fewer.

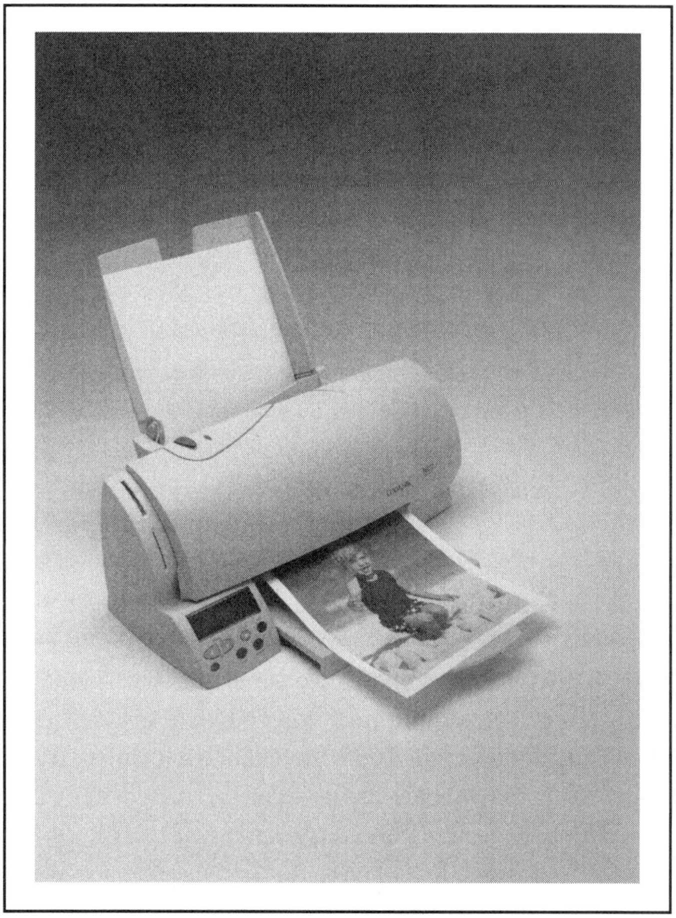

*Figure 9-5. Lexmark's Photo Jeptprinter 5770 (photo courtesy Lexmark).*

### THE FUTURE WILL SEE PRINTERS ON THE INTERNET

The Printer Working Group of the Internet Engineering Task Force, which manages the standards of the Internet, has proposed Version 1.0 of the Internet Printing Protocol (IPP). This standard will allow printers to have their own IP addresses on the Internet. You will then be able to send a file to be printed directly to a printer connected to the Internet. The protocol would also allow you to get the status of a print job or cancel a print job. This is an obvious alternative to faxes, though it could also be used by printer service bureaus. To work, this new protocol requires printer manufacturers to develop printers that support it, and operating systems must also be updated to support these printers so that when you are in your word processor you can choose a printer on the Internet to print to.

But what will happen when printers can be attached directly to the Internet? Will they replace FAX machines? Will it make it easier to use service bureaus? Do users really want their printers to be available to the hackers of the Internet?

## SUMMING UP

Printers are improving in quality and coming down in price. However, a hidden cost is the consumables; the cost of ink and special papers. Whenever considering color printers, you must always include these costs. Print longevity for most inkjets remains a problem and should be a concern; though you can always reprint a photograph, do you want to have to worry about reprinting a photograph ten years from now? For special work that you know you want to last, your best bet might remain with traditional reproduction (using a service bureau that produces a "real" photograph).

And soon you may find that the picture you hang on the wall isn't printed anymore; Sony has announced the Digital Photo Frame, a 5-1/2-inch digital display with a decorative frame. The Digital Photo Frame can display MPEG Movie Mode video or digital still images stored on Sony's Memory Stick media. It includes a slide show mode.

According to the press release, “a wave of one’s hand turns on the Digital Photo Frame with its touchless sensor, and a sleep timer turns it off automatically.” The only downside to this imaging innovation; the sticker price is $999. You could buy a computer for that.

# CHAPTER 10
# ACCESSORIES

When you buy your camera it will probably come with several accessories. At a minimum it should come with cables and software for transferring the images from the camera to the computer. Some cameras include power adapters, rechargeable batteries and a memory card, while others offer these only as optional accessories. Which accessories you will need depends primarily upon how you use your camera. The more frequently you use it the more you need to consider faster methods for transferring images, larger memory cards, and rechargeable batteries and AC adapters. For the casual user, the camera right out of the box may be all that's needed.

This chapter covers memory (including options for reading the cards), batteries, flash and lenses. More information about how to reduce the time it takes to transfer images to the computer is covered in Chapter 5.

## MEMORY CARDS

While the early cameras included only internal memory, most cameras now support memory cards. This allows you to buy additional cards and change them while in the field. These cards are hardly accessories for some cameras—those that don't have memory included in the camera. However, since they can be purchase separately they are included here.

There are three common types of card. PCMCIA is used in high-end cameras and are large enough to hold a hard disk with several hundred megabytes of space. The cards are expensive, but if you have a notebook that supports the PCMCIA cards then you can transfer images very easily from the card to the computer.

CompactFlash and SmartMedia cards are the popular memory cards for cameras under $2000. Sony has also released the Memory Stick which they are planning to use in their cameras. These technologies use nonvolatile memory; they do not require a battery to retain their data, and they contain no moving parts, so their ability to withstand physical shock is much higher than hard disks. Probably the greatest danger to any of these devices is the connector getting damaged through careless insertion or other physical mishandling. Though these connectors are designed for frequent insertions, you should always take care to make sure there are no obstructions when inserting them, and not to push too hard just in case something has fallen into the connector and is preventing reinsertion of the card. Memory cards are generally rated to work within a wider range of operating temperatures than mechanical devices (such as disk drives); if the camera will work, then the memory card will probably work.

Since the CompactFlash cards include a controller, the cost of the smaller CompactFlash cards (4 or 8MB) has tended to be higher than comparable SmartMedia cards, though this difference disappears as you move to the larger sizes. With the volume of sales of CompactFlash cards increasing, this price difference could disappear entirely.

Though these cards are reasonably sturdy, you should protect them from static electricity, particularly SmartMedia cards which have exposed contacts. Running the camera with low batteries should also be avoided; if the camera gives out while writing a file to the memory card it could write a corrupted file. It's possible this could corrupt the directory structure of the card and make it unreadable.

Memory chips are rated for speed, and generally the faster the better. Some manufacturers list the speed of their cards. Faster cards can improve the performance of your camera, particularly if the camera is saving a large file. Differences between cards could result in differences of a second or more in saving a single image. If you've ever had to wait while a camera is saving a file to memory you'll appreciate any reduction in time you can get.

## COMPACTFLASH

CompactFlash cards were developed by SanDisk Corporation, but are now manufactured by several different companies, and the CompactFlash name and logo is administered by the CompactFlash Association. The CompactFlash association is a nonprofit corporation that promotes CompactFlash, and makes the logo and specification available without charging royalties. You can find more information at:

*www.compactflash.org*

Vendors using CompactFlash include: Canon, Casio, Epson, Kodak, and Nikon.

The CompactFlash cards measure 1.7" (43mm) x 1.4" (36mm) x .13" (3.3mm), which is half the thickness of a PCMCIA Type II card. The CompactFlash card has a 50-pin connector, but follows the ATA specification for PCMCIA cards, so a 50-to-68-pin adapter (a CompactFlash PCMCIA Adapter) can be used to plug a CompactFlash card into a PCMCIA slot. The adapters cost less than $20.

CompactFlash cards support both 3.3V and 5V operation, and are interchangeable; any CompactFlash card can operate at either voltage.

The CompactFlash card's built-in controller manages the memory on the chip, detecting errors. In theory, since the card manages the memory, a camera will work with any size memory chip, but there have been reports that some of the very early CompactFlash cameras won't work with cards greater than 2MB.

It's claimed that a CompactFlash card can be used for 100 years or more.

## CAN YOU MOVE COMPACTFLASH CARDS FROM ONE CAMERA TO ANOTHER?

Probably not. While the basic CompactFlash card is the same, and CompactFlash uses a standard file structure, a camera may add its own files and directories that are not recognized by another camera and vice versa. Most importantly, the files that cameras write are not always the same, so images on a card captured by another camera may not be readable by another camera. While you can probably move the card between cameras, reformatting it before doing so is recommended.

Some users have had problems with some CompactFlash cards, and switching to another manufacturer's card has been the only solution. However, you should first try initializing the card on your camera. Also, if you have a card reader for your computer, try initializing the card using the computer. Reinitializing can often solve transient problems.

If your camera uses CompactFlash cards and you have a computer with a PCMCIA slot, then you might want to consider buying a CompactFlash PCMCIA Adapter. It will cost you about $10, and lets you plug the CompactFlash card into the PCMCIA slot. The card then appears as a regular disk drive on the computer, and you can copy files from the card simply by dragging and dropping them. Transfer speed is as fast as any other transfer method, and it's cheaper than buying a CompactFlash reader.

If your computer doesn't have a PCMCIA slot, then you can buy a CompactFlash reader which connects to the parallel, SCSI, or USB port. You could also get a PCMCIA reader and use an adapter. Adding a PC card reader allows you to use other PC cards, but really only makes sense if you are going to use it with other cards in addition to the CompactFlash card.

Choosing between parallel, SCSI or USB card readers depends upon your computer and the connections you have available. Speed wise, SCSI should be the fastest, with USB second and parallel the slowest. All three are faster than serial connections. Windows computers usually include parallel ports but not SCSI, while Macs usually include SCSI but not parallel. USB is a recent interface standard. While it's not as fast as SCSI, it's much easier to set up and use, so if you have USB ports it's the one you should consider first.

## SMARTMEDIA

SmartMedia cards do not include a controller (unlike CompactFlash cards), so the camera must manage the memory handling on the chip. Unfortunately, in some cases this has meant that as the size of the cards has increased, some of the earlier cameras and card readers do not work with the larger cards. This has mostly occurred at the 8MB level. Older cameras won't read cards larger than this size. If you have a camera that uses SmartMedia cards manufactured prior to 1998, and you haven't used a larger card in the camera, you should first check with the manufacturer before you buy a larger card. Olympus has offered some upgrade programs for older cameras.

Vendors that use SmartMedia include: Agfa, Fuji, Olympus and Toshiba.

In general, SmartMedia is very similar to the CompactFlash technology. Which you use depends upon the camera that you buy. While CompactFlash has been adopted by more companies than SmartMedia, the only real reason for choosing between two cameras based on the media is that SmartMedia cards have generally been a step behind CompactFlash in terms of largest card size available. If you want to be able to get the biggest card available then CompactFlash currently is the way to go.

## FLASHPATH ADAPTER

For SmartMedia users, the Flashpath floppy adapter offers a reasonably inexpensive high-speed method for transferring images from the card to the computer. The Flashpath floppy adapter resembles a floppy disk, and the card is inserted into the adapter, which is then inserted into a standard floppy disk drive. You must first install special driver software. Unfortunately, the driver software doesn't provide a transparent interface with the Macintosh operating system; the disk doesn't appear as a drive on the desktop. Instead, you must run the software before inserting the disk, and the software provides a rudimentary interface for selecting and copying files from the card. For the PC, the interface is much better, and the disk does appear as a device that you can open in the Explorer and copy from, much like a real disk.

If you can find this device at a discount price then it's worth considering; otherwise, a SmartMedia reader costs the same (or less) and offers a better interface. Also, the Flashpath adapter is slower and consumes batteries. However, a card reader does require connection to the computer while the floppy adapter simply inserts into a floppy drive.

The Flashpath adapter includes a battery; if you are having trouble using the adapter try replacing the battery. It does not work with nonstandard floppy disk drives, for example, an external drive connected using a PC card, or a high-capacity (120MB) floppy disk drive (FDD).

Transfer times are less than 5 minutes to transfer an 8MB card, which is about half the time it would take with a serial connection.

**WARNING**: The Diamond RIO PMP300 Portable Music Player stores and plays MP3 files from SmartMedia cards. Unfortunately, many users have found that after inserting a SmartMedia card in the RIO player it cannot be read by their digital camera and cannot be reformatted. This might be solvable if you have a SmartMedia card reader and can use the computer to mount and reformat the card, but it's an experiment you might want to avoid.

Also, there seems to be an incompatibility between the Flashpath adapter and the EZBIOS installed by Maxtor large hard drives on older systems that don't handle large block addressing (i.e., large disk drives). The new bios intercepts all I/O requests and the Flashpath driver no longer works.

## OTHER DISK OPTIONS

While no other storage options have yet been adopted, several have been talked about, including the Mini-Disc and the 120MB floppy disks. There are some interim disk storage options available that make it possible to download images from your memory card without having to connect to the computer. Two such devices are the Iomega Clik!, which uses cartridge disks, and Disk MGVision Digital Wallet which contains a hard disk. Both let you connect the camera to the unit, download the images, then later transfer the images from the drive to your computer. Because the Clik! drive uses cartridges, the disks can be used to archive your images, though you should always create a duplicate copy in case a disk becomes unreadable.

## IOMEGA CLIK! DRIVE

The Iomega Clik! drive comes in three configurations: for mobile computers, for digital cameras, and the Plus Pack which includes both features.

The digital camera kit includes the Clik! drive, a parallel port connection, a Flash Memory Reader (for CompactFlash and SmartMedia), a NiMH rechargeable battery, desktop docking station and a disk. Additional disks currently retail at about $10 each, and each disk holds 40MB.

The Clik! drive is designed to be used as a cheaper method for storing images while in the field. After filling your CompactFlash or SmartMedia card, you put the card into the adapter connected to the Clik! Drive. Push the ON button and the unit creates an Images directory on the disk and copies the files from the card to the disk. For an 8MB card, this entire process takes about a minute. You can then remove the card, insert it in the camera, erase the card and start taking photographs again.

Once back home the Clik! drive is connected to the computer through the parallel port and it appears like any computer disk. The contents of the disk can be copied to the hard disk. If you are using the Clik! drive to archive your images, you should remember to keep a duplicate copy of the images on a second disk in case of disk loss.

The batteries are recharged using a separate A/C adapter, which means that if you are going on a long trip you'll have to take the adapter as well. The drive itself is 4.6" long and 1.75" wide without the card adapter and battery pack attached.

At $10 for 40MB, a Clik! disk is cheaper than a card. At current prices, a 32MB card costs $129. However, the unit itself costs $250, which is how much a 64MB card costs. If a 64MB card would hold as many pictures as you'd ever need, you'd probably be better off buying a large, or multiple, memory cards. Also, the card manufacturers are predicting that the price of memory cards should be as low as $1 per megabyte in a year.

However, if you're planning a long trip away, and don't want to take a portable computer with you, the Clik! drive would be cheaper than buying many cards.

The Click! drive creates a new directory each time files are saved from a card. This avoids existing images being deleted if the camera numbers new sequences with the same number; some cameras always increment, while others start at zero after memory is cleared.

So far, no camera manufacturer has announced that they will include a Clik! drive in the camera, though clearly that is what Iomega is hoping will happen. Agfa has said they plan to release a camera using the Clik! drive sometime in 1999. For a limited period Agfa bundled the Clik! drive with the el680 camera. At the moment the Clik! only works with Windows computers.

With newer cameras including uncompressed formats which save files that are 4MB or larger, the 40MB Clik! drive, for many, may prove too little, too late.

## MGVISION DIGITAL WALLET

The Digital Wallet contains a 540MB (or greater) hard disk that can be used to store images from your camera. You connect to the drive using the camera serial connection. The Digital Wallet stores digital images in groupings called *virtual film rolls*, and has a capacity of 99 film rolls. Each roll contains the images transferred from the camera in one session, with a maximum of 256 images per session. The connection to the camera is made using a supplied serial cable, and the drive only works with cameras it has been programmed to support. Currently it supports most consumer Kodak and Olympus cameras, and MGVision says it is working on support for other cameras.

Images are then downloaded from the Digital Wallet by connecting it to the serial port of the computer. Images are viewed and managed using the included FotoExplorer. The Digital Wallet is small, 4.49" x 3.59" x 0.92" and weighs 9.6 oz with batteries. It works with Windows 95/98 and NT.

Digital Wallet provides a means for downloading a large number of images while in the field, though its serial connection makes the downloads no faster than if you were connected to your computer. You can get more information on Digital Wallet at:

*www.mgvision.com*

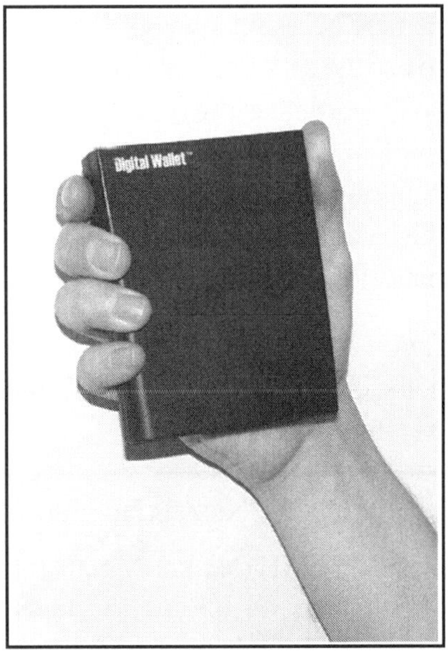

*Figure 10-1. MGVision digital Wallet (photo courtesy MGVision).*

# POWER

## BATTERIES

Digital cameras are high-drain devices. They'll suck the life out of batteries, particularly if you take a lot of pictures in rapid succession. If you're using the flash and the LCD display at the same time, don't be surprised if some cameras start reporting the batteries are low after only ten minutes of use!

The kind of battery you use is important. Many manufacturers are starting to include rechargeable batteries with their cameras because they know that it makes little sense to use regular batteries with their cameras.

There's one exception to this. If you're only taking one picture every three weeks then regular batteries are preferred. They maintain their charge much longer; rechargeable batteries lose charge each day, so if you're only using the camera occasionally you'll probably find that they are always flat when you go to use the camera. Make sure you buy good quality batteries. Brand name isn't so important as the type of battery. Some of the older design batteries don't work well at all with a digital camera. We once inserted some new carbon batteries into a Kodak DC120 and the camera wouldn't even turn on. Perhaps the batteries had been sitting on the shelf for a long time. High-quality alkaline batteries are best

If you are using your camera regularly, and don't already have rechargeable batteries, then you should definitely get them. There are two kinds of batteries widely available:

NiCD (Nickel Cadmium)

NiMH (Nickel-Metal Hydride)

There is a third type of rechargeable battery, **Lithium Ion**, that is becoming popular. A number of cameras come from the factory with Lithium Ion rechargeable batteries. Lithium Ion offers even better performance than NiMH batteries; however, they are more expensive. Also, you cannot currently buy Lithium Ion batteries in the standard AA battery size, so you can't use them with most cameras the way you can with NiCD and NiMH batteries.

Rechargeable batteries lose their charge sitting in the camera. NiMH batteries lose between 2 and 4% of their charge a day. Since many cameras start reporting a low battery at 80% of full charge, the batteries could be reported low in as little as ten days. This number will vary greatly depending upon the age of the battery, temperature, and camera.

Memory effect was a problem with older NiCD battery. This problem has been reduced with newer batteries and chargers, though how you use your batteries may also impact performance. NiMH batteries do not have this problem, though they cost a little more and require a better quality charger. NiCD may perform better in extreme cold than NiMH.

Rechargeable batteries perform a little differently than non-rechargeable batteries. The voltage coming from an alkaline battery decreases steadily as the battery is discharged. NiCD and NiMH have a lower starting voltage than alkaline (typically 1.2-1.3V vs. 1.5V for alkaline), but rechargeables maintain this level until they are almost discharged. So while rechargeables have a starting voltage that's lower than alkaline batteries, this does not mean that the cameras won't work successfully with the batteries. One potential problem; the battery life indicator of the camera may be calibrated for alkaline batteries, which could make it somewhat inaccurate. You may get much less warning that the batteries are getting low on charge. Both NiCD and NiMH can be recharged several hundred times.

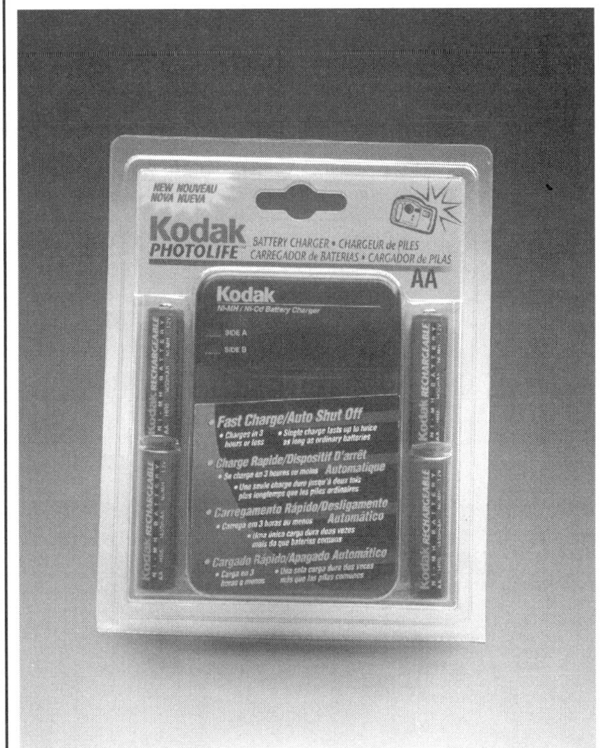


*Figure 10-2. Battery charger and batteries (photo courtesy Eastman Kodak).*

Is it okay to leave batteries in the charger? Depending upon the battery and the type of charger this is either an acceptable practice or something to be avoided. Even if the charger is a trickle charger, or drops to a trickle charge once the battery has been charged, leaving the battery on charge can shorten its lifespan. Most handbooks recommend not leaving a battery on charge for longer than 24 hours.

Those that suggest leaving batteries charging are probably assuming that you will take the pack off within a day or so and not leave them charging for a week or two. NiMH cells are less tolerant of overcharging than NiCd batteries.

Rechargeable batteries are rated by the current they deliver, i.e., 250 mAh, (milliamps per hour), 500 mAh, etc. Chargers are rated by how much current is delivered to the battery in an hour. A slow charge is at a rate of 1/10 the rating of the battery (expressed as .1C) or lower. For a 250 mAh battery that's 25 mAh. A Fast Charge charger may go as high as 1C (i.e., 250 mAh for a 250 mAh battery). While it's probably okay to leave a battery on a slow charge, leaving one at high charge will damage the battery. Expensive chargers include a temperature sensor to detect heat buildup, and will shut off the charge if it occurs. This feature is important for NiMH batteries.

If you're looking to buy a charger, you might want to consider one that includes a 12V adapter for car cigarette lighters.

### WHAT'S THE DIFFERENCE BETWEEN THE TAB AND BUTTON BATTERIES?

Button contacts are the common ones used in spring-contact battery holders. The tab batteries are used by amateur battery-pack builders and allows them to solder the batteries without damaging the cells.

## SOURCES

### BATTERIES

**Thomas Distributing**

1-800-821- 2769 *OR* 1-888-4CBRADIO (1-888-422-72346)

*http://www.thomas-distributing.com/*

NiMH batteries, chargers

**Materials Resources – Battery Products Division**

419 Sherman Drive Scotts Valley, CA 95066-3631
831 430-9650 or 888 502-6611
Fax: 831 430-9651

*http://www.mrbattery.com/*

NiMH batteries, chargers

**The Battery Station, Inc.**

557 S. Milner, Unit A, Ottumwa, Iowa 52501 USA (515) 683-7621

*http://www.batterystation.com/*

NiCd and NiMH batteries, chargers

**GP Batteries**

11235 West Bernardo Court
San Diego, CA 92127
Tel: 1 (619) 674-5620 Fax: 1 (619) 674-7237

*http://www.gpbatteries.com/*

Battery manufacturer. NiCd, NiMH batteries, charger

**Nexcell Battery**

1251, S. Shamrock Ave., Monrovia, CA 91016, U.S.A.
TEL 1-626-3580121
FAX 1-626-3585322

*http://www.nexcell-battery.com/*

Battery manufacturer

NiMH Batteries

### AC ADAPTER

If you're using the camera close to AC power, buying an AC adapter for your camera may make more sense than using rechargeable batteries. An AC adapter is a good reliable way to keep the camera working at its best, though the thin cord trailing from the camera can sometimes be a pain.

## LIGHTING

### FLASH

Flash and flash units are covered in Chapter 4, but in addition to the built-in flash unit you might want to consider getting additional flash units so that you bounce the flash off the wall, add fill-flash, and reduce red-eye.

If your camera has a PC sync jack, then you can use any flash unit that comes with a PC sync cord. If your camera lacks this feature, then a slave unit that fires when the flash on the main camera fires is your only alternative. Slave flash units are available, and you can also get a flash trigger which connects to the flash and fires it when the flash is fired. The slave is a light-sensitive trigger that fires the flash when it detects the flash from the camera.

### ADDITIONAL LIGHTING

Windows are a great source of additional light. A large white card can be used to bounce sunlight into faces or onto objects. Incandescent and fluorescent lights are often the only choices when inside. Incandescent generally gives an orange cast to pictures, while fluorescent gives a blue-green cast. This can be corrected using filters, or by adjusting the color in your editing software.

Photofloods are available at camera stores and are inexpensive lights that you can use, though you need a lamp that will accept a high-wattage bulb (often 250W or more). These lights put out a lot of heat too, and if you use several of them make sure the wiring and circuit breakers at your location can handle the combined wattage.

## TRIPODS

Available as tripods and monopods (a single leg) in various sizes and weights, a tripod has many uses, including panoramic photography (see Chapter 15). With tripods the factors to evaluate are weight and head design. The heavier and larger the tripod the less likely you are to want to carry it long distances; however, a light and flimsy tripod will suffer from camera shake. If you simply want a place to rest a camera now and again, pretty much anything will do, but if you are working with tripods a lot, a smooth head that can be pointed quickly and easily in any direction and locked into position is worth the extra money.

For panoramic shots, a panoramic tripod head makes it a lot easier to set up the orientation of the camera, align the nodal point, and helps you with taking the pictures with click stops that are used to quickly turn the camera to the next position. Two companies make panoramic tripod heads in a variety of styles, including some specifically designed for digital camera models.

## TRIPOD MANUFACTURERS

**Manfrotto**

*www.manfrotto.com*

**Sima Products Corporation**

*www.simacorp.com*

**Slik Corporation**

*www.slik.com*

**Velbon**

*www.velbon.com*

### PANORAMIC TRIPOD HEAD MANUFACTURERS

**Kaidan**

(215) 364-1778

*www.kaidan.com*

**Peace River Studios**

(617) 491-6262

*www.peaceriverstudios.com*

## LENSES

One of the greatest limitations of digital cameras is that they are predominantly rangefinder cameras; there are no interchangeable lenses. Most come with a built-in zoom lens, but these generally fall within the range of 35mm to 105mm (35mm camera equivalent). For this reason, add-on lenses—which fit in front of the regular lens and either extend the lens, convert it to wide-angle, or provide macro capabilities—have been produced. Some are made by third parties, and some are offered by the manufacturer. Unfortunately, some cameras lack a filter thread (a ring with a thread that can be used to attach a lens in front of the camera), and Kodak actually recommends not attaching a lens to the front of the DC260.

For this reason, if you are thinking you'll need a wide-angle lens or teleconverter, then you should check that the camera can be fitted with one before buying the camera. Another consideration is the quality of the lens; in the past, the quality of the images and the resolution produced by most digital cameras was so poor that faults in an add-on lens were likely to be seen only if they were egregious; for example, a wide-angle lens that created vignetting at the edges.

For wide-angle, you might have to accept some vignetting when the zoom lens is at its widest setting (zooming out a little is the only solution). You should definitely try a lens or find a review performed with your camera before buying a lens if it's wide angle. Lenses specifically made for your camera offer some hope that they will work correctly (though this doesn't mean they will be perfect). And a third-party lens that works fine with another camera might not work well with yours.

Don't forget that whatever type of lens you buy, the rangefinder won't work with the lens, and you'll have to use the LCD, or download the images to see how the lens affects your camera.

Close-up lenses allow you to get closer to an object—maybe as close as an inch or two. If your camera has a macro function you may not need a close-up lens unless you need to get very close! These lenses are usually sold by diopter rating (+1, +2, etc.)—the higher the number, the closer you can get. Some are sold in sets and you can use them in combination to get different strengths.

## FILTERS

Filters let you add effects or adjust the image. An *infrared* filter (87c) can be used to create an infrared image, though you'll need a long exposure to make it work. CCDs are sensitive to infrared light.

Cokin offers a *sepia* and sepialight filter which produces results that are similar to black and white. But most image editors offer a function that will convert an image to grayscale (black and white), so it seems pointless to use a lens to produce this effect.

A *polarizing* filter can be very useful for adjusting light effects, but unlike a wide-angle or teleconverter (where it's possible to point the camera and hope the picture turns out), it's very unlikely you can get a polarizing filter to work properly without seeing the results immediately through the camera LCD.

To adjust for tungsten and incandescent lights (which add too much orange to the scene), an 80A blue filter can be used.

### OTHER CONNECTION METHODS

Transferring images to the computer using the serial connection can be tedious. Chapter 5 contains other suggestions for reducing transfer time.

## CONCLUSION

Digital cameras usually come ready to go out of the box, but a few extras can improve your photography, and make the picture taking experience a lot more pleasant.

# CHAPTER 11
# OTHER USES

There's more to digital cameras than just pointing the camera at a subject and snapping a picture. It's possible to use them for digitizing slides, underwater photography, astrophotography and for many other purposes. You'll also find information about using digital cameras for other purposes in Chapter 15, Chapter 18 and Chapter 19.

## CAPTURING SLIDES USING DIGITAL CAMERAS

While a slide scanner will produce much better results, it's possible to use a digital camera to "digitize" a slide or negative provided you have a close focus lens or macro capability. Some users have used a film loupe and simply put the camera lens over the loupe—with a little practice it can work; you'll need some kind of illumination to light the slide (a light table is one alternative). While a loupe might work, if you're going to do this regularly you might want to consider a close-up lens.

One of the major differences between a slide scanner and using a digital camera is that resolution will be lower (unless you have a high-end camera) and the dynamic range the digital camera can capture will usually be lower.

You'll need something to hold the slide or negative while shooting it, and a light source to illuminate the back of the slide. This can be done using almost any type of light, though you'll have to color correct if the light is not a photo light and the camera has no white balance capability. If you have a flash with a PC cord, you may be able to use that to illuminate the slide, though this will take some experimenting. Whatever you do, you'll want to turn the camera flash off.

When digitizing negatives you'll need to use a software editing application to convert the negative to a positive. This will be a problem with color negatives since it's not a true negative; the color negative has a color cast added to it. This can be corrected using color balancing, and some editing tools include a function specifically for converting a color negative to a positive image.

One other suggestion. If you have a slide projector, you can project the image and then use the camera to capture the image. You'll need a matte white surface and the results won't be quite as vivid as photographing the slide directly, but it might be an easy way to quickly copy a lot of images.

## DIGITIZING BOOKS, NEWSPAPER AND PHOTOGRAPHS

Depending on the resolution of the camera, the lighting options, and how you plan to use the image, a digital camera can be a very effective device for copying documents or a page from a book. This is particularly true if you simply want to capture a representation of the page, or digitize an illustration, or headline from a newspaper. The real test, particularly for low-resolution cameras, is OCR (Optical Character Recognition).

OCR works best with high-contrast, high-resolution scans. Usually it requires about a 300 dpi scan. For an 8.5 x 11" page, that's a 2550 x 3300 image. Unless you have a high-end digital camera you're not going to be able to get that kind of resolution with a single image. One solution would be to take two or four shots and then paste them together in an editing program. Unfortunately, it can be difficult to align the images. Alternatively, you can shoot multiple shots of sections that need to be digitized, and then OCR each separate section. Again, this could be a lot of work in OCR— processing, and then cutting and pasting the recognized text together. A regular film camera might be a better solution.

Generally cameras do well with nonglossy paper (newspapers, textbooks etc.), but glossy pages such as magazines can be very difficult to capture accurately. Don't even bother trying the flash with glossy paper. Even for matte paper, most cameras tend to over-flash if the camera is too close to the subject. For this reason, you might want to try using natural lighting or a flash diffuser. Or bounce the light off the wall or ceiling.

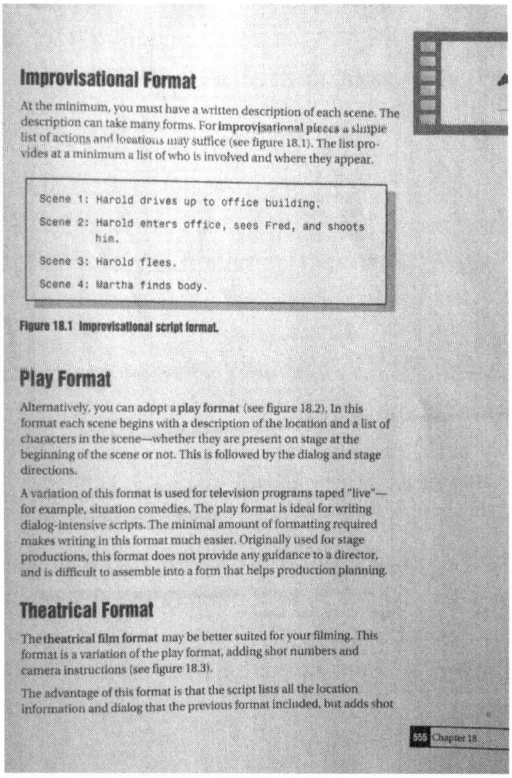

### Improvisational Format

At the minimum, you must have a written description of each scene. The description can take many forms. For **improvisational pieces** a simple list of actions and locations may suffice (see figure 18.1). The list provides at a minimum a list of who is involved and where they appear.

```
Scene 1: Harold drives up to office building.
Scene 2: Harold enters office, sees Fred, and shoots
         him.
Scene 3: Harold flees.
Scene 4: Martha finds body.
```

**Figure 18.1 Improvisational script format.**

### Play Format

Alternatively, you can adopt a **play format** (see figure 18.2). In this format each scene begins with a description of the location and a list of characters in the scene—whether they are present on stage at the beginning of the scene or not. This is followed by the dialog and stage directions.

A variation of this format is used for television programs taped "live"—for example, situation comedies. The play format is ideal for writing dialog-intensive scripts. The minimal amount of formatting required makes writing in this format much easier. Originally used for stage productions, this format does not provide any guidance to a director, and is difficult to assemble into a form that helps production planning.

### Theatrical Format

The **theatrical film format** may be better suited for your filming. This format is a variation of the play format, adding shot numbers and camera instructions (see figure 18.3).

The advantage of this format is that the script lists all the location information and dialog that the previous format included, but adds shot

555 Chapter 18

*Figure 11-1. Book photographed using a digital camera.*

# CAPTURING 3D OBJECTS

Digital cameras can be used for creating the illusion of a 3D image or object using the immersive technologies described in Chapter 15. They can also be used to help create 3D models, using a couple of tools from MetaCreations.

MetaCreations' Canoma is a program which allows you to create 3D models from 2D photos or scanned images. To use the program you use a 2D picture as a base, then add 3D primitives on top of the image. If the picture was of a building, a 3D block is placed "on top" of the building. You have to define the horizontal plane, and complicated shapes will take a bit of work, but once you're finished the program will map the surfaces to the shapes and you can start manipulating the 3D scene. You can also add other surfaces to fill out the scene.

Single or multiple images can be fused together to form 3D models. Incremental modeling makes it possible to add geometry or detail at a later time. Built-in animation tools generate fly-through animations and outputs to QuickTime movie files.

Canoma is available for Windows 95/98/NT and Power Macintosh. Files can be exported in a number of 3D formats.

MetaCreations is also developing MetaFlash, a 3D camera that uses a special flash and converts digital photos into 3D wireframe models. The technology is supposed to be released soon, but there are no example images or good descriptions of what it is or how it does what it does.

There are several other products available that can also be used to convert 2D images into 3D objects. They include:

**Apollo Software Photo3D**

*www.photo3d.com*

**Eos Systems PhotoModeler Pro/Lite**

*www.eossystems.com*

**3D Construction 3D Builder Pro/Lite**

*www.3dconstruction.com*

While PhotoModeler and 3D Builder are the most sophisticated of the three, Photo3D is available for only $38.

# UNDERWATER PHOTOGRAPHY

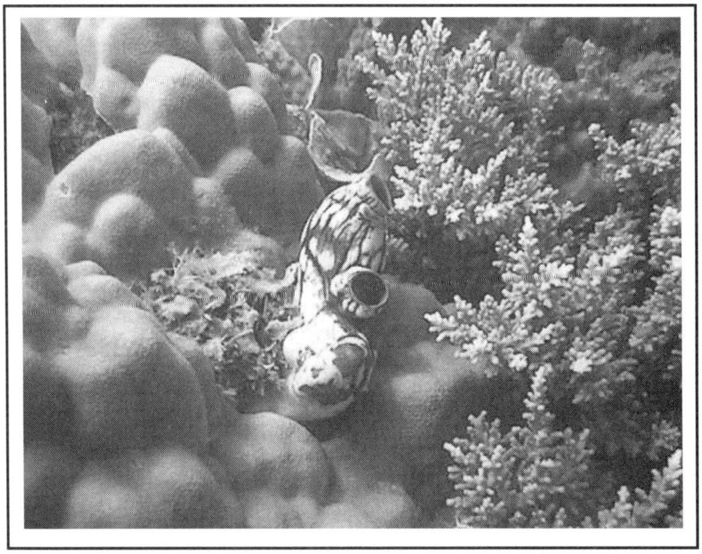

*Figure 11-2. Tinucate and corals image taken with Hitachi MPEG-1A camera using an Ikelite underwater housing (copyright Rob and Robin Burr).*

## INTERVIEW

**Rob and Robin Burr** are website developers based in Coral Gables, Florida. They use digital cameras for capturing images for use in website development. They've also used them for underwater photography and they write for a couple of on-line magazines as diving correspondents.

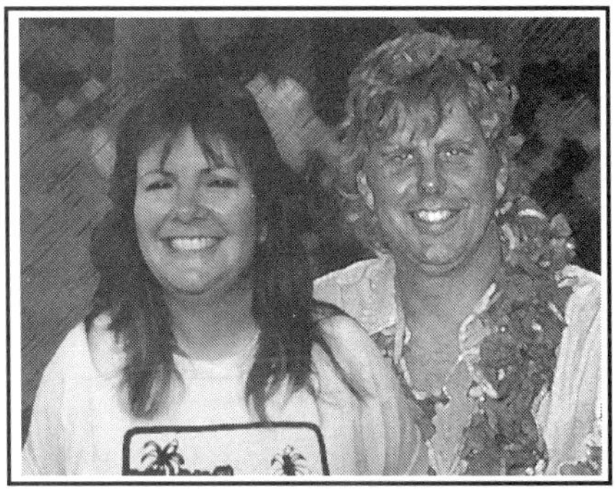

*Figure 11-4. Robin and Rob Burr.*

**Q:** Are you underwater photographers who develop websites, or website developers who are underwater photographers?

**A:** Website development is our primary business and underwater photography is a service we offer. Although the underwater photography is a hobby, it's also part of our profession, as we have many dive resort clients.

**Q:** What do you think of digital cameras?

**A:** We could not be as efficient as we are without a digital camera. It cuts out all the cost of film, processing and then scanning, but with the same end results. The Hitachi MPEG-1A holds 3,000 JPEGs on the 260MB hard drive, or 20 minutes of MPEG video which is ready to go on the World Wide Web.

**Q:** What digital cameras have you used?

**A:** Hitachi MPEG-1A, Canon PowerShot, Kodak DC-260, Sony Mavica, and Sony DX1000.

**Q:** When do you use film cameras?

**A:** We use Nikon N90's and Nikon 8008's for topside and underwater photography that we know will be published in printed matter. For large photos in brochures, etc., there is still not enough resolution in the digital cameras.

**Q:** Can you use any digital camera underwater?

**A:** It is mostly the housing manufacturers that market the cameras—they pick a model they think will be a popular one based on features, etc., develop a housing for it, and then market it to the diving public.

**Q:** Any tips or comments you'd like to share for underwater photography?

**A:** Use a monopod or tripod to ensure steadiness—it is very important to hold "still."

You can find examples of Robin & Rob's work at websites they have developed:

*http://www.provo.net/PNetClients.html*

*http://www.BirdCage-USA.com/*

*http://www. VenetianPool.com/*

As well as two publications they write for:

*http://www.greatoutdoors.com/diving/correspondents/burrs/index.htm*

*http://sofla.com/recreation/outdoors/scubadiving.html*

Email: Robin@QuantumLeap.net

Website: *http://www.QuantumLeap.net/*

Using a digital camera underwater is almost as easy as using a film camera. Housing manufacturers are already selling housings specifically designed for different digital camera models.

Using a digital camera underwater poses some problems, not the least of which is it's impossible to change batteries or memory cards underwater.

When taking photographs underwater, lack of light results in longer exposure times or the necessity for additional lighting. Strobe lights can solve some problems but exaggerate others, particularly the effects of silt in the water. When using longer exposures minimizing camera movement is vital.

A camera in a housing will affect your buoyancy, so practice using it before you have to use it "in the field." Just as when taking long exposures on land, you should hold your breath when pressing the shutter underwater; breath control while underwater will help improve the sharpness of your pictures. Tripods are difficult underwater, but leaning against something (particularly in a wreck) can aid stability.

Remember that many cameras feature an autofocus function, which you'll either have to disable, or trick. Placing a piece of tape over the infrared receiver (if the camera uses infrared for focusing) will force the camera to focus at infinity. Depending upon the type of underwater housing, and the camera, you may or may not be able to access certain controls on the camera. Obviously, you won't be able to change memory cartridges, so using as large a card as possible makes sense.

Underwater photography tends to exaggerate the effect that small flash units produce—brightly lit subjects with very dark backgrounds. Also, if there is sediment or silt in the water, a flash pointing forward will exaggerate the sediment. Rather than using the built-in flash, use an external flash and move it away from the camera.

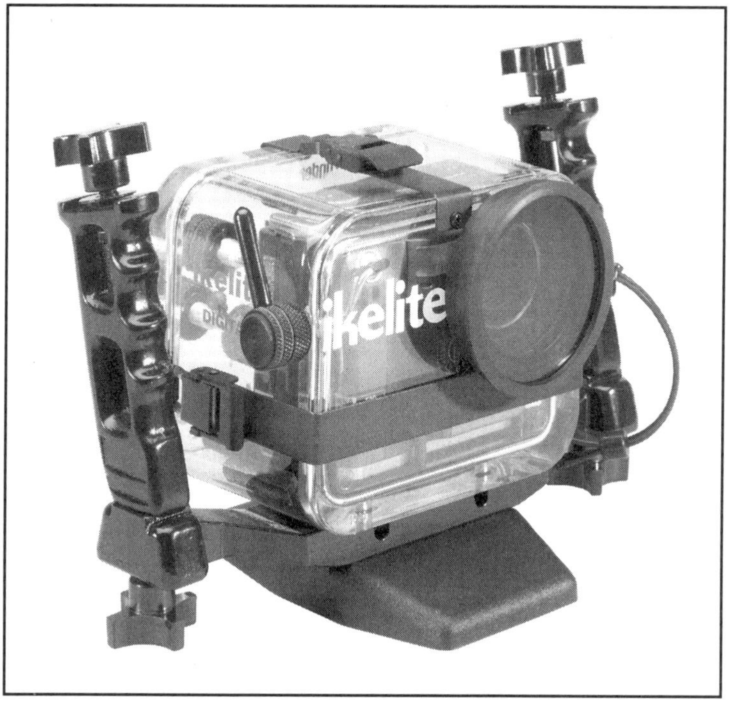


*Figure 11-4. Ikelite housing for Mavica FD-81 (courtesy Ikelite).*

## ASTROPHOTOGRAPHY

A number of people have experimented with using consumer digital cameras with telescopes to take astrophotographs. While these can work well for photos of the moon, or for photographs of terrestrial subjects, other astronomical subjects are much harder to photograph. The long exposure times needed for many subjects makes for a big strike against consumer digital cameras.

For those interested in using a digital camera to capture astronomical subjects, a digital camera built just for that purpose may be the best bet. You can even build one yourself. The Cookbook 211 and 245 are CCD cameras for this purpose that are described in the book *The CCD Camera Cookbook* by Viekko Kanto, John Munger and Richard Berry. The cameras are based on two chips from Texas Instruments; the TC211 CCD and the TC245 CCD chip. The book contains step-by-step instructions as well as the software needed to run the camera. You need a PC to run the software, and the camera must be attached to the computer to work. These cameras cost between $250 and $350 to build. You can also buy the camera as a kit, which raises the price.

The camera is grayscale, but you can produce color images by taking three exposures through different colored filters and then using software to combine the three images. A noticeable difference between this and a consumer camera is that it includes thermoelectric cooling, which allows exposures of up to 60 minutes.

The book costs $29.95 from Wilmann-Bell, Inc. (*http://www.willbell.com*)

The Cookbook CCD Camera Home Page:

*http://www.willbell.com/cookbook/cookbook.htm*

Mike Chaney is the author of Qimage, an image viewer and printing application for Windows. He's also taken some astrophotographs, including the moon photograph reproduced below. Says Mike, "The shot was taken with an Olympus D-600L digital camera. The method was typical *afocal* astrophotography. The camera was on a tripod with the lens very close to the eyepiece on the telescope. The moon was lined up while looking through the camera viewfinder and the self-timer was used to snap the shot so that camera shake was reduced."

The resulting photograph is extremely dramatic despite the random noise in the black space (which appear as light gray pixels) and the odd patterning at the bottom of the moon where it goes into shadow. The yellow halo is reduced if the photograph is changed to grayscale, but this also reduces the dramatic colors of the moon surface that appear in the image (the image is included on the CD-ROM so that you can see it in color).

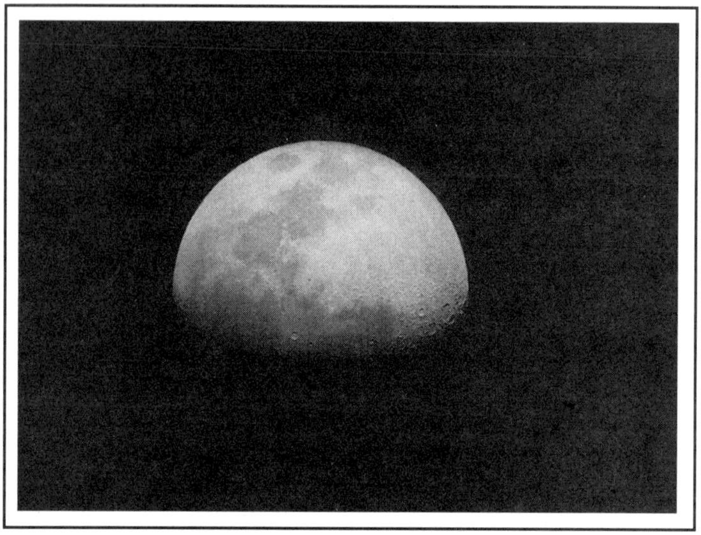

*Figure 11-5. Moon taken with Olympus D-600L (courtesy Mike Chaney). (*http://www.charm.net/~mchaney/*)*

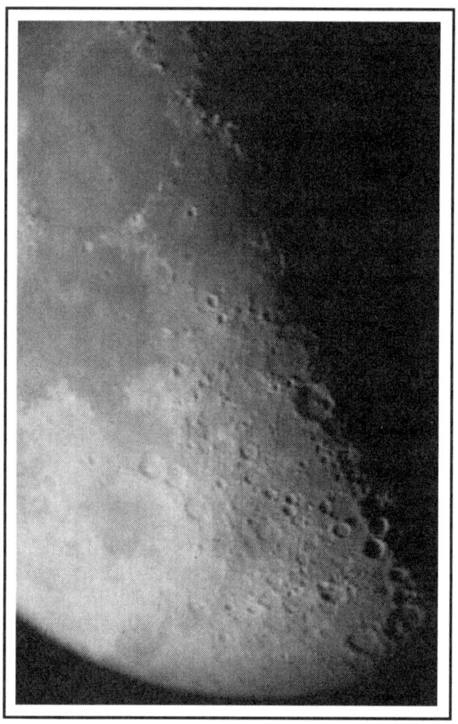

Another photographer, James Shaffer, Jr., has used a Casio QV-770 and a Celestron 8" f/10 Schmidt-Cassegrain telescope to photograph sunspots, the moon, and some other astronomical subjects. Since the camera lacks a threaded lens, he also is limited to aiming the camera through the eyepiece and taking a photograph. He uses a variety of eyepieces—26mm, 20mm, and 15mm—and variety of camera settings.

*Figure 11-6. Moon taken with Casio QV-770 (courtesy James Shaffer, Jr.). (*http://woodstock.csrlink.net/~jshaffer/casio.htm*)*

# CONCLUSION

Digital cameras can be used for a variety of tasks, but be prepared to experiment with settings and your setup to get the best results. Fortunately, you don't have to pay for film, so you are better able to experiment. Unfortunately, each camera is different in the way the CCD works, the exposure controls, and in many other details. For that reason, for many tasks, the best settings and procedures for one camera may completely fail for another model camera.

CHAPTER 12

# INTERVIEWS

**Moose Peterson** is a wildlife photographer who has been shooting wildlife (figuratively!) since 1982. He's also written for a number of photography magazines about wildlife photography, and his website contains a wealth of information for those interested in the subject. There are past articles he's written and a breakdown of what he carries in his camera bag, as well as suggestions for learning about exposure. This website is worth a visit. Moose conducts seminars around the country and also offers safari expeditions where you can go out shooting with Moose. Find out more at:

*www.moose395.net*

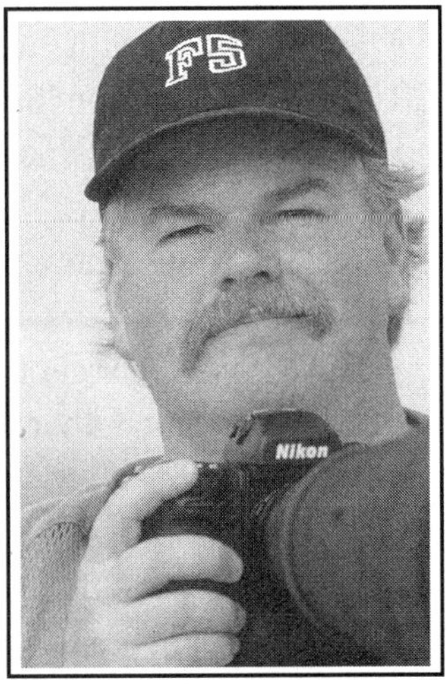


*Figure 12-1. Moose Peterson.*

**Q.** How long have you been a wildlife photographer?

**A.** I've been shooting wildlife since 1982. How I got started, well there was no plan or design, the cards just laid themselves out in such a way that photography was really my only obvious choice. I guess it was just meant to be.

**Q.** Do you have a specialty?

**A.** I'm particularly interested in photographing the life history of California's rare, threatened and endangered wildlife.

**Q.** How were you introduced to digital cameras?

**A.** Nikon asked me to produce some images to help launch the Coolpix 900 in March '98. Prior to that I had looked at a couple of digital cameras just for the sake of knowing a little about them, but the Coolpix 900 was the first time I shot a lot of digital images.

**Q.** So what digital camera are you using now?

**A.** At the moment I only own a Coolpix 900 which I use for many, many things. Shooting products shots for our *BT Journal* and website are the major uses. I also take it with me on projects where I take various images to help better learn the technology and build up a library of digital images. Recently I've been shooting with the Coolpix 950 and the Kodak DCS 620. The DCS 620 is a big step up from the Coolpix; it's based on a Nikon F5 camera body and I can use all of my Nikon lenses.

**Q.** Does the DCS 620 feel and operate just like a Nikon F5?

**A.** The DCS 620 is larger than the F5, which for someone with smaller hands might be a problem. Other than the firing rate, one wouldn't know the difference from the two—quite amazing when you think what Kodak shoved in that body!

**Q.** So if you put the same lens on the DCS620 and an F5, would the resulting image be exactly the same (framing, field of view, exposure) or are there noticeable differences?

**A.** The image size captured by the DCS 620 is smaller than that captured by the F5 with the same focal length lens. For example, if shooting with a 600mm lens on the F5, it would be equal to a 800mm on the DCS 620. But as far as performance, DOF and everything else, it's the same on both.

**Q.** What computer/software do you use?

**A.** I have a hot-rodded PC clone. I use Adobe Photoshop 5.0 and Smart Pix Manager for digital images

**Q.** What is the biggest problem you have with the technology?

**A.** The limited firing rates are a problem, I'm use to being able to shoot eight frames per second. Even the DCS 620 has limitations. You can fire off 12 frames at about 3 fps, but then you have to wait because the buffer is full. This makes it hard to capture action.

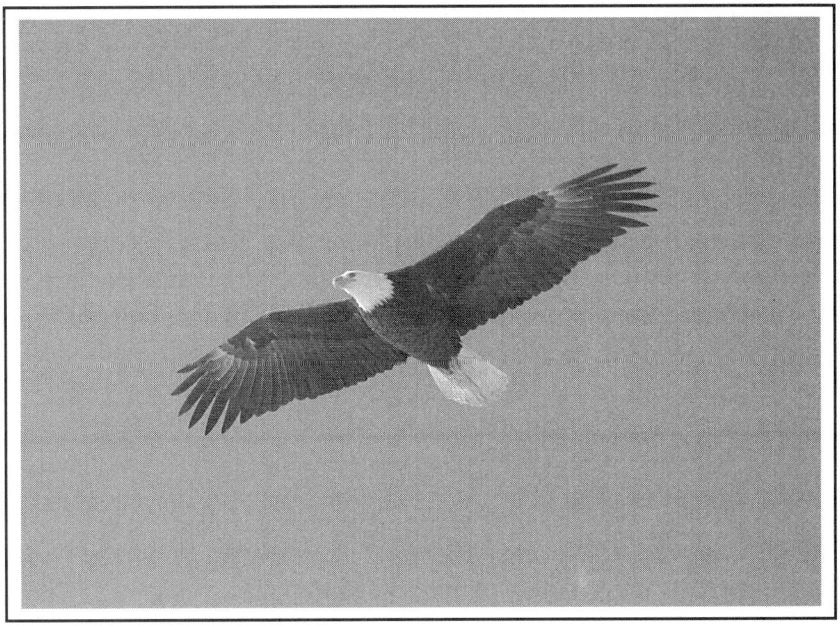

*Figure 12-2. Photo captured by Coolpix 950 on Lexar DigitalFilm (courtesy Moose Peterson).*

I also don't like dealing with file names and the limited image space on the memory cards. But this is a new technology—I know these problems will be solved in a short time.

**Q.** So when do you use a digital camera instead of a film camera?

**A.** I use 35mm film cameras 100% of the time and digital to complement the 35mm. In wildlife/outdoor photography, the industry isn't quite set up yet to accept digital images. They still want to see a sheet of slides. In the future that will all change and I'm using and learning digital photography now so that when that time comes I can step right in and not have to catch up.

**Q.** So what's the biggest differences between a digital camera and a film camera?

**A.** Oh, there's tons of differences, good and bad. Environmentally, the digital camera is great because we're not filling up landfills with film canisters and spent spools or our waterways with spent chemicals from processing the film.

On the other hand, the initial cost of DigitalFilm hurts. On the other hand, it's renewable, takes up little space and can be checked in at the airport rather than hand-checked. On the other hand, it holds limited amount of images and can only capture so many images before the buffer is full and you have to stop to let it catch up.

So, there's a lot of trade-offs. With a digital camera you don't have to wait for the pictures to come back from the lab—you can see your images and file them instantly. But you need lots of disk space to store the images and must make constant backups to insure you don't lose any.

**Q.** So what's the future hold?

**A.** I could go on and on about the differences but they really aren't that important. Digital cameras and images are here to stay, I have no doubt, and I for one am very excited by the possibilities and potential.

**Susan Murie** is an artist who specializes in hand-colored photography. She's also worked in digital editing and has published a number of handmade books, all of which were designed using the computer. She used a digital camera for the images in the book *The Days Flew On Wings*.

Lately she's used digital cameras—as well as film cameras—to photograph objects that she's been selling on the eBay auction site.

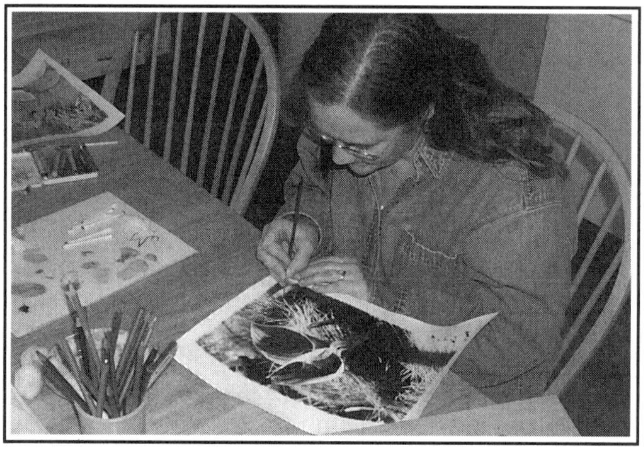

*Figure 12-3. Susan Murie.*

**Q.** How did you get involved in photography?

**A.** In college. There was a darkroom in my dorm and somebody taught me how to use the darkroom. I already had a camera, and that was pretty much it!

**Q.** What is hand-coloring?

**A.** It's done with black-and-white prints on which you apply light layers of oil paints to create color.

**Q.** As well as hand-coloring you've done a lot of image editing work. How do you compare hand-coloring with image editing?

**A.** It's like comparing apples and oranges. The outcome is similar, but the process is not the same. The materials are very different. With computer editing it's more abstract—you don't have your hand on it. For the books I use the computer a lot because of its availability, and because it's more flexible. I like them both equally, it just depends on what I'm doing.

**Q.** How did you become interested in hand-coloring?

**A.** I can remember reading an article in a magazine about how to do it, and I liked the look of it. The article made it seem simple and that was that. I've been hand-coloring for about ten years and now I even teach a workshop on how to do it.

**Q.** What do you use a digital camera for?

**A.** Mainly to photograph items that I sell on the Internet. Occasionally, I use it for personal pictures. One of the things I do is make books and I have used digital cameras to take pictures for one of the books. I've used a Kodak DC120 a fair bit, and then had a DC260 for evaluation, which was really nice. It's a much better camera.

**Q.** What computer and software do you use?

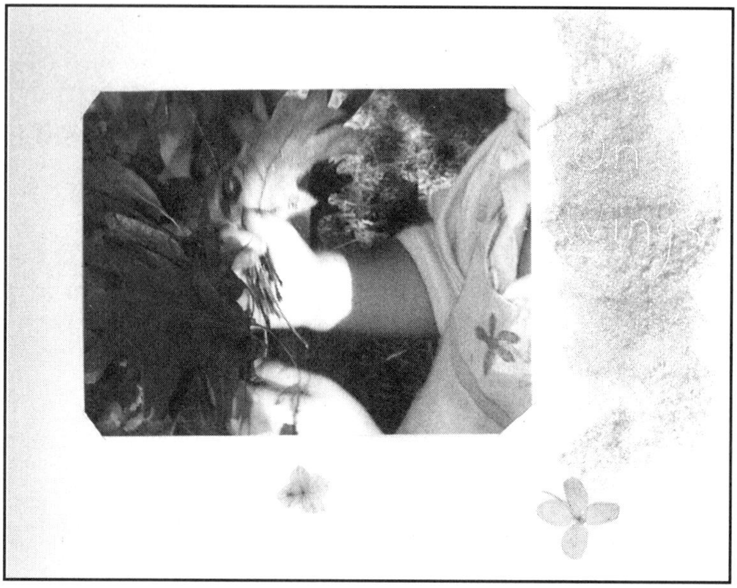

*Figure 12-4. Page of book* The Days Flew On Wings *(courtesy Susan Murie).*

**A.** I have a Macintosh with 64MB of RAM and I use Photoshop most of the time. Some images are scanned in using a flatbed scanner. The book layouts I do in PageMaker. That's pretty much it, though I use a program called Fetch to put images up on the Internet.

**Q.** What do you like about digital cameras?

**A.** They're quick and you can see what you've done right away. Unfortunately the DC260 went back, so at the moment I've been using a film camera to take a lot of pictures for eBay, which means taking the film down to be developed, then scanning it.

My biggest complaint with the earlier camera was the poor resolution. The DC260 was nice.

**Q.** You'd been using the digital camera mainly for eBay, what advice do you have for people using a digital camera for that?

**A.** You've got to shoot the objects in as much light as possible. The flash often washes things out, so I sometimes take things outside to photograph them.

**Q.** Do you edit the pictures much?

**A.** I find I generally have to de-saturate the image a bit. I can't remember whether I had to do that with the digital camera, but I know with regular photographs I have to de-saturate. Generally I found the color to be very good with the DC 260, but with regular film and scanning, getting the color to match the color for a painting, for example, can be a real problem. You also usually have to trim the images a bit to cut out unwanted backgrounds. Sometimes I see pictures and people haven't bothered to do that.

**Q.** How do you decide what size to make the photographs?

**A.** Generally you're concerned about the size of the pictures, and the time it takes to load. Some people size the picture way too big, and it's off the screen and you have to scroll around to see it. You want to size the image to fit small screens (640 x 480). It's better to do two smaller photographs than one really large one. Sometimes you have to show different angles, or a close-up for details. The images are then sharpened and color corrected to give a fair representation of the item. The final step is saving to JPEG.

**Q.** What's the most common mistakes you see on eBay?

**A.** Blurry pictures is the most common mistake. They probably are not taken with a digital camera but with a regular camera that was unable to focus on an object that close.

I bought something once, and the picture on eBay looked a lot better—the colors were a lot brighter—than the actual thing, which is not a good idea. Though you want to sell an item, it's better to have your picture look a little worse than having it look too good!

**Q.** What are you currently working on?

**A.** I was just in an exhibition of handmade books and am now working on an exhibition of new hand-colored photographs.

**Barry Fogarty** is President of DigiNiche Inc., a specialty shop whose expertise is in creating interactive images, such as 360 degree "panoramas" and multiposition "object movies"—what they refer to as "images that work."

**Q.** How did you get started in photography?

**A.** I've been interested in photography since I was a kid and when I sold my first image when I was around 19—a shot of the musician George Thorogood—I realized that it might be possible to make a living out of this medium. By the way, my accountant thinks I was wrong about this.

Shortly afterwards I enrolled in an audio/visual production program at Humber College. While in the A/V program I became interested in going beyond static photography and incorporating images into multiple-projector slide shows, and I guess that my work now is a logical extension of that.

After college I spent five years working in Africa producing communication materials (posters, flipcharts, comic books, slideshows, etc.) for development projects.

**Q.** So what kind of work do you do now?

**A.** In our work we try to marry the art of photography with the science of interactive technologies.

**Q.** But you're still involved in photography?

**A.** Truth be told, I'm not a particularly good studio photographer and I'll generally collaborate with other shooters on these projects, which is something I really enjoy. In these situations I tend to be more of a director during the shoot, and then I handle the post-production.

**Q.** What do you use a digital camera for?

**A.** We use digital cameras for creating both panoramas and object movies. It's really just in the last six months that using 35mm gear has become the exception rather than the norm, though it still is decided project-by-project.

**Q.** What model cameras do you use?

**A.** I have a Canon PowerShot 600, a Nikon CoolPix 900, and a Canon Optura DV video camera. I've had the PowerShot for about eighteen months. Prior to that I hadn't been impressed with the image quality of the Apple QuickTakes and other digital cameras I'd experimented with.

About four months ago we got the CoolPix, which has become our workhorse, and I continue to be surprised at what it's capable of. We're looking forward to getting the new 950 within the next couple weeks.

I've had the Optura for about 9 months, and it's pretty amazing—it's a DV (Digital Video) camera that has a sophisticated still photo mode in which you can control aperture and shutter. I've also used the new Canon Pro70 and I've been pleased with the results.

**Q.** What computer/software do you use?

**A.** The usual suspects....desktop and portable Mac G3's with 128MB of memory. For software I use Adobe Photoshop, Acrobat, Illustrator, Macromedia Director, Apple QuickTime VR Authoring Studio, Live Picture Reality Studio, BBEdit, and Electrifier Pro

**Q.** What do you like about the cameras?

**A.** Because the Coolpix is our workhorse, I'll refer only to it. I like the optics, which are surprisingly—at least to me—clean and sharp. The metering provides the standard Nikon options of matrix, center weighted and spot. The ability to white balance and lock exposure is absolutely critical in order to get good panoramic images, along with the availability of a 24mm adapter. There's also a fisheye, though I haven't tried that yet.

**Q.** Anything you dislike?

**A.** The design takes some getting used to, I call it the "Rubiks Cube" of cameras. The interface is convoluted with way too many buttons to push in order to make the camera do what I want. As an example, to lock exposure I have to hit the *Menu* button, scroll down to and select *Extra Functions*, then scroll down to and select *Exposure Lock*, and then scroll up and select *Execute*. That last step eluded me for a while. And you have to have the LCD on to do this and the LCD really burns batteries.

**Q.** You've talked about the Nikon, what about digital cameras in general?

**A.** I really like the immediacy of results. For example, yesterday I shot a theater with the CoolPix. I took the shots, ejected the CompactFlash card and plugged it into the PCMCIA slot on the PowerBook. I did a quick stitch in QTVRAS and discovered that one area was a little hot. So I reshot, did another stitch, and called the client over and let them have a look. They approved the image right there, so I struck my equipment and headed back to my studio to do the final retouching and programming. I left the location knowing I had what I needed. And the client knew too.

I can't speak for other photographers, but I've very often gone to the lab with my fingers crossed—I generally keep my hands in my pocket so people won't notice—and I figure that digital capture has added 10 years to my life expectancy!

The biggest drawback is in the lack of complete creative control that we take for granted on film cameras. With the CoolPix I can't alter the shutter and aperture, can't switch lenses, can't gel and filter. To be fair, some of these functions can be had with the Nikon E3 and the Canon D2000—both of which are great cameras—but for my work it's hard to justify the cost of these cameras, especially given that the technology is moving so quickly.

**Q.** When do you use a regular film camera rather than a digital camera?

**A.** Essentially there are 4 situations when I'll use a 35mm: if we need a wider focal length, if we need to fire strobes, if lighting is funky, or if we're going to print the image.

The CoolPix, with adapter, gives me a 24mm equivalence, and the Canon Pro70 has a 28mm equivalence. These focal lengths are adequate for most jobs, but when we shoot car interiors we need an 18mm or even a 14mm field of view so we have to use film.

If the light is really funky, say a mixed environment with fluorescent, incandescent, daylight, mercury or vapor lighting then a 35mm with a hand-held color meter and a wide selection of filters is indispensable in situations like that.

For print work we'll shoot on 35mm and stitch from the highest resolution PhotoCD scan. We've been able to get great results from the CoolPix when we print to 8" x 22" paper on our Epson Stylus 700, but on occasion we need much higher resolution.

**Q.** Is using a digital camera for shooting panoramas different from using a film camera in other ways?

**A.** The exposure latitude of digital cameras is much more limited than film, especially some of the latest films. So we don't get the same detail in shadows and highlights. In order to deal with this I'll shoot different exposures with the digital camera. Because there's no film, processing, or scanning costs involved this is just time consuming, but I can then fine tune the images in Photoshop.

**Q.** What problems do you have shooting panoramas no matter what kind of camera you use?

**A.** Situations with lots of people are a challenge, as they'll often be distorted, ghosted, sliced, or blurred, when the images are stitched, but there are ways of dealing with this.

More generally, the fact that you have to compose and light the entire 360-degree view takes some getting used to. I've found that I haven't been able to get adequate results by lighting one section, exposing, and then lighting the next section.

**Q.** How do you solve that?

**A.** When we shoot 35mm we manage to hide small lights—we use Lowell Tota's and Omni's—around corners and behind the furniture. When we shoot digital we're patient and work with available light, bracket extensively, and give thanks to the people who invented Photoshop!

CHAPTER 13

# FILE FORMATS

This chapter covers most of the file formats that you are likely to encounter. First it explains why there are different file formats, and the most common file formats. The cross platform section deals with issues you might face taking files from one operating system to another, focusing primarily on Mac-PC conversions issues.

The table of extensions presents a list of file formats by extension, while the rest of the chapter presents individual information about the file formats. These later sections are presented in alphabetical order, not in order of importance. Finally, the chapter concludes with a note about moving files back to the camera.

## WHY DIFFERENT FILES?

If only there was one file format. Life would be so much easier for users and for software developers. And in a sense, there is. If you only want to read about one file format, then check out the information about JPEG and ignore the rest. Chances are that it will cover the lion's share of what you need to know!

However, the complete truth is that there are many files formats. Some are more common than others, and they exist for a number of reasons. Some are old formats that remain popular even though more efficient formats should have replaced them. Others are specific to software applications, or are used for particular purposes, while still other formats are "new" formats that promise many features not found in older formats but haven't yet completely replaced the existing formats.

Digital cameras tend to use either TIFF or JPEG formats for saving their files, with JPEG being the most popular. JPEG offers one great advantage; it compresses the files by a large amount while retaining the general appearance of the image. Once the image has been downloaded from the camera, you'll most probably keep it in JPEG format unless you need to edit it or use it for some other purpose.

## COMMON FORMATS

**JPEG.** Most cameras use this format to save their files. Compression provides dramatic reduction in file size. Common on the Web as one of only two standard graphic formats supported by Web browsers (see also GIF).

**TIFF.** Used for digital files when actual file size is not the primary consideration. Very common in publishing. Compression is lossless (there's no change in the appearance of the image).

**GIF.** Also known as GIF89A after the most recent variant of the format which supports animations as well as single images. Developed for compressing computer graphics to a small size, specifically for sending them over modems. While it does a good job with color graphics that contain solid areas of color, it is not good for photographs because it is limited to an 8-bit palette.

### APPLICATION-SPECIFIC

**BMP.** The "native" file format for the Windows operating system. Often used simply because most programs support it.

**PICT.** The "native" file format for the Macintosh operating system. Again, this format is often used mainly because nearly all programs on Macintosh computers support it.

**PSD.** PhotoShop file. One of the most popular graphics programs, the file is now imported by a number of other applications.

**EPS.** PostScript is the most common printer language. Because of this, graphics are often prepared for printers in EPS (Encapsulated PostSxcript) format.

It is also used as the file format by some graphics programs (Adobe Illustrator, MacroMedia FreeHand). You should use this format for images only if you are doing high-end printing. If you are simply printing to a PostScript printer connected to your computer then you should print your graphics file from an application that supports the printer, and let the application (or the print driver) do the conversion during printing.

**PDF.** The file format for Adobe Acrobat. Images can be saved within Acrobat (and you can specify the kind of compression you want it to use for images).

### NEW FORMATS

**FlashPix.** Offers some performance improvements over the other formats. Actually uses the JPEG compression algorithm as an optional lossy compression. Only supported by some graphics programs.

**PNG.** Developed as a replacement for the GIF format, it's now being used by some graphics programs, but is not yet widely accepted.

## FILE COMPRESSION: LOSSY VERSUS LOSSLESS

Because of the amount of information a digital image can contain, it is usually necessary to use compression to reduce the size of the files. This is particularly important when saving the image in the camera and when transferring the image to the computer. Camera memory remains comparatively expensive (particularly compared to film!), and transferring images at serial speeds is very slow. Since a 24-bit image with a resolution of 1500 x 1200 pixels is larger than 5MB, even users with large hard disks on their computers will be concerned about conserving space.

There are a number of compression techniques that have been adopted for use with different file formats. The techniques can be divided into two categories: lossless compression and lossy compression, which are described below.

### LOSSLESS COMPRESSION

Compression techniques that reduce the size of the stored file, yet do not throw away information from the original image, are called "lossless." In other words, after passing through compression and then being decompressed, the final file contains exactly the same picture data as the original.

Examples include LZW (Lempel, Ziv, Welch) and Run Length Encoding. Depending upon the source image, the amount of space saved will typically be less that 50% of the original file size, and are often much less.

### LOSSY COMPRESSION

These techniques reduce the size of the compressed file dramatically, but the decompressed image will not contain the exact same picture data as the original file. Often there's a parameter you can adjust to vary the amount of compression applied to the image. The more compression, the less the decompressed image resembles the original. However, when the compression ratio is not high, the differences in image appearance are so minor that it's unlikely you would notice the difference between the original and the compressed file.

Examples of lossy compression include JPEG and Wavelet.

## WHICH COMPRESSION FORMAT IS BEST?

For display on the Web or for emailing, a lossy compressor is practically required. During editing it's best to use lossless compression, many editing programs such as PhotoShop use their own file formats that are not compressed. After you've finished editing a file, saving it in JPEG format (but at the highest quality setting) saves considerable space, yet arguably any loss can't be seen. This is particularly true given that most original camera files are JPEG.

But for some this issue becomes almost religious in significance. One camp says you should never save your files using lossy compression, while the other says that you can't tell the difference, so why worry? Personally, I come down on the side of hedging my bets. Most digital cameras save files using JPEG compression. Converting those files to a non-lossy format (unless you're going to edit them) makes no sense at all. Save the files as they came from the camera. Even if I'm only going to make a minor change I always keep an untouched copy of the original.

When editing files I usually use Adobe PhotoShop. If I'm only doing simple editing (say adjusting brightness and contrast) then I'll save the new file in JPEG as well.

If a lot of editing is done to an image, Photoshop actually saves layers containing different parts of the edited image. If the file is converted to JPEG, this information is lost. For that reason, it makes a lot of sense to save the work file should you later decide you want to change the file again. I do this even though I then will create a JPEG file for use.

If I'm creating a photograph that will be printed using a desktop publishing program such as Adobe PageMaker or Quark XPress, I usually use the TIFF file format. That's probably as much out of habit as anything else because for a long time these programs only worked well with TIFF files.

## WHY NOT SAVE IN TIFF LZW?

Some cameras will save files as TIFF rather than JPEG. The TIFF format supports LZW compression, which is lossless. This option is available in image editing applications, but current cameras that support the TIFF format don't use LZW compression. This because it takes processing time to compress a file, and this would increase the time to save pictures considerably. Saving the larger file to memory takes less time than compressing the file and saving the smaller, compressed file. The problem with using the TIFF format is that the files size is dramatically larger, and despite the falling prices of CompactFlash and SmartMedia cards, saving 4MB files rather than 500K ones really takes a bite out of the number of images you can capture. While it's worth experimenting with the highest resolution captures, for most people the highest JPEG quality will be good enough.

On the other hand, if you do capture as a TIFF file, don't save it in JPEG format and throw away the TIFF file. Always save the original computer file, even if you compress the file for sending to others.

### DOESN'T COMPRESSION THROW AWAY PIXELS?

No. None of these compression techniques change the number of pixels in the image, though they may change the information saved for each pixel. Some cameras offer multiple resolution files, as well as different compression settings, but that's downsampling the file (see Chapter 7). In down-sampling, pixels are "thrown away" and a smaller file is the result. While throwing away pixels does reduce file size, and could be considered a compression technique, it's not usually considered a compression technique in the way JPEG or LZW are.

### WHY ARE THE FILE SIZES DIFFERENT?

Both JPEG and TIFF produce variable sized files. If the image contains a lot of repetitive data then it will compress very efficiently, but if there's a lot of detail then the file won't compress as well.

## LONG-TERM ARCHIVING

One of the problems with computers is that they keep changing. The files created on a microcomputer 15 years ago would be very difficult to read on a modern PC. You may not even be able to read the media on which the data is stored...which poses a problem—will the files I save today be readable 20 years from now when I come to look at them again?

The short answer to that question is; maybe. Certainly they will be if you keep them active; that is, if you store them in a readily accessible format and actively update the files as technology changes.

Alternatively, maybe it's a good idea to archive your images on stone tablets or as cave paintings; they're two of the few media available that are proven to have a long shelf life!

## COMPARING FILE SIZES

The size of graphics files vary depending upon the subject of the file. A single file was saved using a number of different formats. The resulting file sizes are listed below.

| | |
|---|---|
| FlashPix[1] (uncompressed) | 6,260 KB |
| TIF[2] | 4,616 |
| BMP | 4,609 |
| PhotoShop | 4,604 |
| PICT | 4,597 |
| IFF | 3,743 |
| PNG | 3,588 |
| LZW.TIF | 3,090 |
| JPEG 10 | 1,200 |
| GIF[3] | 939 |
| JPEG 5 | 205 |
| JPEG 0 | 89 |

[1] FlashPix adds additional copies of the file at different sizes to improve file load times. This accounts for the file size increase.

[2] Without compression.

[3] GIF requires converting the file to an 8-bit palette, which throws away a considerable amount of information.

# CROSS-PLATFORM ISSUES

Converting files between platforms (Macintosh and Windows) is much easier today then it was a few years ago. Many of the graphics programs are able to read files from either platform.

Some files are virtually cross-platform by definition (JPEG, GIF for example).

TIFF is virtually cross-platform, but it's not really a general-purpose file format and some people may not be able to open it. For the casual user, a JPEG file should be just fine. Remember to save the file with the ".jpg" extension in the file name, even if it's a Macintosh disk.

## RGB VERSUS CMYK VERSUS 8-BIT VERSUS LAB

CMYK uses four channels at 8 bits per channel, while RGB stores three channels at 8 bits per channel. Which stores the most information? In reality, that is a somewhat meaningless question because the color spaces are different; RGB can represent colors that CMYK cannot, and vice versa. CMYK files are 1/3 larger than RGB simply because they have an extra channel, but that doesn't mean they have a third more color information (especially since nearly all digital cameras capture in RGB). The black channel is really only a replication of information that could be stored in the CMY channels. CMY with the same value should produce gray or black. In general, CMYK represents fewer colors than RGB. The LAB file format is also available for storing image data. While it is able to store a wider color spectrum than CMYK or RGB, it's not a common format and is only supported by high-end color editing applications.

While converting between one color space and another is pretty harmless (the converted file should appear very similar to the original), one thing you don't want to do is convert between color spaces many times. The conversion between each color space is not necessarily lossless. During conversion there can be roundoff errors because colors that are exact integers in one color space will not be so in another. The errors won't be large, and aren't worth worrying about if you're doing it once, or are going to do some kind of effect which will manipulate the image anyway. But they can accumulate.

# TABLE OF FILE EXTENSIONS

The following table lists the most common graphics formats. Most of these are covered in more depth below.

| | |
|---|---|
| BMP | Windows Bitmap (Also RLE) |
| CRW | CCDRAW Canon Proprietary Format |
| EPS | Encapsulated Postscript |
| FPX | FlashPix |
| GIF | GIF89A |
| HAM | Amiga file format |
| ICB | See Targa |
| IFF | Amiga file format |
| JPG | JPEG |
| KDC | Kodak proprietary file format |
| MIX | File format used by Microsoft PhotoDraw and PictureIt! |
| PCT | Apple PICT format. |
| PCX | PC Paintbrush format. Now supported by a number of applications. |
| PDD | Adobe Photo Deluxe file format |
| PDF | Adobe Acrobat |
| PSD | Adobe PhotoShop [version 2.0 and 3.0(layers)] |
| RLE | Run Length Encoding, Windows Bitmap, (see BMP) |
| TGA | Targa is a video production board and software combination. |
| TIF | TIFF |
| VDA | See Targa |
| VST | See Targa |

LZW : Dictionary-based compression.

# FILE FORMATS

## BMP

This is the standard bitmap file format for the Windows Operating system. Most Windows editing applications are able to open this file format, as well as some Macintosh applications. RLE (Run Length Encoding) is a compression option available for use with BMP files. Not all applications that can read a BMP file can read a file save in RLE.

## CRW

CCD_RAW format. Canon's PowerShot 600, A5, Pro70 and A5 Zoom use this file format. Rather than converting the data from the CCD and compressing the image using a compressor such as JPEG, these cameras save the 10-bit data from each CCD element. Canon's software later converts the data to RGB, after it is downloaded from the camera.

The data is not compressed, yet it is much smaller than a 24-bit RGB file simply because each single pixel is stored as 10 bits rather than 24 bits (remember, each photosite only produces a value for a single color).

The downside of this method of saving files is that you have to use Canon's software to convert the files; you can't just open it using any editing application. You might want to save the resulting files in TIFF or some other compressed format rather than CRW so that it is easier to manage and work with the files.

Note that some of the Canon cameras have a different pixel count, so the software for one camera may not be able to read the files from another. Also, there is at least one third-party application that can read CRW files. It appears that the conversion of the files requires some tricks to get really good results; Canon's software is the preferred transfer method. It comes as a plug-in TWAIN driver that you can use with many applications.

## EPS

PostScript is the most common printer rasterizer software. PostScript is really a computer programming language that was designed for printing. A PostScript "program" is written and then sent to the PostScript interpreter (called a RIP) in the printer which interprets the PostScript code and translates it into the individual dots that the printer outputs.

When PostScript first appeared, some people actually wrote PostScript files by hand, but this is rarely done now (or if it is, I don't know of anyone who is doing it!). PostScript is great for text, and includes vector mathematics for storing line artwork very efficiently. The graphics programs Adobe Illustrator and MacroMedia FreeHand both save their graphics files in EPS format.

EPS can also be used to save digital images, but this should only be done if you are sending the graphic files to a PostScript printer; and even then, I would encourage you only to do it when sending the file to a printer at a service bureau. And only do it if you know what you are doing. See "Working with a Service Bureau" in Chapter 9.

If you are simply printing to a PostScript printer connected to your computer then let the application (or the print driver) perform the conversion during printing. Don't change your TIFF file to EPS, simply print it from an application that supports the file format.

**WARNING:** EPS files also contain a low-resolution bitmap file that represents the PostScript data. This is included primarily for use in page layout programs such as PageMaker or Xpress so that these programs can show the user what the graphic looks like and makes it easier to layout the rest of the page. The bitmap image included in the EPS file is at a much lower resolution (72 dpi) than the PostScript file. Be sure, when opening an EPS file, that your program is opening the PostScript information and not the preview image; otherwise you'll end up using a low-resolution image. If your program can't open the EPS file, or can only open the preview, then you'll somehow need to get the original file.

## FLASHPIX

FlashPix was developed by Kodak, Hewlett-Packard, Microsoft and Live Picture Inc. Based on Live Picture's IVUE format, Flashpix was designed to be a universal, royalty-free format for simple, resolution-independent imaging.

The Flashpix format contains a hierarchical set of images at different resolutions with the lowest resolution at the top and the highest at the bottom. Looking kind of like a pyramid, the image is further divided into tiles of 64 x 64 pixels. When the file is opened, the viewing application only has to read the data for the resolution it needs to fill the screen; it doesn't need to read all the data in the file and then scale that data to fit the screen. Similarly, if the user zooms in on the file, then the viewer only has to read the appropriate slices needed to update the screen, not all of the data. When a file is edited, the changes are saved in a separate area of the file and applied during viewing or when printing. Live Pictures' Live Picture application, and the graphic editor xRes use this technique. Though both programs are supposed to be much faster at editing large files than other applications, neither have been widely adopted.

This technique has advantages for both viewing as well as for editing. For viewing very large files, only a comparatively small amount of information needs to be read in at one time.

Unfortunately, this feature doesn't necessarily translate to performance savings when viewing over the Web because the Web server typically must send the whole file and can't provide a way for the browser to read and download just part of the file.

Live Picture has released a Web server that, when used with a special plug-in for the browser, provides a way of serving these files so that the Web browser doesn't have to download the whole file, just the parts the user wants to see.

This format is not compressed, though it can use the JPEG compression format as an option. Flashpix supports RGB and Kodak's YCC colorspace, but not CMYK. This format is supported by a number of applications, including

Microsoft's PictureIt! and Live Picture's LivePix, MGI Software Inc's MGI PhotoSuite 8.0, Adobe PhotoDeluxe.

See also MIX, JPEG.

## GIF

Pronounced with a soft *g*, like "giraffe," GIF (Graphics Interchange Format) has become one of the two most important graphics formats for the Web (along with JPEG). It's a lossless compressed format; however, it only supports a maximum of 8 bits of color per pixel (see Chapter 6). So if you are going to save a photograph in GIF format, then in most cases you will have to first convert it from a 24-bit image to an 8-bit palette before saving it as GIF, which means that you are throwing away information. So even though GIF is described as lossless, for photographs it essentially is lossy.

Generally, it's better to use JPEG for photographs on the Web or when emailing to others. Nearly all browsers that support GIF support JPEG, and JPEG files are usually smaller than GIF files for photographic images.

Instead, GIF files should be used for computer graphics containing solid color regions, logos and other artwork that does not compress well using JPEG.

The other reason you might want to use GIF is that it supports transparency colors and animation. While the JPEG file format also can support transparency, most browsers do not support transparency for JPEG files, and you have to use GIF if you want to make a short animated sequence of images. GIF animation capabilities make it possible to create simple sequences consisting of frames that play one after the other. Since JPEG does not support multiple frames, your only other alternative is to use a video format such as QuickTime or RealVideo (neither of which is preferred when you have long sequences to animate). If you want to learn more about animating sequences, read Chapter 19, *Animation*.

**WARNING**: Because GIF files are only 8-bit, one thing you should never do is convert your images to GIF and throw away the original files.

### *Converting a Photograph to GIF*

Just in case you do have to convert a graphic to GIF, here's what you need to know. Most graphics applications will save in GIF format. Some will ask you to convert to an 8-bit palette when you go to save the file, while others will only let you save the file as a GIF if you first convert it to an 8-bit or less palette.

Chapter 6 features a discussion of palette-based graphics. When you convert to an 8-bit palette the program should give you at least two options. One is to create a custom adaptive palette which contains a set of colors which most closely matches the colors within the image. If you're lucky, the resulting image will look almost as good as the original. The other option is to convert the graphic to a standard system palette; there are several, including Macintosh and Windows system palettes. If you're really lucky, the program will also provide an option of the browser palette, which contains the standard 216 colors used in Web browsers.

For photographs I recommend using an adaptive palette. It produces the best looking image. Assuming that the image is being viewed in a browser, if the user is looking at the photo on a monitor set to thousands or millions of colors, then the image will look almost as good as the original. If the monitor is running in 8-bit color, then the image will be converted to the browser palette. This results in dithering, and the graphic won't look as good as the original, but it won't look too different from an image converted to the browser palette using your image editor. Using your graphics program to do the conversion—rather than having the browser perform it on the fly—results in a slightly better looking image, but it's not drastically different, whereas for users running in more than 8-bit, the color improvement is dramatic.

NOTE: The Mac and Windows System palette is not the same.

## JPEG

Named after the committee that created it, JPEG (Joint Photographic Experts Group) was developed for the compression of photographic images. It's a lossy compression format, and it can do tremendous work with photographs, reducing them smaller than a comparable GIF file. JPEG takes advantage of the fact that the human eye is less sensitive to small changes of color than to changes in brightness.

JPEG is not good for compressing nonphotographic artwork that contains extreme changes in contrast. Artifacts (pixels that aren't the 'correct' color) can occur in those situations.

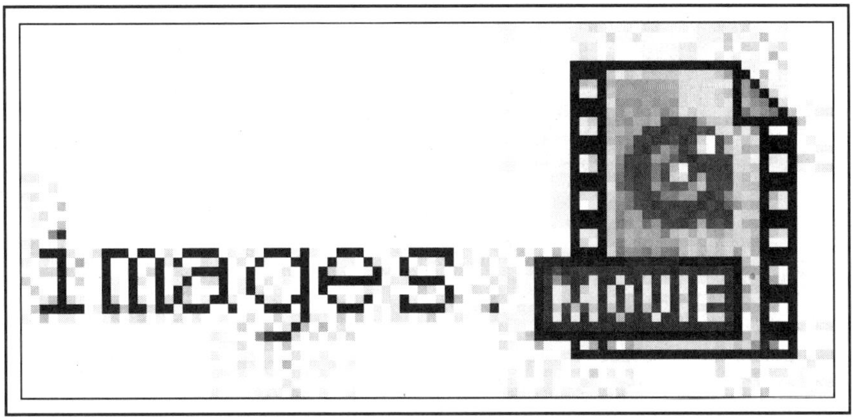


*Figure 13-1. JPEG aliasing. Artifacts (incorrect pixels) can be seen in this highly compressed version of a computer graphic. While such errors will also start to occur in photographic images, they appear earlier, and are more objectionable in this kind of graphic.*

While it can be used for either color or grayscale images, it cannot compress 1-bit (black-and-white) images, nor is it good for sequences of images. There are similar algorithms available for compressing sequences, notably MPEG.

When you save using JPEG, you can choose the compression factor; the greater the compression, the more information is thrown away. This appears mainly as artifacts; the image will start to appear "blocky," as though it is made of larger pixels than the actual resolution of the file. This is because one of the things the compression routine does is to save information for 8-pixel-square blocks of the file. With higher compression, the accuracy of the individual pixels within these blocks gets thrown away, until the block becomes a single value. Artifacts will be most visible in high-contrast and high-detail regions. In high-contrast regions artifacts can create a halo or outline effect.

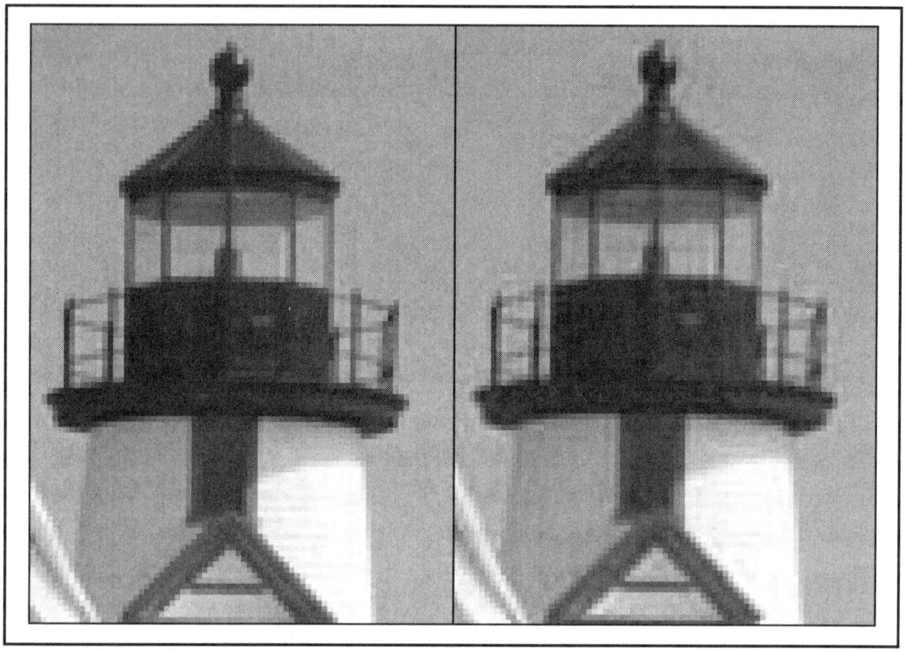

*Figure 13-2. An enlargement of an image before (left) and after (right) heavy JPEG compression. The halo effect caused by the compression algorithm at high settings can be seen in and around the dark areas of the lighthouse.*

Which compression factor to use? This depends upon the image and the purpose. Often, when saving for a Web page, you want to get an image that is as small as possible, yet looks as good as is acceptable. The latter is a subjective opinion; that's why if you are often saving images in JPEG you should consider a tool that let's you preview the effect of the compression factor you choose. Fireworks from Macromedia and ImageReady from Adobe are two programs that show you the effect of applying different compression ratios to the same image. This can be really helpful when you want to get the smallest possible file, while still maintaining acceptable image quality.

NOTE: "Acceptable image quality," as used here, is a completely subjective term. What is acceptable to one person may not be acceptable to another.

Even using the highest quality and least amount of compression, the JPEG algorithm (at least as available in most image editing programs) produces images that are slightly changed from the original. It's doubtful, however, that anyone would be able to tell the difference between the files. Compression ratios up to

10:1 are virtually undetectable. *Figure 13-3* shows two copies of the same image before and after compression. The image is not heavily compressed; it's not possible to pick out the difference without zooming in on the image. The third image shows the differences between the two images. This was created by subtracting the channels of one graphic from the other (channels are described in Chapter 6). While the differences are small, you can see how they are clumped around the areas where the color values in the scene change dramatically, creating an outline of the lighthouse. The sky, since is it mostly the same color, contains much fewer differences. Note that in this example the differences have been exaggerated by adjusting the contrast of the difference image.

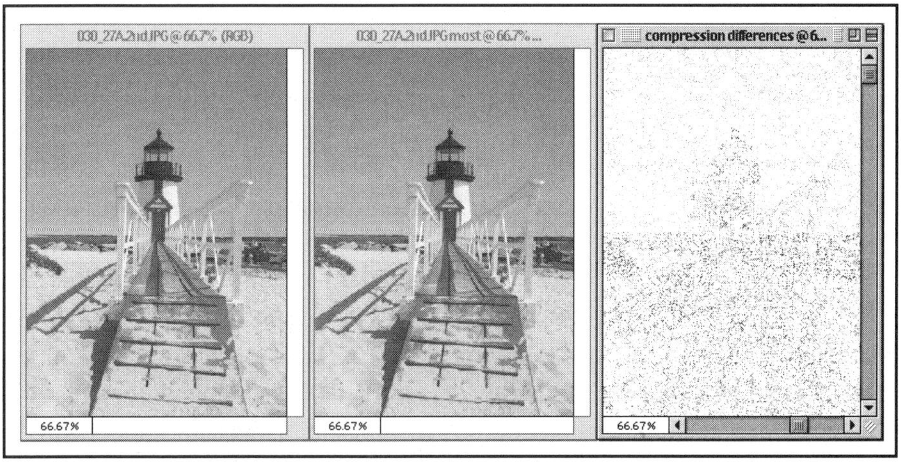


*Figure 13-3. Image before (left) and after (center) compression. Frame on right shows areas of difference between the two.*

Just to confuse things, the JPEG specification has been updated to support a lossless compression algorithm (which is a completely different algorithm from the lossy JPEG compression algorithm!). This option is not widely supported by current editing applications.

**WARNINGS:** If you are saving a JPEG file for use in a Web page you should not use the progressive option that some graphics programs offer: most browsers don't support that option. Also, while Photoshop lets you save a CMYK file as JPEG, you should only use RGB, as browsers don't support that either.

Some programs don't preview the effect of applying compression—you have to save the file then open the saved file to see the effect. If you save using JPEG compression in Adobe Photoshop the program continues to display the file as it is in the program's memory buffer, not how the file has been saved. Only by closing the file after saving it and reopening the saved file can you see the effect of the compression.

It's not recommended that you crop a JPEG on anything other than a 16-pixel boundary because JPEG processes images in 8 x 8 blocks. The color accuracy of the edge pixels will be impacted, but you probably won't notice this.

## WHAT'S JFIF AND EXIF?

Most people refer to files saved using JPEG compression as JPEG files. Strictly speaking, the compression routine is JPEG, while the common file format used to save JPEG compressed data is JFIF. So few people use this distinction that JPEG is used to refer to both the compression format and the file format.

EXIF is a variant of the file format which includes header information. This can be used to save a large amount of Meta data (data that describes the file). Most commonly, digital cameras that support the EXIF format store date, shutter and aperture information in the file. This information can be viewed using a number of viewer applications. If the viewer application does not support EXIF, it can still read the JPEG file. The file appears to the application to be a standard JPEG file, and the reader will read the JPEG data and ignore the EXIF information.

## KDC

A proprietary file format developed by Kodak and used by some of their older cameras (such as the DC 120), Kodak has since moved to more "standard file formats" such as JPEG. KDC files contain a thumbnail image and a YUV 4:2:2 image. The images from the DC 120 are at a resolution of 848 x 976, but Kodak's routines take the images and scale them to the 1280 x 960 size. Most image editing applications can't open this file format, but Kodak has licensed these routines to ThumbsPlus and VuePrint, which can open these files.

NOTE: If you have a lot of KDC images and want to convert them to a manageable format, then the ThumbsPlus utility is worth getting (see Chapter 17, *Useful Software*). It will batch convert all the files, whereas the software that comes with the camera is more cumbersome.

## MIX

MIX is a variation of the FlashPix format that can contain multiple images stored in a single file. Used by Microsoft PhotoDraw and PictureIt!

Note: There's also a sound format with the same name (*Ouch!*).

## PDD PHOTODELUXE

This is the native file format of Adobe PhotoDeluxe. The PhotoDeluxe format is nonlossy, and supports layers.

Note: When PhotoDeluxe exports an image to JPEG format, it attaches a thumbnail preview to the JPEG. America Online 3.0 and PhotoPaint 5.0 do not support JPEG images with thumbnail previews. It isn't possible to disable the thumbnail preview when exporting a photo as a JPEG file from PhotoDeluxe 1.0.

## PDF ADOBE ACROBAT

Adobe Acrobat was developed as an electronic document distribution system. It consists of three parts:

The PDF file format

Acrobat Reader

Acrobat Exchange

PDF is the file format that contains the electronic documents. These might be brochures, newsletters, or practically any other document you can create on the computer. Adobe Acrobat Reader is an application that can open and display PDF files. It is available free from Adobe and runs on a number of different platforms. It provides features such as searching and printing.

PDF documents are created using Acrobat Exchange. This is a tool that consists of two parts; a print driver and an editor.

The purpose of Acrobat is that while computer documents are created in a number of different applications, it's difficult to distribute them to other people because not everyone has every application available. So Acrobat provides a print driver that will capture the information from any application and turn it into a standard electronic document. After creating the document in an application (such as Word or PageMaker) you choose the print driver and "print." The Adobe Acrobat print driver creates a PDF file.

When saving PDF files you can choose a number of options about the compression of images within the files.

PDF is great for brochures or other documents that contain images and text. It might be a great way to distribute a collection of photos too; you can restrict whether users can print or copy from a file. However, to use it to distribute a single image makes little sense compared to JPEG.

## PCD PHOTO CD

This is Kodak's proprietary file format used by Photo CDs. The disk holds 550MB of data, and each image is digitized at 2048 x 3072 resolution. The files themselves use compression and each file is about 4.5MB each. When opened at the largest size in RGB, the file is 18MB. The PCD file is not in RGB format—rather it is in the PhotoYCC luma-chroma space.

For Windows users, nearly all graphics applications will open a Photo CD image directly from the CD. The Macintosh operating system includes extensions that recognize the Photo CD disc and display an interface that includes a set of thumbnails for the images on the disk, and separate folders containing copies of each image at different resolutions. These files appear to Macintosh applications as though they are PICT files, so any graphics application can open them (the Macintosh operating system converts the information to PICT format on the fly when the file is opened).

The scanner used to create Photo CDs from 35mm film has an effective resolution of 2200 pixels per inch in both dimensions. A scanner used for pro film can scan at 4400 pixels per inch.

Because Photo CD is proprietary, only Kodak equipment can be used to write a Photo CD.

While the basic scan is as 2048 x 3072, the Photo CD plug-ins and software will provide a copy of the graphic in any one of five different sizes. The software asks you which size you want before it opens the file. The sizes are:

| | |
|---|---|
| Base/16 | 128 x 192 |
| Base/4 | 256 x 384 |
| Base | 512 x 768 |
| 4 base | 1024 x 1536 |
| 16 Base | 2048 x 3072 |

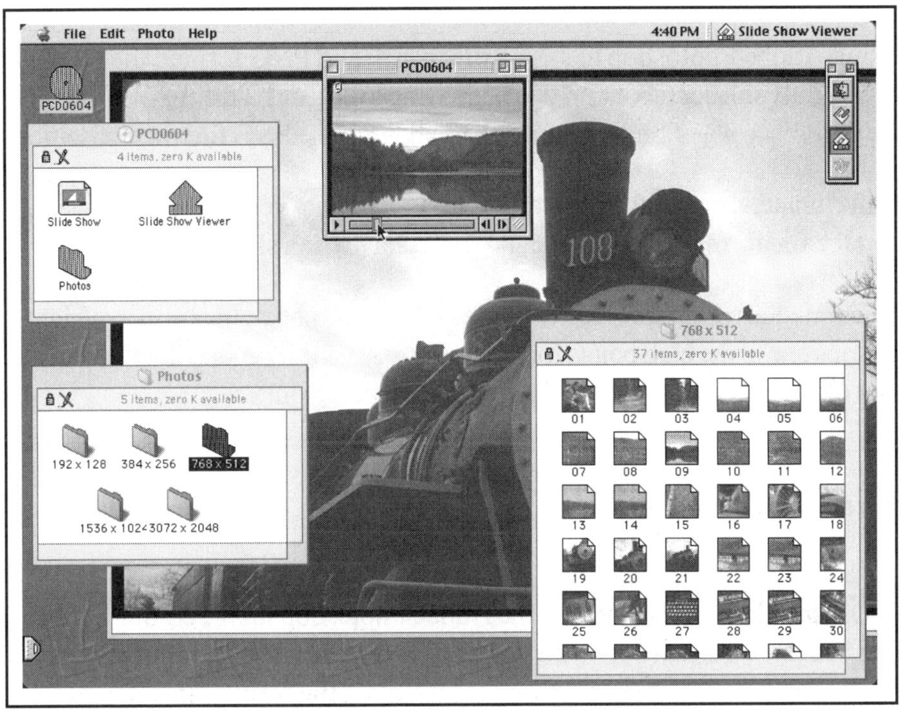


*Figure 13-4. Viewing a Photo CD on a Macintosh.*

## PICT

The standard format for Macintosh graphics, it supports vector as well as bitmap data, though it's mostly used for bitmapped data. Most Macintosh applications can read this format. Optionally, JPEG compression can be used within the file format to compress the bitmapped data.

## PNG

Pronounced "ping," this is a lossless compression format that was developed to replace GIF. The developers of this standard claim that it does a better job at compressing files than GIF or TIFF/LZW. My casual tests don't agree with this, though it's not much of a difference (see chart above). It's also claimed to be better than the lossless JPEG algorithm, but so few programs support lossless JPEG I have not bothered comparing it.

Using Adobe Photoshop to save a file in GIF and PNG format, the GIF file was slightly smaller (about 2%). When saving a file using TIFF/LZW and PNG, the difference was significant (see chart above). The PNG file was 1/6 larger.

Unfortunately, one of the problems with PNG is that very few programs support the file format. In particular, browsers have been slow to adopt it.

Macromedia's program Fireworks uses PNG as its native file format, which is perhaps one of the high points for the file format at the moment (if you do create a lot of Web graphics, Fireworks is worth checking out).

## PSD PHOTOSHOP FILE FORMAT

PSD is the native file format of the Adobe Photoshop image editing application. Because Photoshop is so popular, several other programs now open Photoshop files. It differs from most of the other file formats mentioned here because it supports layers. These can be useful when editing an image.

There are at least two versions of the file format. The original Photoshop file format does not support layers.

### TIFF (TAG IMAGE FILE FORMAT)

This format, pronounced "tiff" (rhymes with "riff"), was one of the first widely used formats for scanned images on desktop computers. Though it is a "standard," there are many 'optional' features of the standard. Because of this, in the early days many applications claimed that they opened and saved TIFF (Tag Image File Format) files, yet often they could not open files from other applications. Over time this problem went away, but it was a serious problem in the early days, and is a problem that file format standards committees seem to have learned from.

There are two ways that a TIFF file can be saved; in Macintosh or PC (IBM) byte order. This can be a problem when moving from a Macintosh to a PC, though many programs now support both versions and let you choose what byte order to use when saving the file.

### WAVELET

Not yet widely used, Wavelet compression is supposed to produce greater compression of images than JPEG, with smaller loss of information. However, the extra amount of processing required to compress an image (and compression time is greater than decompression time for most algorithms) means that using Wavelet compression inside the camera requires faster processors.

## PLACING IMAGES BACK IN YOUR CAMERA

Most cameras feature some kind of review function that lets you view the images contained in the camera's memory. You can also use the camera as a presentation device when connected to a television (if the camera has a video-out capability).

With many cameras you can take the images from the camera, edit them on your computer, and then place them back in the camera, provided you keep the files in the right format.

The first requirement is that you have a camera that lets you place photos back in its memory. Most early (pre-1997) cameras don't allow you to do this. But newer cameras do, particularly the cameras that support removable media (CompactFlash and SmartMedia cards) and use the EXIF (JPEG) file format.

The important things to do are:

**1.** Save the file in the correct file format.

**2.** Use the correct dimensions.

**3.** Use the same naming scheme the camera uses.

Remember that EXIF is a variation of JPEG that includes a special header. Many cameras require that header or they will ignore the file. Most editing applications will open an EXIF file, but they'll save it as a JPEG file without the EXIF header. Many cameras require the EXIF header, yet the header doesn't need to contain any information.

Olympus provides an application called Camedia with their cameras which can be used to add the EXIF header to a JPEG file. Choose *Image>Convert Format* and a dialog will appear that lets you select a format to save a file, including EXIF-JPEG.

If the application software that comes with the camera doesn't let you transfer images back to the camera, then your best bet is to use a CompactFlash or SmartMedia card reader to connect the card to the computer and transfer the files back to the card.

Some cameras, including the Fuji MX-700 and Kodak DC260 and 220, will display regular JPEG files without the EXIF header.

# STORING FILES

Dealing with the many different file formats can be problem enough, but as your collection of images grows you'll find that you'll need some way to archive these files. This issue is explored in the next chapter.

# CHAPTER 14
# SAVING AND ARCHIVING

Once you've had your camera for a while, you'll start accumulating images. You'll probably want to get some kind of cataloging software (see Chapter 17), but you should also start thinking about how to archive images. Even though the cost of hard disks has come down dramatically, no one ever seems to have enough hard space to hold all of their images. If you are fortunate enough to have a 10-gigabyte drive that you can devote to storing your images, then you're probably set for a while —depending upon file size and compression this could translate into anywhere from 2,500 4MB files to 20,000 half-megabyte files. But you'll still need a method of backing up these images just in case the disk drive develops an error. Either tape or CD makes the most sense, and for most people CDs are more convenient than tape, and have the added advantage that they are easily read by others. Tape drives aren't nearly as common as CD drives. Removable drives are also an option, though the media is more costly.

Whichever medium you use, there are three important points that you should come away with:

**1.** Back up regularly.

**2.** Keep at least two copies of any valuable item on separate media.

**3.** Keep them in different locations.

It's important to back up (make a copy that is stored off your computer's hard disk) regularly. Unfortunately, you never know when—or if—your hard disk is going to have a problem and become unreadable. You might be lucky and never suffer a disk failure, but you always need to plan in case it does happen—particularly if your thesis is due next week and you haven't made a backup copy in three months.

For this reason, always make a backup copy of an item that would be difficult to replace "just in case." Even placing critical files on a floppy disk may save you weeks of lost work.

The easier you make it to be able to make copies and backups, the more likely it is that you will do it every now and again. A CD burner is inexpensive and is a great backup device for those who only need to back up every few weeks (most home users). The cost of disks is so low that you could burn a new disk once a month and spend less than $20 a year on backups. A CDRW burner is most practical for doing regular backups, though if you need to back up every day a tape system might be less expensive.

## BACK UP AND ARCHIVE

*Back up* means to make a copy that is stored on another media other than your hard drive. It's usually recommended that backups are kept in generations; grandfather, father, son. As an example, the first backup is made on Monday. On Tuesday a new backup is made on another tape or disc. On Wednesday another new backup is made. On Thursday the Monday backup tape is re-used, and the cycle continues. This process can be done daily or weekly (or some other appropriate time frame). The advantage of keeping generations of backup is that you can go back and get older versions of a file that you have been working on if you suddenly realize that you need the file you threw away two days ago. Of course, this only works if your oldest backup goes back far enough. You should try to keep a backup "offsite" just in case your house is broken into or burns down.

*Archive* means to remove from the hard disk. You've decided that it's unlikely you'll need the file in the near future and you want to remove it so that it doesn't take up space on your hard drive. When you archive a file you should make two copies on separate tapes or discs and keep them in separate locations.

**TIP:** First rule of restoring backups and archives: If you insert media that was previously readable into your drive and the drive says it's unreadable, it may be that the drive is malfunctioning. Try another piece of media that is not important (i.e., not your other archive or backup) before trying to read your other backup or archive. It would be more than unfortunate if you destroyed your backups or archives in a drive that was malfunctioning.

## SOFTWARE

Depending upon how frequently, and what equipment you are using, you may need to use backup software to back up and archive your files. If you have a tape drive you will have to use the software that comes with the unit. If it's some kind of removable media such as a Zip disk, then you can either manually drag the files you want to back up to the disk, or use a backup application to automate the process.

Backup applications are good if you have a lot of files that change frequently and you want to back up regularly. They will automate the task for you and can save some aggravation. But if you have a small number of files and only need to back up occasionally, making a copy by hand may be perfectly acceptable.

## STORAGE OPTIONS

The major options are floppy disks, tape, removable disks, and CDR and CDRW.

### FLOPPY DISK

Inexpensive and plentiful, floppy disks are a good medium for backing up small files, but not so good for backing up 8MB of pictures. The Sony line of cameras actually use floppy disks as their storage medium, so maybe it makes sense to save those disks as one of the backups. Just remember to make a second copy if you're not planning on keeping the images on your hard drive.

## TAPE

There are several "types" of tape drive, from low-priced Travan drives, medium-priced DAT drives, and high-priced cartridge-based systems. The primary differences between the different types of units is the speed of data transfer of the unit and the amount of data that can be stored on a single tape.

Tape systems generally work out to be the least expensive to operate when you are backing up a lot of data because of the low cost of the tape. The software that comes with them will usually make backing up a very simple process.

However, tape is the least flexible option; retrieving a single file takes time because tape does not provide random access the way a disk drive does. Instead, the tape must be read from the beginning to find the file you want to retrieve. Also, you have to use special software to operate the unit; you can't just drag files from the desktop.

## REMOVABLE STORAGE

Removable storage ranges from Iomega's Zip and Clik! disks, to the Magneto Optical drives which are available in a range of sizes, from 128MB up to several Gigabytes. Each has different advantages and disadvantages, mainly concerning cost of the mechanism and cost of media. All share the advantage that they are disks, and information stored on them can be read randomly, which makes access much more flexible.

Most of these technologies are much slower than a hard disk, so you don't want to use them as your primary data source. Also, you should always have at least two copies of any content; if you're using Clik! drives to transfer images between the camera and your computer, don't forget to make a second copy of any images you don't copy to the computer hard drive.

Magneto Optical disks resemble CDR and CDRW discs. They contain an aluminum reflective layer and a crystalline layer. A laser is used to read and write to the disk; the laser heats the surface and a magnetic head is used to change the alignment of the crystals. The alignment of the crystals represents the data value (1 or 0) recorded.

Magneto Optical drives are generally slower to write than they are to read because multiple passes must be made (at least one to erase and a second to write).

The speed of all removable disks is also dependent on the connection to the computer. Parallel connections are the slowest, while SCSI is the fastest. Zip drives, which are available in both parallel and SCSI versions, are noticeably faster in the SCSI version.

## CD-R

It's hard to believe that six years ago CD burners cost about $5,000. They now cost about $200 and computer manufacturers are starting to include them as options, or even as standard equipment. CD-R and CD-RW— their slightly more expensive cousins—provide a reasonably inexpensive storage medium for large amounts of data; each disc can hold about 640MB.

For digital photographers the CD is the cheapest and, I think, best method for storing and distributing images. The manufacturers of the CD discs claim that the lifespan of the discs is as much as 100 years, providing they are stored correctly. If you are archiving (i.e., making a copy and then removing from your hard disk), then you should always verify the copies that you make, and make at least two copies just in case one of them develops problems. Store discs in CD jewel cases; they provide better protection than cardboard covers or simply stacking them in a pile.

When writing a disc you have the choice of saving in different disc formats. The most common of these is ISO-9660, which can be read by most computer operating systems. ISO-9660 is a file system for storing information on CD-ROMs. It supports multi-session discs, and has two "levels." At the lowest level, it is compatible with the MS-DOS file system, with file names in upper case in the format 8 + 3 (i.e., MYFILE.JPG). Sub-directories up to eight levels are supported. Level 2 of the format supports longer file names of up to 32 characters, but such a disc is not readable by MS-DOS.

Saving a disc in ISO-9660 has the advantage of creating a disk that is readable by both Macintosh and Windows systems. The disadvantage, for Macintosh users, is that an ISO-9660 disc lacks the Macintosh file system, which includes a resource fork and file creator and type codes. While most files can be saved in an ISO-9660 format, those with a resource fork will be corrupted. Most importantly, applications cannot be placed on an ISO-9660 disc. You could, however, use a compression program such as StuffIt to compress an application file and place that compressed file on the disc. The user would already have to have the decompression program (unStuffIt) and then decompress the application to their hard drive.

Most Macintosh CD burning software includes the ability to create a Macintosh disc. This means that the CD has a Macintosh file structure, so it can contain any kind of file the Macintosh file system understands (including applications). This is fine if you have a Macintosh and the person you are sending the disc to has a Mac, but Windows users will be unable to read the disc.

You can have your cake and eat it too. A Hybrid CD provides a solution to the problem. It creates an ISO 9660-compliant disc that can be read by Windows systems, but also places a Macintosh file system on the disc as well (thus it is a hybrid). The best part of the whole thing is that while the same data can be read by both discs, you can also make files for the Macintosh (such as applications) visible only to the Macintosh user, while other files are only visible to the Windows user.

This feature is only available in Macintosh CD burning applications, since a Windows system couldn't recognize the Macintosh resource fork files.

## CD-RW

This is the more recent variation on the CD drive. It offers rewriteable functionality; a disc that can be rewritten. To rewrite a disc, the entire contents are erased, so this is not as convenient as a hard disk where files can be deleted and added at will. A rewriteable drive can also write regular CD-R discs, which are much cheaper than their CD-RW cousins. Also, writing to a CD-RW disk is usually done at half the speed of a CD-R disc. Note too that CD-RW disks use a slightly different process from writeable disks, and may not have as long a shelf life as CDs.

Which is not to say the CD-RW drives are all bad. One of their advantages is that they can write CD-R discs as well, so even if you don't use the CD-RW capabilities very often, if you are considering a CD-R drive you might want to consider spending more for this capability.

As with CD-R drives, you should never entrust a single copy of any valuable media to these disks. You should make at least two copies if you are archiving material.

## DVD

The new big brother to the CD, DVD discs are identical in size to CDs, but can store much more data than a single CD. They are also double-sided, which also increases the size of the discs. Transfer rate (speed at which data can be read) is a maximum of 9.8 megabits per second.

A single-layer side can hold 4.7GB, while a dual-layer side can hold 8.5GB. A single-layer, double-sided disc could hold 9.4GB, while a dual-layer, double-sided disc can hold 17GB, but in both cases you have to flip the disc. Mastering and pressing dual-layer discs is much more complicated too.

At the moment there are several "types" of DVD drive:

**DVD-ROM**. A DVD drive that can read DVD discs.

**DVD-R**. Units that burn a DVD disc that can be read by a DVD-ROM drive. They are currently very expensive; $15,000 (though the price may fall to $5,000 soon!). The current drives can't write a dual-layer disc.

**DVD-RAM**. A rewriteable format that holds 5.2GB on a double-sided disk that's stored in a cartridge much like a magneto optical disc. Drives can read regular DVD media, and the transfer rate supports playing DVD-Video files from the disks.

**DVD-RW**. Just to confuse things, there are two additional rewriteable formats under development that are supposed to work with DVD-ROM players.

While it seems likely that DVD will replace CD, at the moment the only acceptable writeable format are the DVD-RAM drives, which cost several hundred dollars and write to a disc that can only be read on other DVD-RAM drives.

Disc or Disk?

At the beginning of the CD revolution, the developers of the CDs (Phillips and Sony) ordained that Disc would be used for a non-rewriteable unit (i.e., CD) to distinguish it from rewriteable disks (i.e., hard drives). Many people aren't even aware of this distinction: a search of Altavista returned the following number of entries:

| | |
|---|---|
| "CD Disk" | 2528 |
| "CD Disc" | 2075 |

Since CD-RW is a rewriteable format, it's unclear whether these discs should be called disks or discs.

## CONCLUSION

What technology you use to backup and archive your files isn't as important as the fact that you are backing up your files.

CHAPTER 15

# IMMERSIVE IMAGING AND 360-DEGREE PANORAMAS

This chapter covers immersive imaging using digital cameras. The simplest description of immersive imaging is that it is the simulation of virtual reality using 360-degree panoramic photographs. Those interested in 3D modeling might also check out MetaCreations' Canoma, which is described in Chapter 11.

There are several immersive technologies available, including QuickTime VR, SmoothMove and IPIX. All are based on a similar theme; a camera is used to photograph an entire 360 degrees around a central point. This can be done with a single shot, or multiple photographs (see below). If multiple images were taken, they must be "stitched" together to create a single panoramic image. The panoramic image is then converted into the file format of the playback technology. A viewer application or browser plug-in is then used to open and display the file. The viewer provides a single window that appears to the user as though they are looking at part of the scene. As they click and drag, the viewer displays different parts of the scene, as though you were looking around, creating the illusion of standing at the point where the original photograph was taken.

The different technologies each have different strengths and weaknesses. While each of the companies that have developed these technologies offer stitching and creation tools that can be used to create the panoramas, there are also some third-party authoring tools available for creating panoramas.

Even if you don't want to create a computer-based simulation, a 360-degree panoramic image can be visually stunning and useful in other applications. The first part of this chapter covers techniques for capturing a 360-degree panorama. The second part covers turning those images into a single flat image, while the last part explores the different playback technologies that are available. Note that IPIX differs from the other authoring technologies in that it only

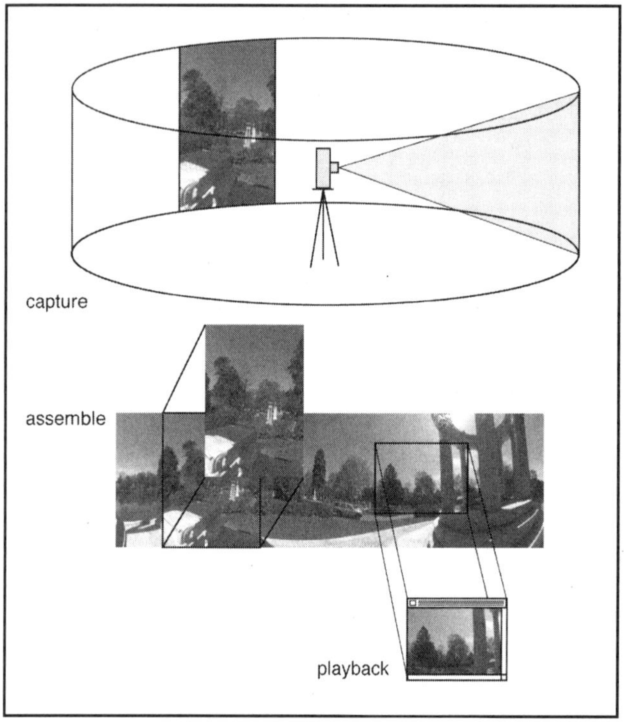


*Figure 15-1. Creating an immersive panorama (sketch).*

works with a fisheye lens used to take two half-hemisphere images. Their stitcher is the only one that can be used to assemble the two images and create the file.

## OBJECTS

While panoramas provide an experience of being in a place, many of these technologies also support *objects*; that is, something that you can view from any (or many) angles. This is done in a very different way from creating a panorama. Photographs are taken of an object from every angle that you want the user to be able to see. These are then linked together and as the user clicks and drags the software displays the correct frame, creating the illusion of rotating the object while looking at it.

It's possible to make very simple object movies using a tripod and rotating an object on a table. If you want to create really smooth presentations then you might want to get an object rig—a device with a rotating platform and an arm on which the camera is mounted.

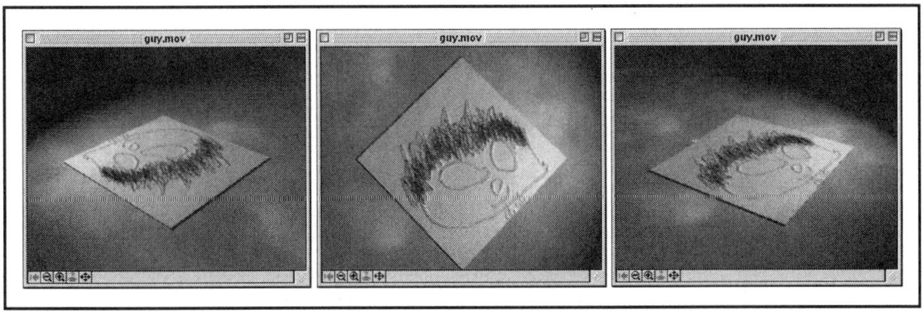


*Figure 15-2. Object movie.*

# IMMERSIVE PLAYBACK TECHNOLOGIES

If you want to create a photographic immersive experience for a multimedia presentation, or to add to your website, you first need to decide which technology to use. That choice will impact the options you have for creating the panorama. The major choices are:

## QUICKTIME VR

Based on Apple's QuickTime technology, QuickTime VR has been in existence the longest of any technology, and is probably the most widely used. When initially introduced, authoring tools were only available for Macintosh users, but that has now changed and there are several tools available for Windows users that create QuickTime VR movies.

QuickTime VR is based on a cylindrical projection; it does not support spherical projection, so it will not display the "ceiling." It supports object movies, linking between movies and hotspots within scenes and objects. There are browser plug-ins for both Macintosh and Windows; however, it does require QuickTime to work, which is a 6-megabyte download for the user.

For multimedia authors, Macromedia Director includes support for QuickTime VR, which makes it easy to integrate into interactive presentations. There's no licensing fee. You can find it at:

*www.apple.com*

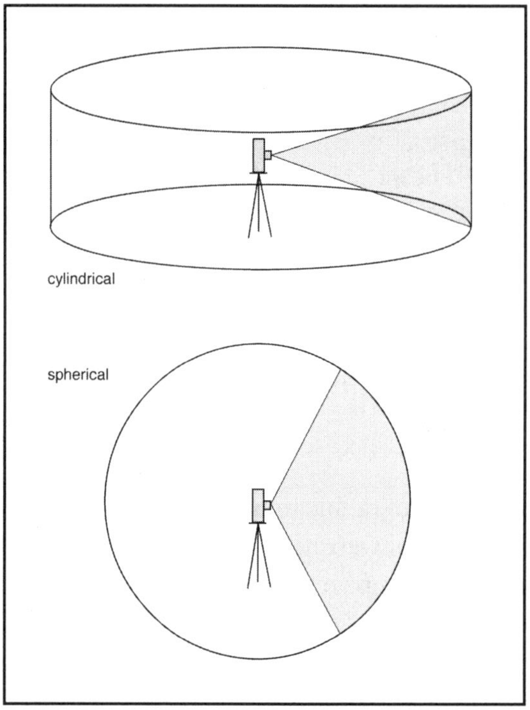


*Figure 15-3. Spherical and cylindrical projections.*

## SMOOTHMOVE

The SmoothMove technology from Infinite Pictures provides support for spherical projection (the roof!) but has been limited until recently because IPIX has a patent on using a wide-angle lens to create immersive environments. They just announced a new tripod from Kaidan, and special software, that takes multiple images and then stitches the pictures together to create a spherical panorama.

Infinite Pictures offers some exporting tools for creating panoramas in 3D programs (3D Studio Max and Lightwave 3D) as well as a Photo Stitcher, that creates panoramas. There is no support for object movies, but it does support hotspots within a movie.

An ActiveX control for Visual basic and Director is available for $99 from Infinite Pictures, but the Web plug-in only works for Windows users. Get more information at:

*www.smoothmove.com*

## IPIX

IPIX, from Interactive Pictures Corporation, uses a spherical projection based upon a stitcher which takes two half-hemispherical images (captured using a 180-degree fisheye lens) and stitches them together. Only IPIX's stitcher can be used to create these scenes, and IPIX charges a license fee per scene (starting at $25 per scene). IPIX has a patent on using a fisheye lens to create an immersive image, and has successfully prevented other companies from using this method to create panoramas. IPIX does not support objects, but there is an Xtra for Macromedia Director. It does offer plug-ins for Windows and Macintosh systems.

IPIX has shown a 360-degree video technology, but there's no announced shipping date. They are selling kits that include a digital camera, lens and software.

Unfortunately, with a digital camera and only two images, the resolution is going to be low; though probably acceptable for Web applications. Exposing the two hemispheres correctly can also be difficult. Because there's so little overlap between the two halves, the stitcher can't do much to minimize the transition at the seam. However, the big advantage of this technology is that you only have to take two photographs to capture the entire scene.

### OTHERS

There are at least two other technologies available that can be used for immersive imaging, though they haven't received wide industry support and it's unclear if they can survive:

**Reality Studio**. Developed by Live Picture, Reality Studio grew out of Real Space, which itself was developed by one of the original QuickTime VR developers. It supports photographic panoramas and objects, as well as VRML (3D models) within scenes. The authoring tool was Windows only. Unfortunately, Live Picture no longer seems to be developing the product.

**Surround Video**. Originally developed for Microsoft, it was then licensed back to its developer Black Diamond. The latest release supports objects as well as movies, and also supports hotspots. The authoring tool is Windows only, but they offered a fairly complete Active X control, and there were Mac and PC plug-ins.

Unfortunately Black Diamond is a small company and the product has not been widely promoted.

## CAPTURING 360-DEGREE PANORAMAS

### PANORAMIC CAMERAS

There are four ways to capture a panorama. The first method involves using a panoramic camera. These are currently only available as film cameras. The camera rotates while exposing and advancing the film. The film is exposed through a small slit so that only a small portion of the film is exposed as the camera rotates. In a sense, imagine a sheet of film being formed in a cylinder, and then the film is exposed to the scene by a slit that rotates about that film.

One disadvantage of this system (apart form the price of the cameras themselves) is that the negative is not a standard size, which can make it difficult to get the image scanned for use on the computer. A flatbed scanner that can be used to digitize film may be the best solution. Alternatively, have the panorama printed and scan that.

In theory it would be possible to create a digital camera that does the same thing using a linear array CCD and rotating that while capturing the information. No one has developed such a device yet.

## 360-DEGREE PARABOLIC LENS

The second method is to use a parabolic lens to capture the entire 360-degree image. A hole is cut in the top of the parabolic mirror, and a secondary reflecting mirror is placed on a rod above the parabolic mirror. Light from the main mirror bounces to the small mirror and is then reflected down the hole in the main lens to a camera underneath. The resulting image looks like a doughnut. Special software has to be used to dewarp the image to produce the flattened final image.

The advantage of these lenses is that they take only a single frame to capture the entire scene; you don't have to worry about part of the scene moving while the picture is being taken. It would even be possible to create a video sequence using such a lens (though no company has yet formally released a playback engine that will play such a sequence).

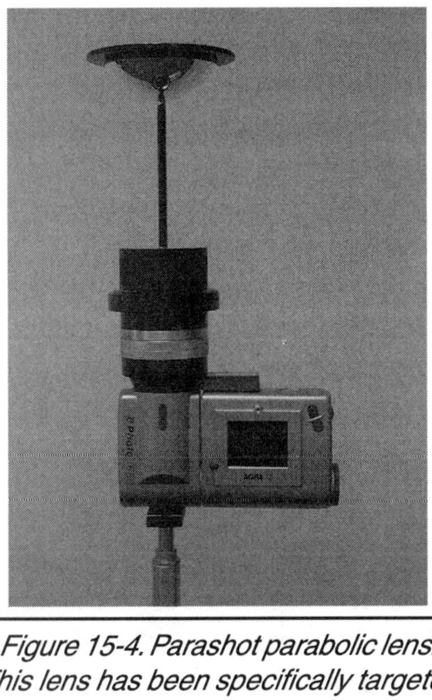

*Figure 15-4. Parashot parabolic lens. This lens has been specifically targeted for use with digital cameras such as the Agfa e1680 (photo courtesy CycloVision Technologies Inc).*

This advantage is also its main disadvantage; the resolution of the final image is greatly limited by the resolution of your capture device. If you are using a digital camera then this is a serious consideration; a film camera might be a better choice unless

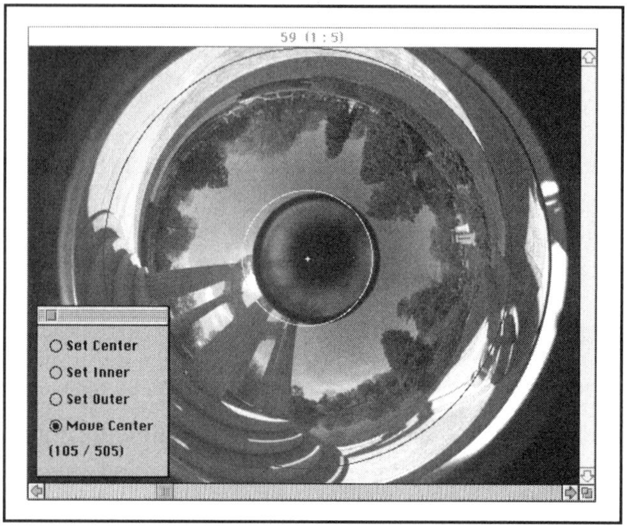


*Figure 15-5. Unaltered image captured by the BeHere Portal Parabolic lens system.*

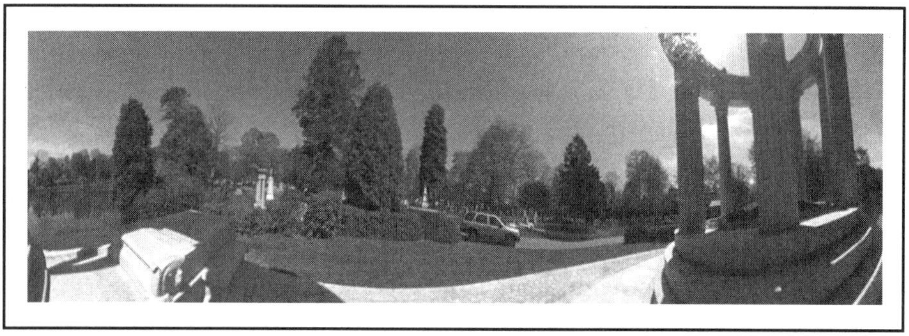

*Figure 15-6. Dewarped image.*

you can afford a high-end digital camera. Another disadvantage is that the actual resolution in the final dewarped image is affected by the distortion of the lens and how it is mapped to the captured image. The top region of the image is mapped to a smaller area of the capture image (the inner circle of the doughnut) than the bottom area, so there is less actual resolution in the top part of the picture than in the rest.

Companies offering this technology include BeHere and Cyclovision.

## 180-DEGREE LENS

The next solution is to use a 180-degree fisheye lens. In a single picture this takes an image which covers the entire half of a hemisphere. Taking two such images provides a complete 360-degree scene. Again, if you are using digital cameras then the resolution will be low; but probably perfectly accessible for Web-based scenes. Exposing the two scenes evenly can be a problem, making it difficult to seamlessly match up the two halves of the hemisphere. IPIX is the only commercial company and technology available that supports this method.

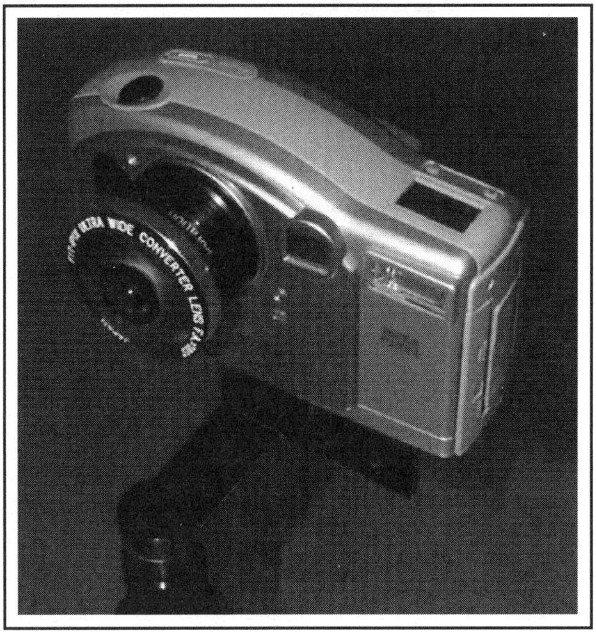


*Figure 15-7. A 180 degree lens attached to a digital camera.*

## MULTIPLE-IMAGE PANORAMAS

The last technique is to use a wide-angle lens and a tripod to shoot a sequence of images that are then stitched together to create the final image. You don't even need a tripod to do this, providing the scene you are shooting does not include objects that are very close to you. A panorama of a landscape can be shot by hand.

If you're taking an image inside a room, then you will need to use a tripod. If you're taking a lot of these images, then you'll want to consider getting a special tripod head which makes taking these sequences much easier.

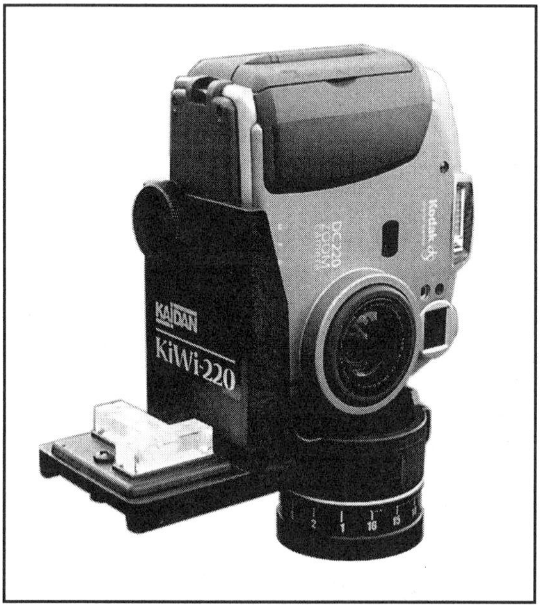


*Figure 15-8. Panoramic tripod head.*

The other piece of equipment you'll want to consider is a wide-angle lens adapter. Without it you'll have a very restricted field of view, and you'll also need to take a lot of photographs to get a complete panorama. It's generally recommended that there be plenty of overlap between one photograph and the next. This provides plenty of room for the software to stitch the images together, and creates a gradual transition from one image to the other, which is especially important if the exposure changes dramatically in your photographs. With the 35mm film equivalent of a 28mm lens, eighteen pictures are usually taken, while with an 18mm lens only twelve pictures are needed.

The chief advantage of this technique is that special equipment is not really required; you can do it with a regular tripod and a standard lens if you want to. A special tripod head and a wide-angle lens can help if you're taking lots of scenes.

The multiple images can be both an advantage and a disadvantage. Producing consistent exposures so that one image isn't wildly brighter than the other will produce the best results. The stitching software will create a transition from one image to the next, but best results are obtained by using the same exposure setting across all of the exposures. Some cameras feature an exposure lock precisely for this reason.

The other problem with multiple-exposure shots is that objects and people can move while you are moving the camera to take the next picture. This may mean having the same person appear multiple times in the final panorama, or it might mean you have to perform some image editing to fix this problem. You might even find you have a shadow in a picture for a person who isn't visible!

## TAKING THE PICTURES

Whether you have a special tripod or not, most of the software that stitches together a sequence of images requires a rectilinear lens. Most standard lenses are rectilinear; the vertical lines in the image aren't curved. But it is not true for all lenses, specifically fisheye lenses.

The photograph is taken with the long axis of the photograph pointing vertically. This increases the vertical field of view, but also means that you'll need to take more photographs to cover the 360 degrees than would be necessary with the camera oriented horizontally.

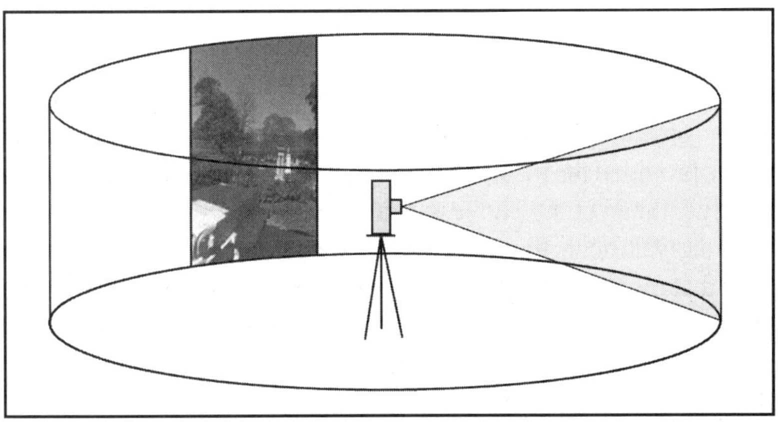

*Figure 15-9. Taking a panorama.*

Using a tripod makes aligning the images much easier, and can also ensure that the orientation of the camera is correct. The camera should rotate around a single point and should "capture" a flat cylinder representing the scene. If the camera is oriented off-center, then stitching together the sequences can be much harder (though some stitching programs have been developed to allow you to take off-axis shots so that you can capture the entire 360 degrees (i.e., the ceiling) and not just a cylinder.

It's also important, if there are objects close to the camera, that the camera is rotated around the nodal point of the lens. The *nodal point* is a point between the front of the lens and the capture plane (for digital cameras, the CCD) where the light rays cross. If you don't rotate around this point then parallax error will occur, which is very noticeable with foreground objects. If all your objects are far away it's not a serious problem.

The parallax problem can be seen readily enough just by looking at a foreground object relative to a background object and turning your head left and right. Chances are very good that the foreground object will appear to move relative to the background object.

If this error is captured in two frames that the stitching software is trying to combine, then the resulting image will contain a blurry area where the foreground object has moved in the first frame relative to the second one.

## CALCULATING THE NODAL POINT OF A DIGITAL CAMERA

If you're really lucky you'll either get one of the Kaiden tripod heads made specifically for your digital camera, or you'll get a camera for which the manufacturer has provided the measurements for the nodal point. If neither of those cases are true, then you will have to figure out the nodal point yourself. This isn't particularly difficult, though it can be tedious. The problem is compounded by the fact that most digital cameras are of a rangefinder type. This means that you can't rely on what you see through the rangefinder to determine whether you have aligned the camera correctly. Instead, you'll need to use either the LCD preview, or actually capture frames and view them. In both cases, it's

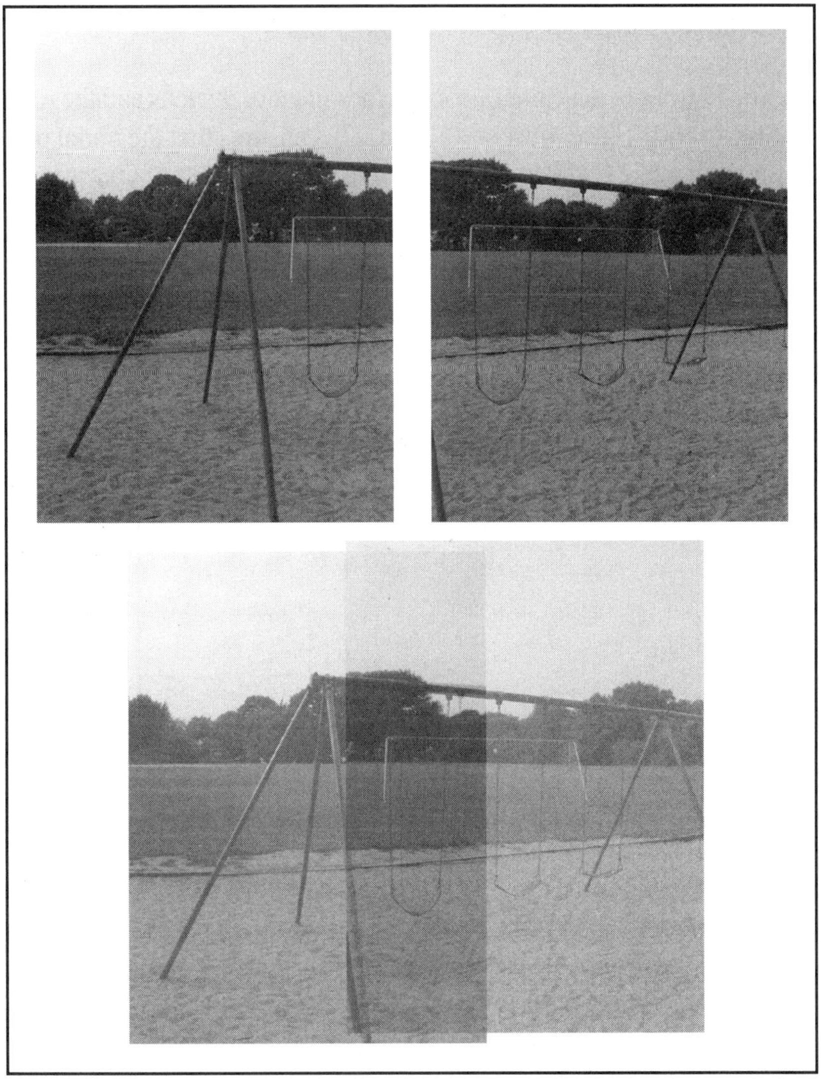

*Figure 15-10. Parallax error caused by not rotating around the nodal point of the light path. Notice that the chain of the swing seat appears to move relative to the goalpost.*

easier to do the initial measurement inside, connected to the AC and your computer, rather than outside in the field. Then measure or mark the position of the camera so that you can set it up again later.

Here's what to do:

**1.** Mount the camera on the tripod head.

**2.** Standing in front of the camera, look at the center of the lens and the rotation point of the tripod. These must be aligned. This ensures that the nodal point is somewhere on a line that passes through the rotation point of the tripod. The next part is the hardest part.

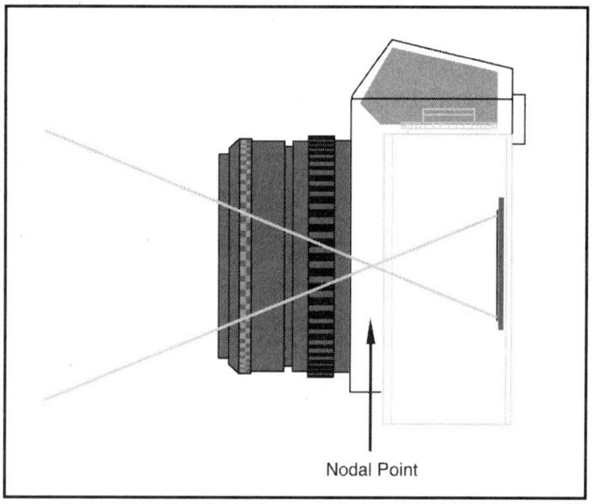


*Figure 15-11. Nodal point.*

**3.** Now it's time to move the camera forward (or backward) until the nodal point is located directly over the center of rotation.

Position an object in the foreground (about three feet from the camera) and position it relative to something as far away as possible (difficult if you have a small room!). It's easiest if you have two vertical objects; the side of a chair back and a door frame, for example.

If you look through the rangefinder as you turn the camera left and right you'll probably see the movement of the foreground object relative to the background one. If you have an SLR camera you can move the camera while adjusting the camera (see *Figure 15-9*).

If your camera has an LCD display that features a live update, then you can accomplish this task using the LCD preview. If your preview is not live, then you can still probably use it for the first few tests. As you get closer to the nodal point you might have to switch to viewing images.

In all cases, what you want to do is position the target point as far left as possible in the frame, then rotate the camera so that it is on the extreme right.

However you do this, you should start with the camera lens as close to the rotation point as possible, then slowly move the camera forward while continuing to take test pictures. Move the camera very small amounts; what you don't want to do is waste time moving the camera back and forth and then getting confused about which direction you should be moving the camera to correct the problem! Continue taking and comparing images until the foreground object no longer appears to move relative to the background object.

## STITCHING

There are several different tools available that will stitch together a sequence of images and turn them into a panorama. Many of these will even save the panoramas into some of the different formats. Just some of the tools include:

Enroutes QuickStitch and QuickStitch 360, Ulead Cool 360, LivePicture's PhotoVista, PanaVue's Visual Stitcher and Apple's QuickTime Authoring Studio. The next couple of pages will focus briefly on just three of these tools.

### QUICKSTITCH

Enroute Imaging offers more than one stitching program, but you need to make sure you understand the differences between them. QuickStitch 360 can be used to stitch together a 360-degree panorama, but QuickStitch is a very different program. Its forte is stitching together several images that overlap. This enables creating a higher resolution image of an object or scene than would be possible with one picture.

First you take your pictures. You can take a grid of up to 6 x 6 pictures. You want to use a wide—yet not too wide—lens; if the images are too distorted the sequence won't be stitched correctly. Similarly, you want to be far enough away from the object that your images overlap as a straight projection (i.e., you're not rotating as you photograph and creating a cylindrical projection). Once you're arranged the images within the software grid, you tell it to stitch the images and it does the best it can with the images you gave it.

Your results really depend on how well you take your images. The software does not handle extreme warping. QuickStitch is available for both Macintosh and Windows.

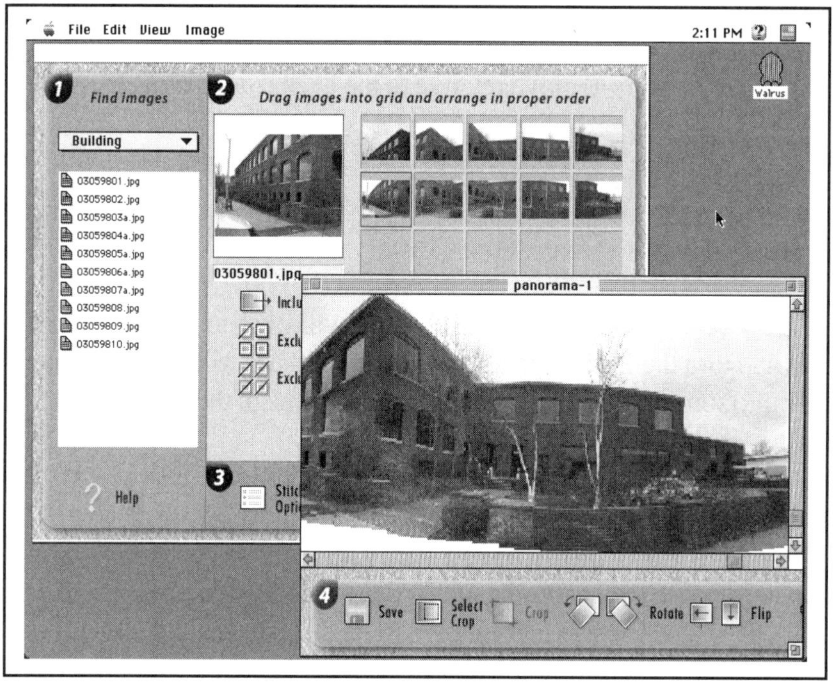


*Figure 15-12. QuickStitch.*

## APPLE QUICKTIME VR AUTHORING TOOL

Though it's a couple of years old, Apple's QuickTime Authoring Tool is still one of the easiest to use, and most polished tools available. Unfortunately, it's a Macintosh-only tool.

Images to be stitched are imported into a single window. The orientation can be changed by selecting and clicking a button. The order of the images can be changed by clicking and dragging. It's important to tell the software what kind of lens was used to take the image; the software expects a certain number of images depending upon the lens setting.

Click a button and the software stitches the software and saves the panorama; you can also go in and change settings for compressor and other options. The program lets you add hotspots to a scene as well.

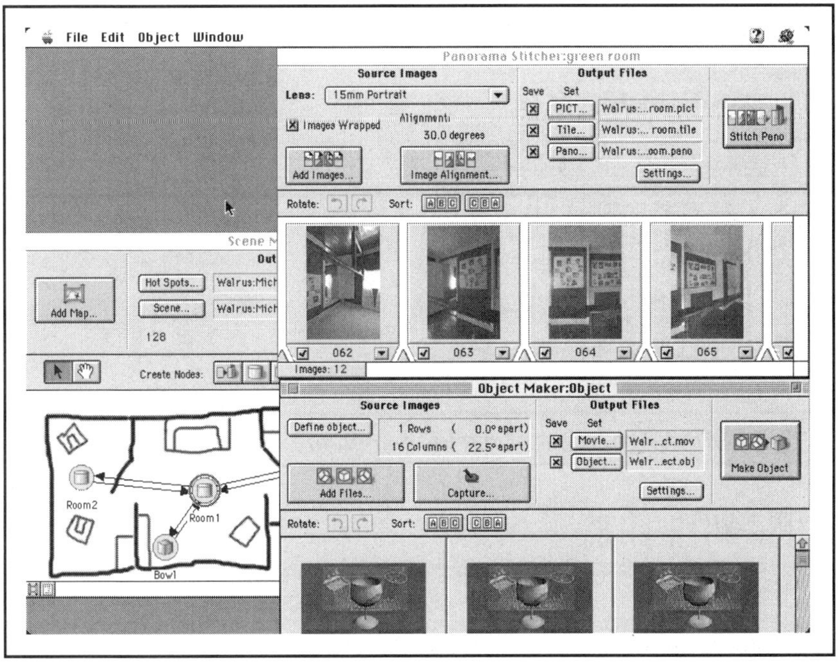


*Figure 15-13. Apple QuickTime VR Authoring Tool setting up a panorama.*

Creating object movies is almost just as easy. If you tell it how many rows and columns are used for the object it will arrange the images as you import them. Then you can go in and edit hotspots (though this can get a little tedious if you have a lot of images).

## PHOTOVISTA

While the Apple QuickTime VR Authoring Studio is one of the best stitching tools because of its combination of object and panorama functions, for stitching panoramas LivePicture's PhotoVista does just as good a job as long as you adjust the settings correctly, and it's available for Windows as well as Macintosh.

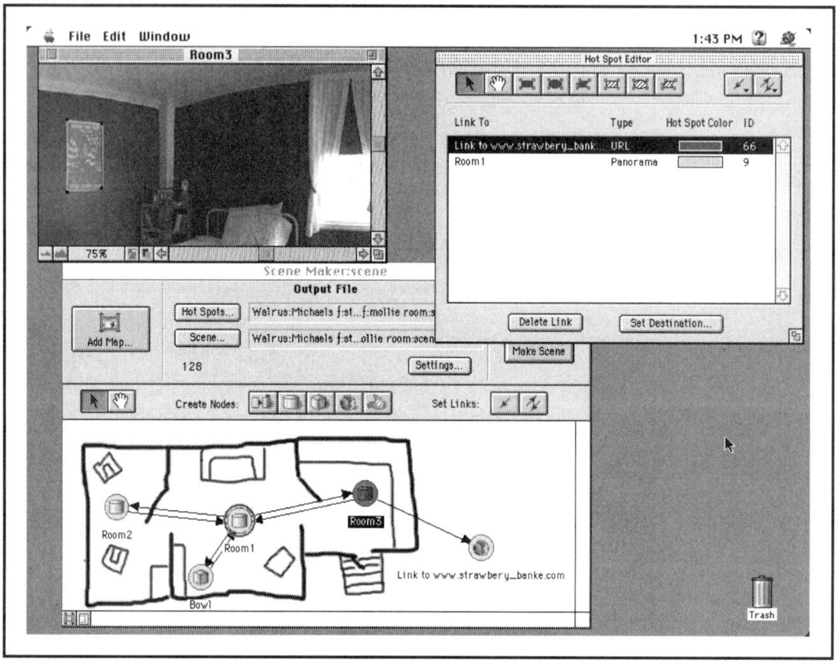


*Figure 15-14. Apple QuickTime VR Authoring Tool setting up an object.*

# WORKING WITH FILES

Before you stitch the images, there are some preprocessing tasks you might need to do.

## MANIPULATING FILES

Generally, if your images are slightly misaligned you do not need to do any cropping; the stitching program should be able to handle the alignment. If your images include some border—this doesn't happen with digital cameras, but can happen with scanned slides—then you must crop the border.

Whether you should perform any image editing on individual images before stitching is debatable. If you have one image that is seriously over- or underexposed then you should try to correct this problem. But in many cases, it's easier and more productive to stitch the sequence and then perform any image adjustment on the final panorama.

Once the panorama is completed, you might have to perform some cleanup tasks, such as removing ghosts or people who appear in the transitions between pictures; a person in the transition between two images who was in one photograph but not the other will appear faded in the final panorama.

### RENAMING FILES

When working with files to be stitched, it's very helpful to have the files correctly named in a sequential order, i.e., PIC 01, PIC 02, PIC 03, etc. Many digital cameras use such a naming scheme, but not all do. If you use a scheme of PIC 1, PIC 2, PIC 3 (no leading zero) then the file PIC 10 will come between PIC 1 and PIC 2.

You can rename the files manually or you can use a utility.

For Macintosh users, a simple AppleScript called "Rename script," written by Jim Anders of Kaidan, renames images and is included on the CD-ROM.

## ADDING THESE TECHNOLOGIES TO YOUR WEB PAGE

Adding these technologies to your website is going to require a little work on your part. For a start, you should always warn your users that they need a special plug-in to view the scenes before they get to the page that contains them. A link to where the user can download the plug-in is also desirable. IPIX offers a Java version of the viewer, though that may not work with all browsers.

### ADDING QUICKTIME VR TO A PAGE

Your ISP must set up the MIME type for QuickTime files (it's likely that they already are):

video/quicktime file extension=mov

The following code will work with the plug-in:

```
<EMBED SRC=”graphics/59high.mov” PLUGINSPAGE=”http:/
quicktime.apple.com” WIDTH=320 HEIGHT=240 CONTROLLER=false
LOOP=false AUTOPLAY=false>
```

## ADDING AN IPIX FILE TO A PAGE

Your ISP must set up the MIME type for IPIX files:

application/x-ipscript file extension=ips

application/x-ipix file extension=ipx

The following code is added to your Web page—just change the FILENAME.ipx to your file name. Width and Height defines the size of the window (you must set it in two places; the OBJECT part is used by Microsoft Explorer, while the EMBED tag is used by Netscape). The codebase points to the IPIX viewer file which will download automatically for Internet Explorer users.

```
<OBJECT ID=”IpixX1" WIDTH=320 HEIGHT=240

CLASSID=”CLSID:11260943-421B-11D0-8EAC-0000C07D88CF”

CODEBASE=”http://www.ipix.com/viewers/ipixx.cab#version=2,2,0,1">

<PARAM NAME=”IPXFILENAME” VALUE=”FILENAME.ipx”>

<embed src=”FILENAME.ipx” border=0 width=320 height=240
palette=”FOREGROUND”>

</embed>

</OBJECT>
```

### WHAT IF VISITORS DON'T HAVE THE PLUG-IN?

CYLPAN VR panorama scroller is a Java applet that plays back a JPEG or GIF panoramic image in much the same way as the immersive technologies do. It works with Microsoft Explorer 4.x and Netscape Navigator 3.02+ on both Macintosh and Windows. Get it at:

*http://www.thrift.com/cylpan/*

There are a number of other similar Java applets available. LivePicture offers several Java viewers and IPIX also has one for their technology.

While the Java applets are often limited in functionality, and performance will suffer if the file is too large, it does provide a way for visitors to get an idea of what you have on your Web page without having to download the plu-in. Use it as a teaser to your actual content.

Similarly, you could create a small linear scrolling thumbnail as a preview of your content. Apple offers QTVR PanoToThumbnail, a free utility which converts a panorama to a thumbnail-sized linear QuickTime movie; users need QuickTime to see this movie so it only provides a low-bandwidth preview to someone who already has the playback technology used by QuickTime VR. However a GIF file of the same thing could be seen by anybody.

## CONCLUSION

Immersive imaging produces fascinating interactive images that a user can explore. It can be a great way to add depth to your website. The next chapter covers Web graphics in more depth.

# CHAPTER 16
# INTRODUCTION TO WEB GRAPHICS AND HTML

This chapter is an introduction to working with graphics on the World Wide Web. It's meant only as an introduction and to get you going; if you want to become an expert in HTML and Web production then this will get you on the road, but you may need to buy another book!

If you already have a Web page with a photograph in it, know what an HTML image tag is, or are making your living as a Web consultant, then you can safely skip this chapter.

For the rest, this chapter provides first an overview of how HTML, browsers and Web servers work and covers how to create a simple page and insert a graphic into it. As a specific illustration, adding a picture to an eBay page is described, and there are tips for preparing graphics for the Web. There's also a look at PhotoNet; a website for displaying and sharing images.

Finally, some suggestions on how to learn more, and what other tools to get, are included at the end of the chapter, along with a little advice for sending images via email.

## HOW HTML WORKS

HTML is a tagging system for adding formatting and links to a text file that is displayed by a Web browser application. An HTML file is simply a plain text file that contains the HTML tags. Tags are contained within a pair of "<" (less than) and ">" (greater than) symbols. An example is the tag:

<BOLD>

...which indicates that all the text following the tag should be displayed bold. Most tags have an ON and OFF form and appear in pairs within the document. For example, the <BOLD> tag indicates that all text after the tag is bold, while the </BOLD> tag turns off the bolding. There are some exceptions, notably the IMG tags used to insert graphics.

An HTML line that contains the following...

    The quick <BOLD>brown</BOLD> fox jumps over

...would be displayed as:

The quick **brown** fox jumps over

## CODING A SIMPLE WEB PAGE

You can edit HTML files with just about any text editor (i.e. Notepad on Windows, SimpleText on the Macintosh) or even a word processor such as Microsoft Word or Corel WordPerfect. If you do use a word processor you have to be careful not to save the file in the native format of those applications. An HTML file must be a text format file; if the file isn't a standard text file the web browser won't be able to deal with it.

While HTML is fairly free format, there is some basic formatting that all pages should follow. This includes an HTML open and close tag, and head and body tags. The following page contains a single line of text, and has the background color of the page set to a light blue color:

```
<HTML>
<HEAD>
</HEAD>
<BODY BGCOLOR="#66CCFF">
<P>The quick brown fox jumps</P>
</BODY>
</HTML>
```

You can then get sophisticated by adding text formatting (bold, paragraph, change type sizes), and add links and graphics to the page. Adding graphics and links are explained below.

One of the advantages of HTML is that you can view the source of any page you look at either by using the view source option of the browser, or saving the source file and then opening it in your text editor. Looking at how things are done in other pages can be a great way to learn!

## INSERTING A GRAPHIC

HTML does not support graphics inside the file in the way that your word processing application does. Instead, you place a link within the HTML file that points to the graphic file. The Web browser first downloads the HTML file and starts to display it. When it encounters the link to a graphics file it downloads that file. A graphic file can be contained on the same Web server as the HTML file, or it could be on a Web server in another country. The browser doesn't really care. The Web browser inserts the graphic into the formatted page it creates.

Most Web browsers support both JPEG (.jpg) and GIF (.gif) files. The link to the graphics file is created using the Image tag. The format of the image tag is:

```
<IMG SRC="picture.jpg" WIDTH=448 HEIGHT=262>
```

The width and height are the pixel dimensions of the graphic and is optional. If you don't specify the height and width the Web browser will download the file and display it at the actual pixel dimensions of the file. However, it is desirable to include the dimensions for graphics even if you want them displayed at their actual size. If you don't, the Web browser will have to download the file before it can figure out how to format the rest of the page. If you provide these dimensions then the browser can format the rest of the page and display it to the user correctly while it is still downloading the graphic.

Width and height are the display dimensions of the graphic, not necessarily the actual dimensions. For example, if you wanted the Web browser to display the graphic much smaller than its actual size, then you could scale the graphic in the browser by adjusting these parameters. However, as a general rule I don't recommend this practice unless you want to later display the graphic enlarged. If you're shrinking the file, then you're sending more information over the network than you need to. You'll improve user performance if you scale the graphic before uploading it to your server. Also, in most cases your graphics program will do a better job of scaling the graphic than the browser will.

In the example above, the graphic is in the same directory, on the same Web server as the Web page. If it were in a subdirectory named "graphics" on the same machine the SRC link would look like this:

SRC="graphics/m2wmain.gif"

If the graphic were on a different Web server, then the entire Web address must be inserted:

SRC="http://www.m2w.net/graphics/m2wmain.gif"

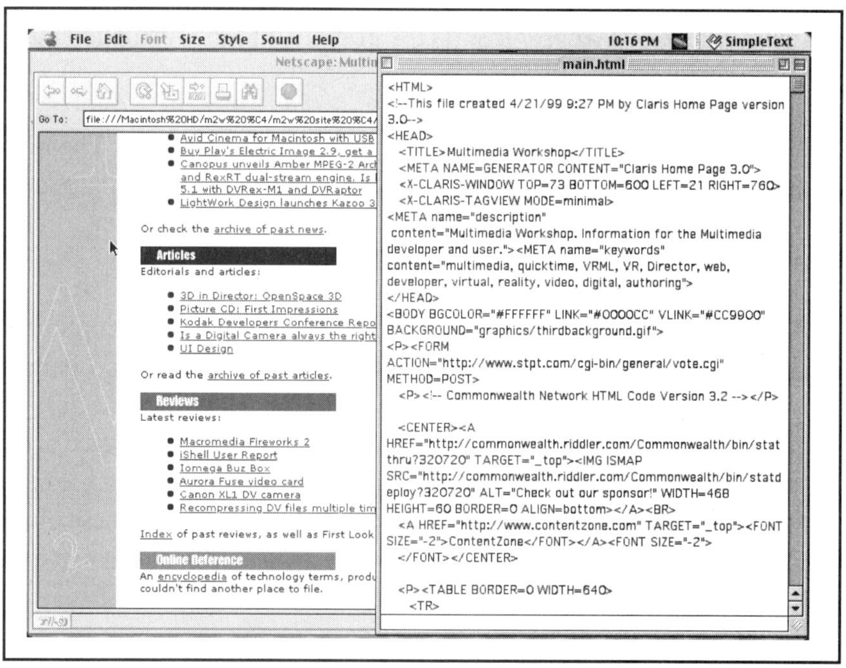


*Figure 16-1. Page displayed in browser along with source of page being viewed.*

The following simple page displays a graphic in the center of the page.

```
<HTML>
<!—This file created by me—>
<HEAD>
</HEAD>
<BODY BGCOLOR="#FFFFFF">
<CENTER><IMG SRC="Nasa.jpg" WIDTH=317 HEIGHT=241
ALIGN=bottom></CENTER>
</BODY>
</HTML>
```

## ADDING LINKS

Adding links is reasonably easy. A link can be attached to a graphic, text, or both. A *link* is a piece of text or graphic that when clicked will cause another HTML page to be loaded. Links are formatted in the following way:

```
<a href="http address"> [HTML content]</a>
```

Where the HTML content can be any valid HTML, including text or an IMG tag. Here's an actual example:

```
<a href="another.html">Click this to go to another page</a>
```

## HOW BROWSERS AND SERVERS WORK

If you want to set up a Web page you need a Web server (also called an HTTP server) to host the pages and graphics. Internet Service Providers (ISPs) provide Web page facilities, as do several free services such as Tripod and Geocities. They let you upload your HTML pages and graphics.

HTTP is the transfer protocol used to send information from the server application to the browser. All servers on the Internet are identified by a unique IP number. This is a four-byte number; and is in the form 255.255.255.255, where each of the values can be a number from 0 to 255. Numbers are assigned by a governing body, so to get a fixed IP number you usually have to get it from the service that provides your connection to the Internet. The Internet is really just a collection of connections.

Rather than having to remember numbers, the domain name function is a directory of unique names mapped to unique IP numbers. When you type in "*www.m2w.net*" the Web browser connects to a domain name server which looks in its database and sends back the IP number that matches the domain name.

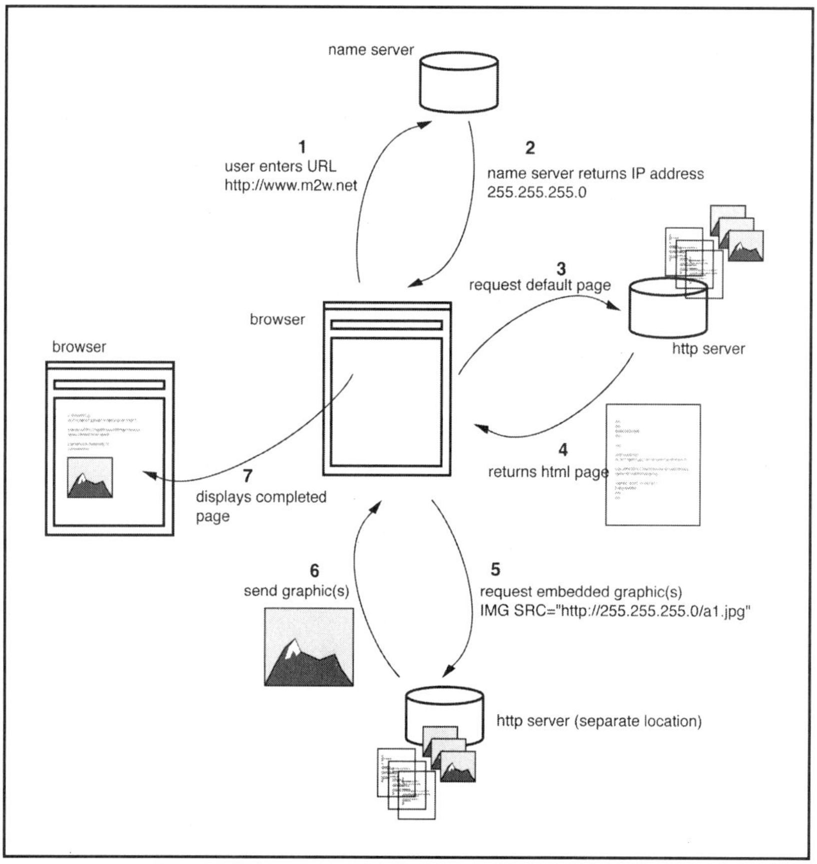


*Figure 16-2. Server connection.*

The browser can then connect to the Web server by sending a request to the IP address. The browser requests a page. The server returns the page, and if the browser needs additional graphs, then it will send additional requests to the Web server.

There are a number of Web serving applications, and the latest Macintosh and Windows operating systems include personal Web servers; basic applications that will work fine on internal networks, or for limited websites. The biggest problem with a Web server is that it really must be connected to the Internet 100% of the time so that users can connect to it at any time. So-called *permanent* connections are expensive to operate, even at phone speeds. It's much cheaper to use the Web server at an ISP. If you don't want to spend any money, you can use one of the free Web hosting services.

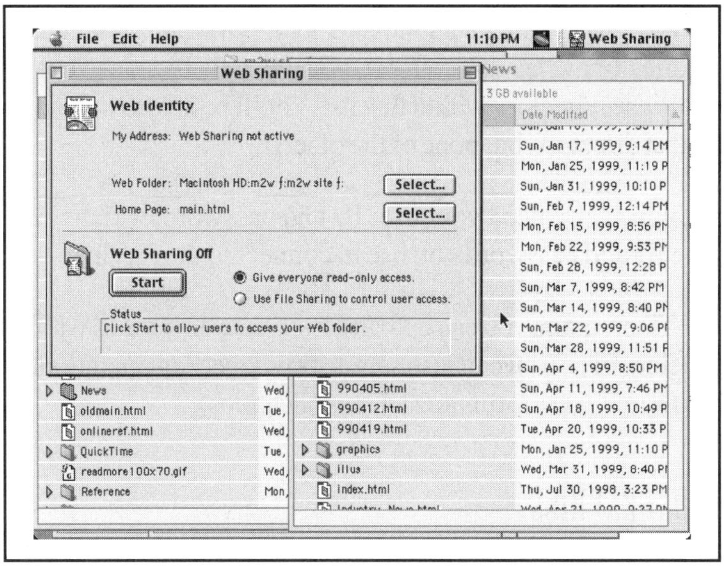


*Figure 16-3. Windows personal Web server.*

## UPLOADING PAGES AND GRAPHICS TO A FILE AREA

Once you've created your Web pages, you'll need to upload them to the Web server. Before doing so, you should test them on your local hard disk first; simply open the files using your Web browser.

To upload the files you'll probably have to use an FTP (File Transfer Protocal) program. The FTP protocol is different from the HTTP protocol, but it is another way to transfer files from one computer to another. There are several shareware applications available for FTP for both Macintosh and Windows. Check out a download site such as *www.download.com* to find some.

Several of the HTML editing applications include FTP transfer features which can make it very easy to update your website.

Your ISP should provide information about how to access the file area where the files must be placed. This usually means supplying an IP address or FTP server name, which is not necessarily the same as the HTTP server. For example, while your site might be accessed at the URL *www.myisp.com*, the FTP name might be *fred.myisp.com* or something else. Many ISPs place all of the user directories in a single public directory. But while you might end up in the public directory when you connect, you will only be able to upload files to your specific user directory within that list. You'll be able to upload and delete files from that directory, but none of the others.

Of course, you'll also need your user ID and password, which is usually the same as the ID and password you use to connect to the service.

Using the FTP program, you can upload files, create directories within your file area, and delete files. You can also set file privileges (an operation you might need to perform if you are running CGI applications).

## ADVANCED GRAPHICS

For the basics about the GIF and JPEG file formats, you should also read the File Formats chapter (13). However, in general, photographs should be saved in JPEG since it supports 24-bit color, while GIF only supports 8-bit color.

### GIF OR JPEG

While JPEG will offer the best quality for photos, there are a few situations where you might find that you want to use GIF files. The two most important reasons are: 1) GIF supports an animated format that displays a sequence of images (for more about this, see Chapter 19); and 2) GIF supports a transparent color, which lets you see the background color or graphic through the transparent color.

## TRANSPARENT COLOR IN GIF

Most programs that save in the GIF file format will let you specify the transparent color.

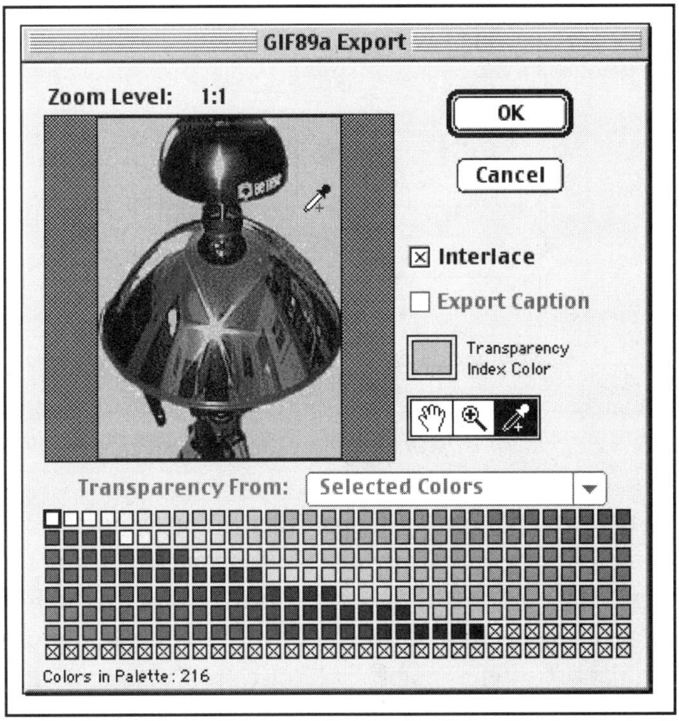


*Figure 16-4. Defining the transparent color in Adobe PhotoShop.*

Some HTML editing applications let you open a GIF file and set the transparency color, which can be a useful time saver.

Once the transparency color is set, browsers will display the graphic with the transparency. One thing to watch out for is keeping the color solid when you

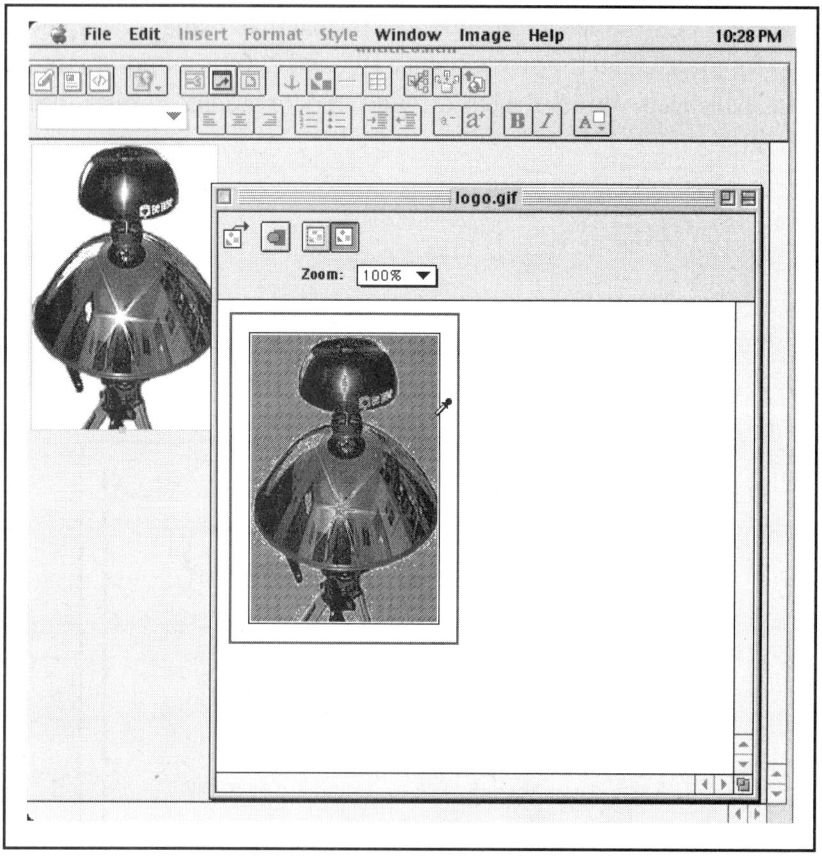


*Figure 16-5. Setting transparent color in Claris Home Page HTML editor.*

convert them to 8-bit. If you have a graphic that you are editing in 24-bit color, and there's a region that you want to make transparent, when you convert it to 8-bit you might want to turn off dithering during the conversion so that the region you want transparent is only one color. Otherwise, the region will have patterns of similar but different colors, and you can only have one transparency color. Of course, you can fix the problem by hand by re-editing the color.

Unfortunately, you can only set a single color to be transparent. You can't define regions with a percentage of transparency, so you can't create soft, anti-aliased transitions. To do this would require an alpha-mask. While there has been talk about adding such features to browsers for support in JPEG, so far it's not possible to do this.

## IMAGE MAPS

Creating an image map (hot spots on graphics that point to other locations) is much more difficult than creating a simple link. The MAP tag lets you define regions that point to other pages, but you have to work out the coordinates to do this. WYSIWYG HTML editors (see below) will do this for you automatically; you draw a circle, square or polygon on the image and define the link and the software works out the coordinates. It's much easier than doing it by hand! The following is an example of a single region defined for a graphic.

<MAP NAME=map1>

<AREA SHAPE=rect COORDS="244,44,365,222" HREF="nextpage .html">

</MAP><IMG USEMAP="#map1" SRC="picture.jpg" WIDTH=448 HEIGHT=262>

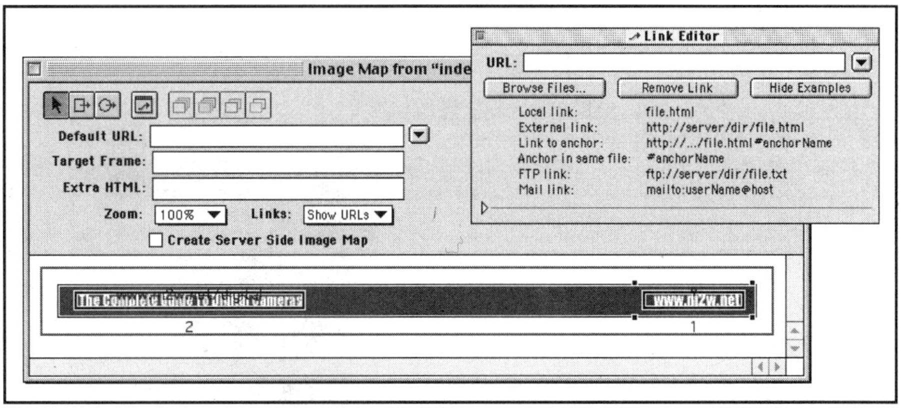


*Figure 16-6. Defining an image map.*

## CREATING THUMBNAIL PAGES

If you're posting a lot of photographs to a Web page, you should think seriously about how best to let a visitor to your site navigate the pages. There are three ways to do this.

**1.** Post all the graphics at actual size on one page.

> **Pro:** Easy to do.
>
> **Con:** Takes forever to download and may cause memory problems if there are a lot of graphics on the pages. Doesn't look very good if you have a lot of large images.

**2.** Create a "thumbnail" page which downloads all the graphics at actual size, but shrinks them to thumbnail size. When you click on a thumbnail, it displays the graphic at actual size.

> **Pro:** Relatively easy to do, looks good, and when the user clicks on a thumbnail the graphic will display immediately.
>
> **Con:** Takes forever to download and may cause memory problems if there are a lot of graphics on the pages.

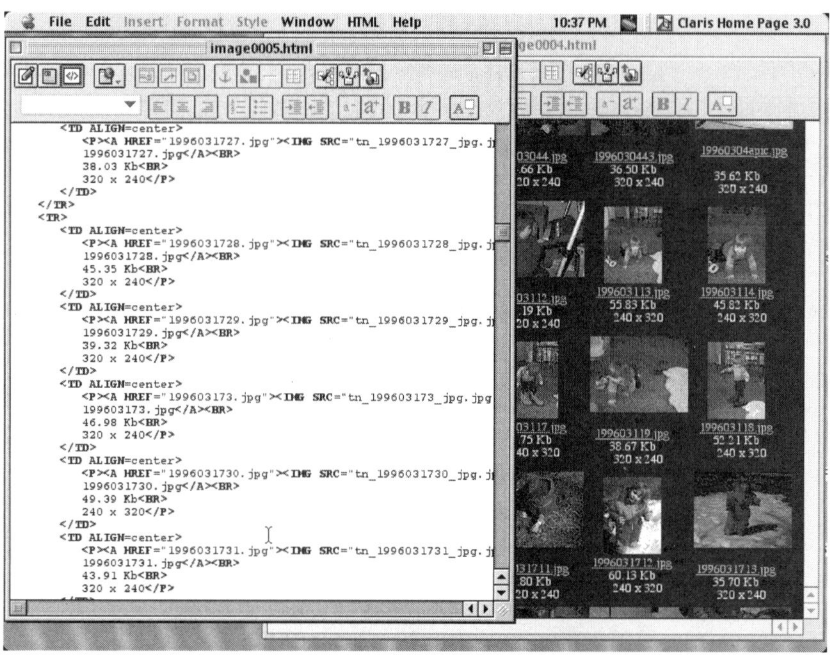


*Figure 16-7. Thumbnail page.*

**3.** Create thumbnails of the graphics which are downloaded on the page, then download the actual graphic if the user wants to see that.

> **Pro:** Very fast and efficient to download the thumbnail page and the thumbnails will look better than scaled images.
>
> **Con:** More work: downloading the full-size graphic will take additional time.

I strongly recommend the last option, as it provides the best performance for users. There are even programs such as ThumbsPlus which will create thumbnail pages and thumbnails automatically (see Chapter 17).

## PROXY GRAPHICS

HTML supports an option called proxy graphics; a low-resolution 1-bit graphic that downloads first before the final graphic. This lets the user see what the page looks like before all of the data has been downloaded. It's relatively easy to do; create the original graphic, then convert it to 1-bit black-and-white and save the copy. They add "lowsrc=filename" to the image tag, where "filename" is the 1-bit file:

<IMG SRC=”sep309802.jpg” border=”0" lowsrc=”lr_sep309802.gif” width=”96" height=”72">

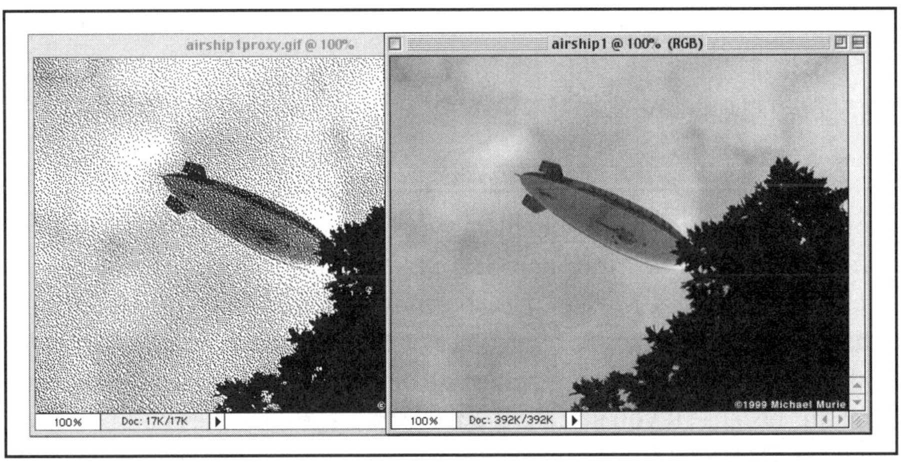


*Figure 16-8. Proxy graphic.*

Proxy graphics are best for large graphics such as image maps used for navigation. They download faster than the color graphic. They are of little point for use with thumbnails.

### GETTING RID OF THE BLUE BORDER

If you've placed a graphic within link tag you've probably noticed a blue border that appears around the graphic to indicate that it can be clicked. If you don't want that to appear around the graphic, just add the parameter "BORDER=0" to the tag:

<IMG SRC="picture.jpg" WIDTH=448 HEIGHT=262 BORDER=0>

## ADDING GRAPHICS TO eBAY PAGES

eBay is one of several websites that let you place items for auction. They let you place a text entry in their database, but you can't place pictures on their servers. However, if you have a place on a Web server where you can place a graphic, then you can place a pointer to that graphic from within the entry that you place on eBay. While the following discussion centers on eBay, it's applicable to the other auction sites.

### THINGS TO WATCH OUT FOR

If you want to add images to your listing you'll need a server to host the graphics files. Services like Tripod and Geocities let you place graphics and Web pages on their servers. Most of these businesses make money by placing advertisements on the pages that you place on their servers. Though you could use these servers to hold your graphics, all of these services have rules against doing this because if you just use them to serve a graphic they can't display their ad and they can't make any money!

If you're only doing it once you'll probably get away with it, but if you're planning to do this regularly then you'll need to get your own file area on an ISP that doesn't have such a limitation. If you're paying for an Internet connection and get a file area with your account you can probably use that file area to hold the images.

## LIMITS ON TRANSFERS

Most ISPs (Internet Server Providers) limit both the amount of disk space that is available to store your graphics (they'll charge you extra if you exceed it), and they often limit the amount of data that can be transferred from your Web area within a month.

For example, Pair.net provides 20MB disk space and 100MB/day traffic for their basic accounts, but will charge extra if you exceed this.

For even the moderate user this is probably not a problem, but if you place a lot of photos on your Web area, and they are very popular at eBay, then you could find that you exceed this quota. How ISPs handle exceeding this limit seems to differ from ISP to ISP.

Some will "look the other way" if you only exceed the quantity by a small amount occasionally. Others will immediately charge you extra for the downloads. Still others simply cut the Web serving facility. This happened to someone I know and it took quite a while to figure out why the graphics weren't being served any longer!

## ADDING THE IMAGE

First, create your image. Then create HTML tags that will point to the graphic. These tags will have the complete URL for the graphic because it is on a different server from eBay's. Once you add the item on eBay, you can simply insert the HTML tag inside the body of the description.

One thing to watch out for: make sure you have the correct URL when you add the tag. It's possible to create a tag that points to the file on your local hard drive. While that works for you, it won't work for anyone else who downloads the page. Also, don't forget to upload the graphic before you test your HTML; it won't work if the graphic isn't there.

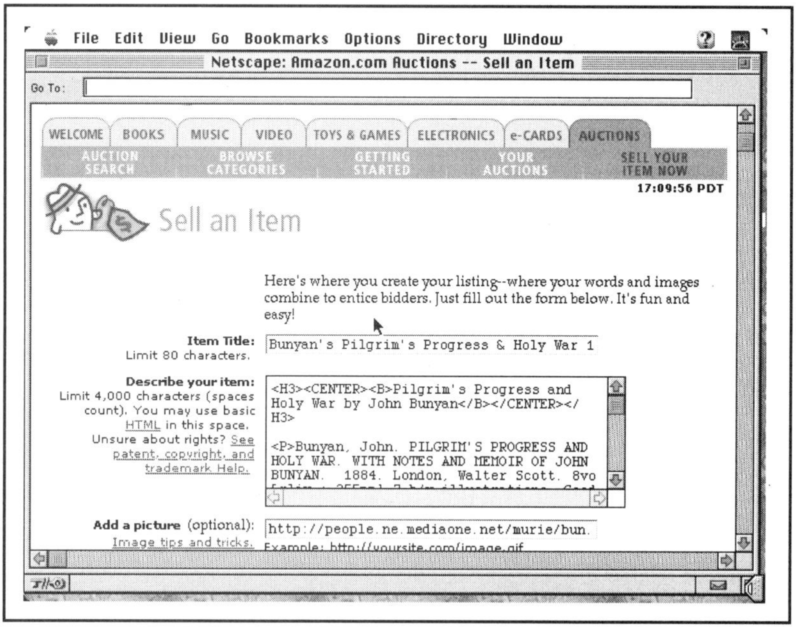


*Figure 16-9. Adding an entry to eBay.*

## Tips:

**1.** Create the best pictures you can; make sure the color is accurate, and crop the images to the object.

**2.** Don't use images that are too large; it'll be a problem for users on small monitors.

**3.** Don't place too many really large graphics on the same page. It takes too long to download and causes problems for users with slow machines or little memory. If you have to place several graphics, use thumbnails that point to the full-size graphics.

# WHAT TO GET IF YOU'RE DOING A LOT OF THIS

As you've already seen, creating simple HTML pages is very easy. If all you want to do is create a page with some text and maybe a graphic or two then you can use any word or text processor to create your pages.

You can continue to do that when you start using advanced features (such as tables) in your pages. But if you're going to be doing a lot of HTML coding then you should look into an HTML WYSIWYG editor. There's a wide range of such tools, including:

Adobe GoLive

Adobe PageMill

Microsoft FrontPage

Claris HomePage

Macromedia DreamWeaver

Any one of these is a good tool for someone maintaining a complex site. The neat thing about these tools is that you can forget everything you learned about HTML coding and tags. The program will do it for you. I couldn't create a table if my life depended upon it, but with one of these programs I just tell it I want a table and specify the number of columns and rows within the table.

Similarly, there are several tools that have been developed just for the preparation of Web graphics. They let you preview a graphic using a number of different compression settings so that you can easily find the right trade-off between file size and image quality.

If you only need to create a couple of graphics every month then you don't need these tools, but if you're doing it every day then look at:

Adobe ImageReady

Macromedia FireWorks

# SENDING IMAGES VIA EMAIL

If you're sending an image to a friend via email, you need to consider what mail program your friend has, as well as what type of computer they use. You might also need to consider whether they have a graphics program or viewer that can open the file you are sending.

## HOW TO IDENTIFY GRAPHICS FILES

The Windows operating system uses an extension to indicate the file type. ".jpg" = JPEG and ".gif" = GIF. When you try and open a file of those types, the operating system looks at a list of programs on the disk with the files that each application can open. It will launch the program that matches the extension.

One interesting thing is that if you change the file extension by accident, then the operating system will assume that the file is of that type and launch the program to open that file. This can result in error messages even though the application trying to open the file is perfectly capable of opening the file. (Go figure!)

The Macintosh operating system doesn't use the file type. Instead, it has a creator and file type code that is associated with the file when the file is created, but are hidden from the user. The Macintosh can handle files that come from another platform and don't have the correct information attached to it—it will attempt to find an application to match the file; otherwise, it will ask you to choose an application to open the file.

If you are sending a file by Internet, you should always use the file extension. It is required for Windows, and makes things easier for Macintosh systems too. Also, JPEG and GIF files are very standard. But send someone a BMP or PICT file and there's no guarantee they'll be able to open the file.

The other thing to watch out for when sending files is the kind of encoding your mail program uses when sending the file. BineHex, Base64, UUEncode and MIME are encoding types supported. Not all mail programs support all encoding types, so you may need to experiment.

# THE FUTURE

Web browsers are adding more features and there are plug-ins that support all kinds of new formats, including Flash and RealMedia. But with JPEG and GIF you know almost anyone can open the file you are sending, whether it's vie email or on a website.

# CHAPTER 17
# USEFUL SOFTWARE APPLICATIONS

There are a number of different applications available for manipulating and managing images. This chapter outlines some different application types, as well as actual software you might find useful. A short introduction to viewers, desktop publishing and printing applications are included. Applications for Image Editing and Immersive Imaging are covered in those respective chapters, and video editors and multimedia authoring tools are covered in Chapter 19.

## OTHER CONNECTION AND IMAGE TRANSFER SOFTWARE

Your camera should come with the software needed to transfer the images from the camera to the computer. The functionality of this software varies widely; some companies provide a dedicated application, while others supply a TWAIN plug-in for use with other editing software. Whatever is supplied, the manufacturer's software may be the easiest and simplest way to transfer images from your camera; however, you might find that a third-party application can do the job better, or offers other features that make it worth using. It may also solve problems encountered using the original software.

I recently had a lot of trouble getting the Nikon CoolPix 950 software to work correctly with an older Macintosh computer. The software worked, but crashed very frequently. At first I suspected the problem was related to the fact that I was using a large memory card (32MB) rather than the 8MB card the camera came with. So I increased the memory for the application, but that didn't fix the problem. Neither did turning off the generation of custom icons, though it did seem to alleviate it a little. Then I saw that the software was recommended for a later version of the operating system so the next couple of hours were spent upgrading the system. Unfortunately, the software still didn't work. I then downloaded *Cameraid*, a shareware application for the Macintosh that works with Nikon Cameras. Not only did it work without a hitch, but it works faster than Nikon's own software!

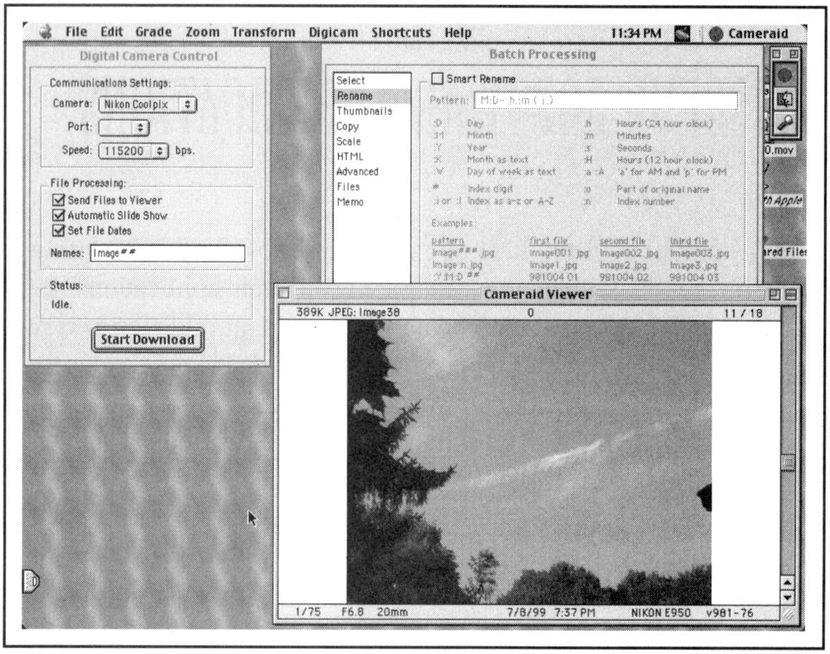


*Figure 17-1. Cameraid application.*

So, if you have a camera that's supported by one of the following applications you might want to download the demo and check it out.

## CAME

OS: Windows95/98/NT

Cameras Supported: Agfa, EPSON, Olympus, Sanyo

Cost: Freeware

URL: *http://www.butaman.ne.jp:8000/~tsuruzoh/index-e.html*

Came works with digital cameras that use the Fujitsu SparcLite Microprocessor. In addition to downloading images you can use it to change the camera's configuration as well as take pictures. It's ideal for webcam and time-lapse applications. While earlier versions worked with Nikon's digital cameras, the current version no longer supports these cameras.

There is also CameDOS for the HP 200LX portable computer.

## CAMERAID

OS: Macintosh 7+

Cameras Supported: Agfa, Apple QT200, Epson, Fujifilm, Leica, Nikon, Olympus, Sanyo and Toshiba

Cost: Shareware $15

URL: *http://www.cameraid.com/*

As well as being usable as a download utility that is faster than some of the applications provided by the manufacturer, Cameraid includes a number of other functions. A viewer provides a slideshow of any selected images, and will display EXIF information (if available). You can rotate images and save them without recompressing the JPEG data. The *Add Thumbnail Icons* function adds custom icons to selected images. A batch processing function can be used to copy and rescale images, as well as create HTML pages.

## EZ-AUTOCAM

OS: Windows95/98/NT

Cameras Supported: Fuji MX, DX, DS or FinePix series, Toshiba PDR-M1/M3, Leica DigiLux

Cost: $39.95

URL: *http://www.beausoft.com/ezindex.html*

This software works with Fuji cameras and cameras manufactured by Fuji for other companies. It will download and save images, as well as let you browse them. It also can be used to print multiple pictures to a single page. The software can also be used to remotely control the camera and take an image at a fixed interval, which makes it useful for webcams and time-lapse photography. Sequences can be saved in AVI movie format.

## GPHOTO

OS: GNU/Linux

Cameras Supported: Range of cameras from different manufacturers. Check website for complete list as is it changing frequently.

Cost: N/A

URL: http://www.gphoto.org/main.php3

Application for transferring images from digital cameras under GNU/Linux. Can "publish" images in HTML templates. Command line interface can be used to control some cameras for webcams and time lapse photography.

## MAVICADABRA!

OS: Macintosh 8.0+

Cameras Supported: Sony Mavica

Cost: $24.95

URL: http://www.blackmagik.com/mavica.html

Most of Sony's Mavica camera write to a floppy disk in standard JPG format, but this application provides some batch processing functionality for those copying the images from the floppy to their hard disk. It will scale images, as well as convert them to PICT format, add thumbnails and change the creator type. It will also split a multi-image photograph into 9 separate images.

## PIE (PICTURE INFORMATION EXTRACTOR)

OS: Windows 95/98/NT

Cameras Supported: EXIF

Cost: Shareware $18.95

URL: http://www.hoju.de/

PIE is a file renaming, information extraction, and manipulation application. It doesn't handle the actual transferring of images from a camera; you have to "mount" the images on the desktop so that the application can access them. A CompactFlash or SmartMedia card reader can be used to do this. PIE will copy and rename files from those devices. It can extract date and time information from EXIF files and rename the files. Lossless rotation and saving of JPEG files is supported.

## VIEWER AND MEDIA MANAGEMENT APPLICATIONS

Once you start accumulating images—and you will, unless you put the camera in a bag and forget about it—you will be faced with the problem of ordering and filing them on your hard drive.

There are a number of applications available that can help with this task.

Many of these applications include functions for resizing and adjusting images. I recommend testing these functions before using them. While they do a pretty good job if you use the right settings, you may get noticeably better results using an image editor to resize images. Do some tests for yourself, then decide whether the savings in time and convenience outweigh any difference in image quality.

### ACDSEE

ACDSee from ACD Systems is primarily a viewing application, though it can also be used to connect to and download images from the Kodak DC200 and 210 cameras. ACDSee costs just $40 to register, though you can download and try it out for free. Its chief advantage is that it can be used to quickly view images in a folder using its thumbnail view. There are two primary modes of viewing images: Browser mode resembles the Windows Explorer interface, though it can also display large thumbnails of the images; in Slide Show mode the image is displayed within the ACDSee window, or can fill the entire screen if desired. You can then page through the images automatically or manually.

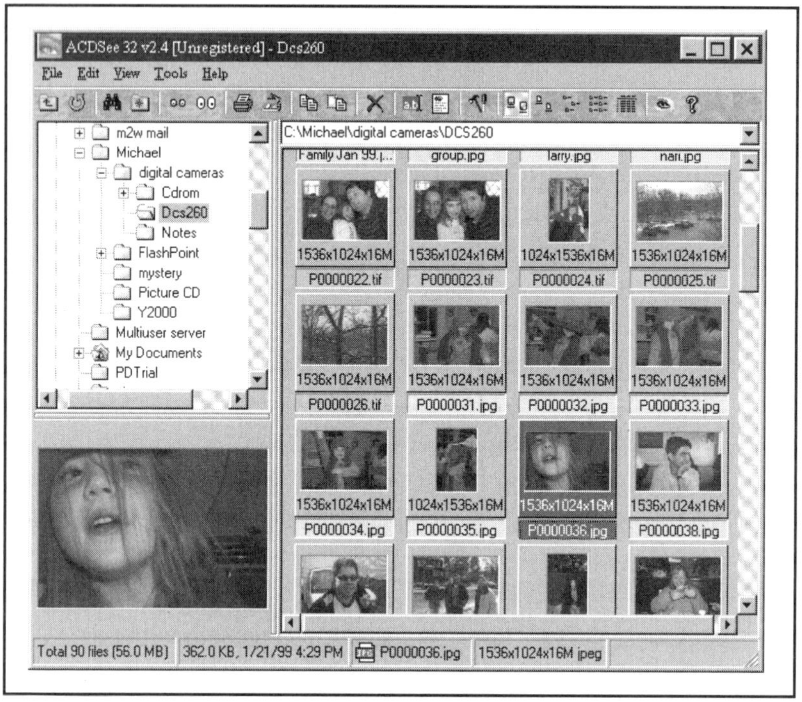


*Figure 17-2. ACDSee browser window.*

The program opens and displays images very quickly, which is an advantage when you have a lot of images to maintain, and can be used to save the images in another format. It can also be used to print files. You can also crop a section of the image by clicking and dragging and then saving the changed file.

By default, when you first run the program in slideshow mode, the window will resize as images change. This makes the window jump around if all your images aren't the same size, which is very annoying. However, you can easily turn this off in the Options window under Viewer options. You can also set the program to scale the image to fit within the window if it is larger than the window size.

From the browser window it's possible to select multiple files and then convert them all to another file format using the Convert function. It can replace or rename the new files. Unfortunately, while you can set compression settings, you can't resize the graphic, which could be useful for creating thumbnails. The program includes a Print Thumbnails option which prints out a sheet containing thumbnails of selected files (you can specify the size of the thumbnails too).

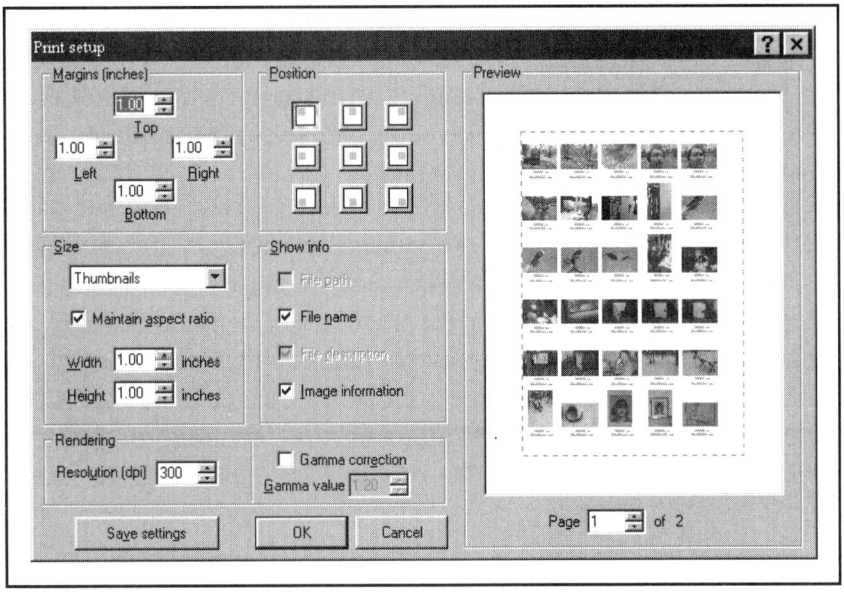


*Figure 17-3. ACDSee Previewing the Thumbnails page.*

A Description field can be added to each image, though this requires actual work on your part!

Other unique features include the ability to read KDC files from the Kodak DC40, 50 and 120 cameras, and the program can also create a text file listing of the contents of a folder. The program also lets you adjust gamma correction for viewing the images (this is not used when printing unless specified in the Print dialog of the program). Since the program doesn't include image adjustment controls, it's unclear how useful this feature will be for most people. You can find out more about this program at:

*www.acdnet.com*

## THUMBSPLUS

Like ACDSee, ThumbsPlus provides image indexing capabilities as well as image processing options. ThumbsPlus is available for both Macintosh and Windows, though the Macintosh version is in beta (and has been for some time with seemingly no change in its status). The Macintosh beta version works very well, however, there are a few menu and dialog box quirks. This could be a beta problem, or a problem with running with the latest version of the OS. Still, none of the problems stop you from being able to use the software, and you can download a trial version and try it out.

Something I really like about this program is the function that creates HTML pages and thumbnails for a selected directory. You can specify the size and number of thumbnails per page, and the program will then automatically create the pages and thumbnail files. I used it to "index" 700 images, and it worked well, though it took quite some time to process the whole job. Like ACDSee, this program can read Kodak KDC files. Find the Cerious Software site at:

*www.cerious.com*

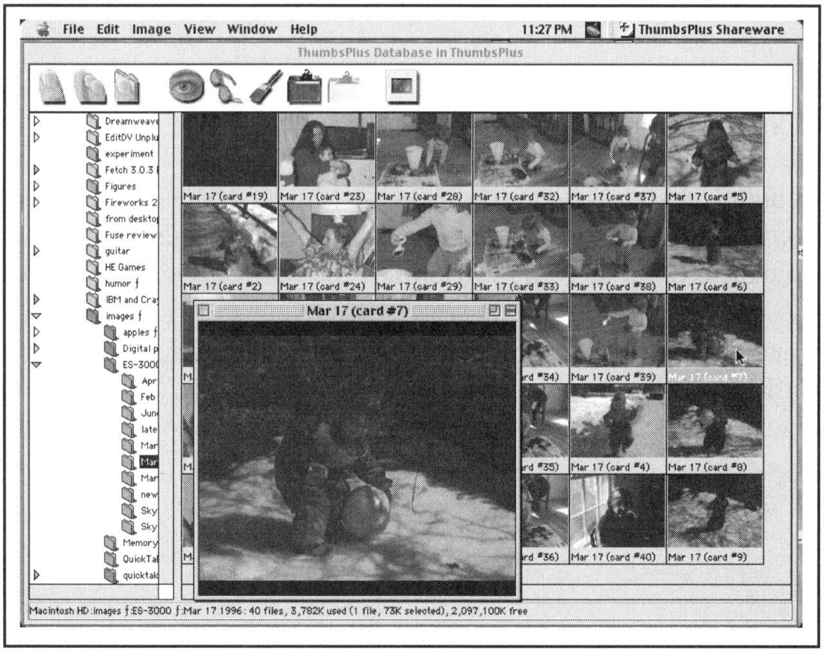


*Figure 17-4. ThumbsPlus.*

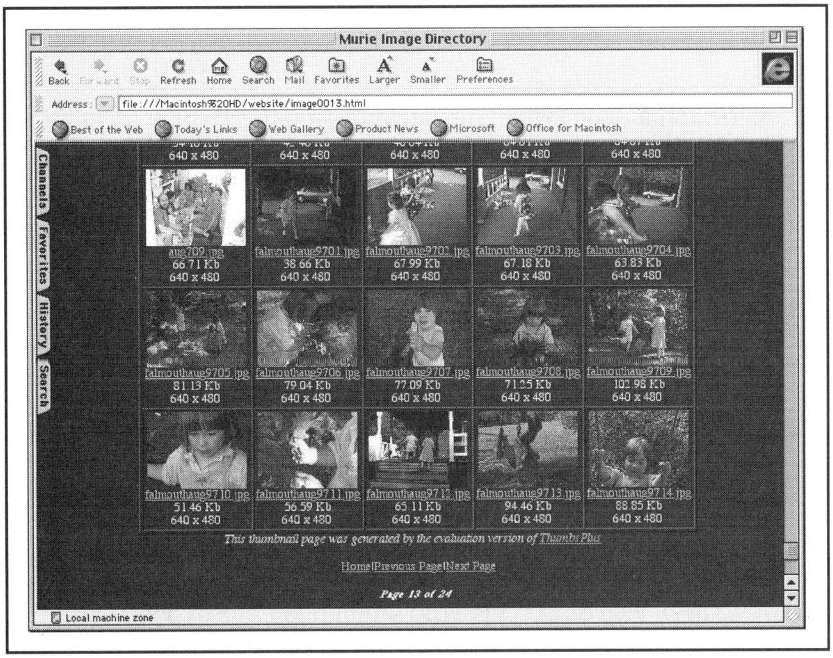


*Figure 17-5. A Web page created using ThumbsPlus.*

## POWERBROWSER

For Windows, this program from Pegasus Imaging is a multimedia organizer that can create folders with thumbnails for images, video and audio files. It can also print images, print thumbnails of images in a folder, and includes filter effects, color adjustment and recompression for the World Wide Web. There's even a slideshow option and it costs about $40. Find it at:

*www.pegasusimaging.com*

## HTMLPIX

This product, for Windows, takes a folder and creates a thumbnail view of all the pictures in that folder. Clicking the thumbnail displays the actual image. The thumbnails are really the resized versions of the actual files, so this function should only be used for relatively small files. It's freeware and available at:

*www.hail.icestorm.com/freed/*

## THUMBER

This freeware application extracts information from EXIF files, including shutter speed, aperature, etc. It also displays thumbnails, though only if the file contains a thumbnail; the program won't create a thumbnail for a file. It can be used for file renaming tasks as well as for creating Web page indexes. The developer also offers some other interesting software, including EXIFread. Check it out at:

*members.tripod.com/~tawba/index.html*

## EXIFREAD

A very simple viewer application which displays all EXIF information in a selected EXIF file. It also displays the file's thumbnail. It's freeware, located at:

*members.tripod.com/~tawba/index.html*

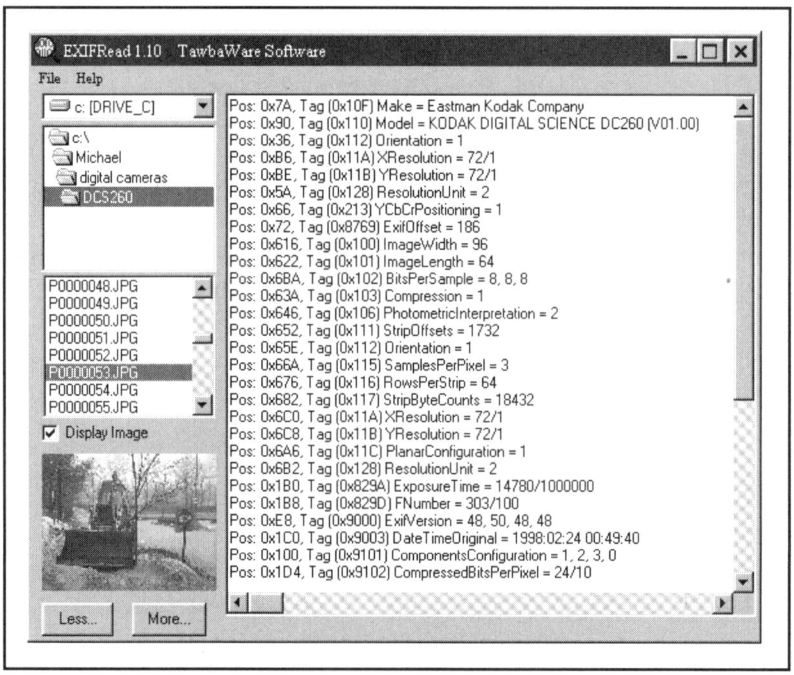


*Figure 17-6. EXIFRead.*

# MEDIA CATALOGING

While several of the viewer applications provide cataloging, and even store thumbnails and information in a catalog file, if you have large numbers of images and are sharing them across a network with others then you might want to look at a more powerful solution. Media Cataloging tools provide more features; the best let you share a catalog across a network.

## EXTENSIS PORTFOLIO

Extensis Portfolio is a client and server combination for the management of large amounts of digital media. The individual client version can be used on a single machine, where it creates a visual catalog of media assets. Images are represented by thumbnails, and can be searched and viewed in customizable views. It also has a slideshow option. Running on a single machine, Portfolio functions in a similar way to some of the other cataloging and viewing applications. Portfolio's real advantage comes when multiple users are running the client software on the same network. A server version is also available for larger installations.

Portfolio is available for Macintosh and Windows, and there's even a Web server plug-in available that makes the Portfolio catalogs available over the Internet.

A free viewer is also available so you can distribute it along with a catalog of images for others to view. The standalone version of Portfolio costs $200; the server costs about $2000. It's available for both Macintosh and Windows. The website is:

*www.extensis.com*

## CANTO CUMULUS NETWORK

Canto Software's Cumulus Network Media Management System is another asset management system for managing large image databases. It's cross-platform (Unix, Windows NT, and Macintosh servers) and a CGI is provided for use over the Web. Surf to it at:

*http://www.canto-software.com/*

### IMAGEAXS

ImageAXS and ImageAXS Pro from Digital Arts & Sciences also offers cataloging and a large set of features, including printing. It can even acquire images from a digital camera. While it's available in both Macintosh and Windows version, you can't open catalogs moved from one operating system to another. This is not strictly a network solution, though the PRO version can export to HTML. Learn more at:

*www.dascorp.com*

### ULEAD WEB ALBUM

A media cataloging and management tool, it lets you organize files into albums with custom fields for annotating information relating to the images. Includes multiple-search capabilities as well as a slide show capability. Software can create thumbnail image pages in HTML. A 15-day trial version is available. Find it at:

*www.ulead.com*

## PRINTING SOFTWARE

Often when printing multiple images you want to arrange several on a single large sheet of paper. The programs below offer features that makes it easier to print multiple images on the same page.

## EASYTEK PHOTO ASSISTANT 98 LIGHT

A sophisticated program that is especially good at printing multiple pictures on a page. The interface lets you drag the images you want to print onto a page and then rotate and arrange them by clicking and dragging. There's also an option that will print a page of thumbnails. Its website is:

*http://www.easytek.com/eng/produkte/light/index.htm*

## PRINCE DIGITAL

Another program that can handle multiple images on the same page. The program also lets you enlarge parts of an image or choose the part you want to print. Check out their site at:

*http://www.erithacus.com/princedigital*

## QIMAGE PRO

This program prints multiple images and also offers image processing features. It boasts a special algorithm for removing random pixels caused by dark current in digital cameras. You can see more at:

*http://www.charm.net/~mchaney/imaging/index.html*

## QUICKTIME 3

QuickTime was originally developed as an architecture for time-based media, specifically for digital video. Over time it has added other components (including QuickTime VR—see Chapter 15).

In addition to its support for digital video and animated movies (Chapter 19), when you install QuickTime you also get two applications: PictureViewer and MoviePlayer. PictureViewer can be used to view single images in a wide variety of formats (including JPEG, GIF and Photoshop's PSD), and MoviePlayer can be used to play different video formats. In addition, if you upgrade from the free version to the Pro version, you can perform simple editing tasks, including recompressing images. It can also be used to perform simple edits on video sequences.

For its combination of features, QuickTime is worth downloading.

## DESKTOP PUBLISHING

It's hard to believe now, but desktop publishing was the Internet of the mid '80s (it just wasn't quite as big!). There was a lot of buzz about it, and it seemed everyone wanted to learn about it. Desktop publishing applications remain different from word processing in that they provide greater control for arranging images and graphics in different formats. You can also use a desktop publishing program to arrange multiple images on a page. It's not the fastest and most efficient way of doing it (there are programs listed above that will do it semiautomatically) but it can get the job done if you don't have anything else. Desktop publishing programs can also be useful for placing text on top of images, and of course for creating brochures and other documents.

While these programs will import many different types of graphic files, when working with images the TIFF file format is likely to give the least problems. If you're printing to a consumer inkjet, you should use RGB files, not CMYK files. Even if you are sending a layout to be separated (the process of creating separate plates that represent the different CMYK colors for us on a printing press), if you don't have experience in color separations it's better to use an RGB file in your layout and leave the separation up to the service bureau rather than creating a CMYK separation in your image editor. You should always talk to your service bureau before you start work on a project like that.

An advantage of using a desktop publishing program to add text on top of a graphic rather than add the text in an image editor is that it's much easier to go back and edit the text; also, the text will be sent at the best resolution for the printer. If you place text in your image editor, then the text is limited to the resolution of the graphic; also, the dithering used to render the image will degrade the sharpness of the text.

With Adobe Acrobat (see *Electronic Publishing* below) you can use any desktop publishing program (as well as word processor) to create electronic documents that are more complex in layout than HTML pages.

Two of the most common desktop publishing programs are:

Adobe PageMaker

Quark XPress

Each has advantages and disadvantages, but QuarkXPress is popular in large production environments and for creating long, very structured documents, while PageMaker is better suited to free-format projects. The reality is you can use either for most tasks.

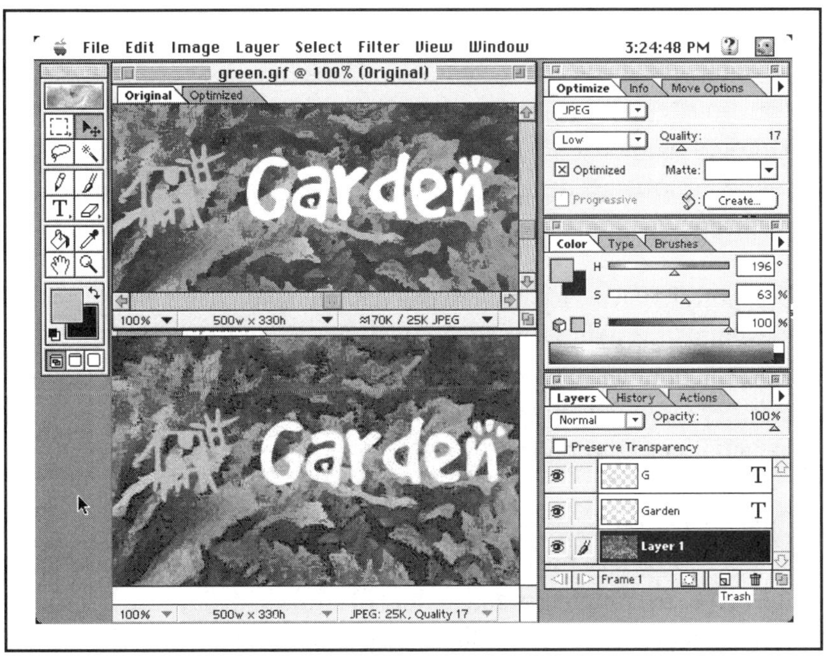


*Figure 17-7. Adobe ImageReady.*

## WEB GRAPHICS PREPARATION

If you're creating a lot of graphics for the Web you might want to look at two programs that offer features specifically for Web designers. Both Adobe ImageReady and Macromedia Fireworks provide options for choosing different compression options and previewing the effect before saving the file. Both offer features that can save a lot of time, including tools for creating GIF animations, and slicing tools for cutting up large graphics into small sections that can be compressed at different settings.

## PAINT GRAPHICS

Chapter 7 lists some of the software that works well for editing images. Most of these applications grew out of "paint" programs—graphics tools originally created for drawing and painting using pixels on the screen. When scanners became available for desktop computers, a new class of software—image editing programs—were released, which offered many of the tools found in paint programs but added color and contrast balance functions and sharpen filters which previously had not been found in paint programs. Over time these image editing programs,such as Photoshop, became so popular and feature-rich that the category of software designed simply for painting has almost disappeared. There are a couple of notable exceptions however, both from MetaCreations.

MetaCreations Painter is a paint tool that has been available for many years. Its forte is in emulating natural media—brush strokes, crayons etc. While it could be used to edit photographs, it's real strength is in the brush and paint tools. It's most effective if you want to paint on top of a photograph. The most recent version has enhanced support for exporting graphics in Web formats.

MetaCreations Expression is another paint program, but it works slightly differently from Painter. Expression paints with brushes that look like natural media—they have the roughness and texture of strokes made with a real brush—but the strokes aren't stored as individual pixels in the document. Instead, the program uses vector graphics with very complicated line shapes to create its images. The advantage of this technique is that the lines can be reshaped very easily; something that can't be done with a pixel editing application.

# VECTOR GRAPHICS

The three most popular vector drawing applications are:

Adobe Illustrator

Macromedia FreeHand

Corel CorelDraw.

They provide pen tools for drawing as well as shape tools. They're excellent for graphic design purposes—logos and artwork—but aren't useful for editing pixel-based graphics (though the CorelDraw package includes CorelPaint which is an image editor). These programs will let you import a bitmapped graphic and draw on top of it.

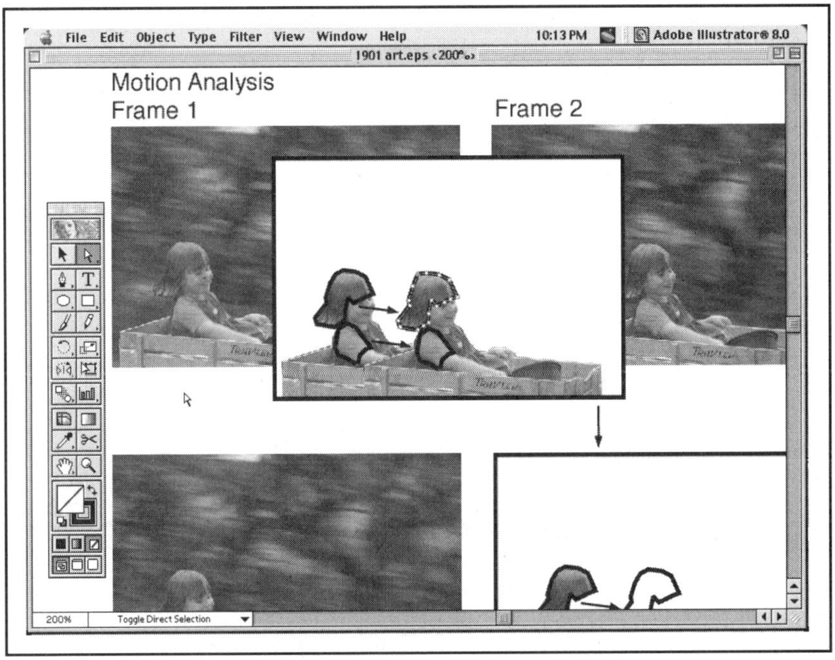


*Figure 17-8. Drawing in Adobe Illustrator.*

# ELECTRONIC PUBLISHING

Adobe Acrobat is an electronic document distribution system. It's designed for creating electronic documents that can be read by anyone who has the Adobe Acrobat Reader. The Reader is available for free at *www.adobe.com* for several different operating systems.

Acrobat consists of three parts:

The PDF file format

Acrobat Reader

Acrobat Exchange

PDF is the file format that contains the electronic documents. The documents can be created in any program that can be used to print a document. Acrobat Exchange includes a printer driver which takes the information that is sent to the "printer" and turns it into an electronic file. The file can be edited (links added, pages added or removed) using the Acrobat Exchange application.

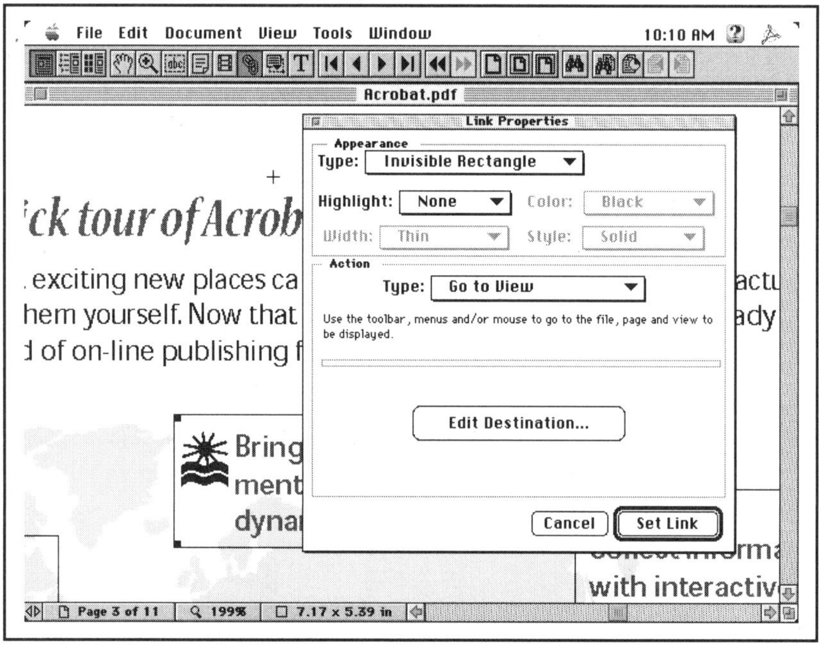


*Figure 17-9. Creating a link in an Acrobat document.*

When saving PDF files, Acrobat provides several different compression algorithms for compressing graphics. The choices include: LZW, RLE, CCITT group 3, CCITT group 4, and JPEG. You can also create hyperlinks within a document. Clicking on a hyperlink brings up another page. You can even specify which part of that page appears in the Reader window.

PDF is great for brochures, or other documents that contain images and text. It might be a good way to distribute a collection of photos too; you can restrict whether users can print or copy from a file. However, to distribute a single-image using PDF makes little sense.

## OTHER TOOLS

### JPEG WIZARD 1.0

A tool developed specifically for compressing graphics in the JPEG format, JPEG Wizard claims an improved algorithm that is more efficient than many other JPEG compression algorithms. A WYSIWYG (What You See Is What You Get) interface shows the effects of compression. It's from Pegasus Imaging at:

*www.jpegwizard.com*

### PIC*SAVER*

A screen saver product for displaying JPEG and PIC images. It's free, and you can find it at:

*www.jpegwizard.com*

### ARCSOFT PHOTOMONTAGE

This very easy-to-use application lets you take one of your own photographs and turn it into a photomontage; a picture made up of many little pictures. If you saw the movie poster for The Truman Show you'll be familiar with the effect.

The software lets you select a photo and specify how much of the picture you want to use as the basis of your montage (as well as the size of the final output), but it doesn't let you use your own photographs as the tiny thumbnail images. Instead, it uses a library of images contained on the CD that comes with the software.

Also, the software license prohibits you from using the images created by the application for commercial purposes. The software costs less than $100, and the results are a lot of fun, but the fact that you can't use your own images as the thumbnails in the print makes the program a little less appealing. Visit the website:

*www.arcsoft.com*

## WHERE TO FIND MORE

While there are dozens of commercial applications that may be of interest to digital camera users, there are literally hundreds of freeware and share applications, ranging from image management to tools for controlling cameras.

Here are two good sources for finding such software:

Perhaps the largest and oldest index of software for Macs and PCs can be found at:

*http://www.tucows.com*

An impressive index to all kinds of shareware and freeware utilities, all categorized by use, can be found at:

*http://www.davecentral.com*

# CHAPTER 18
# WEBCAMS

Webcams—cameras that produce constantly updated images on a Web page—make for an exciting addition to any website. As an application for a camera it's one of the most interesting. There are thousands of webcams live on the Internet today. From the mildly interesting to thc strange, office cams to bedrooms cams, they're all out there. Some update every few minutes, while others update every few seconds. Some cameras are even controllable; users can point the camera in different directions by clicking buttons on the Web page.

Webcams, as a category, are usually considered separate to video servers. Video servers broadcast streams of video at several frames per second, along with an audio channel. They require special server software and use proprietary file formats. Examples include RealNetworks RealVideo, Microsoft's NetShow, and Apple's QuickTime. Streaming video is made complicated by the requirement of being able to maintain synchronization of the frames with the audio. For best results, this means adjusting the frame rate, and the client software has to be able to drop frames so that it does not get hopelessly behind. They also often require specialized hardware and software at the server end.

Webcams, on the other hand, are not as constrained by network performance, and can even be set up using existing software. They typically use a much lower frame rate than a video server, updating the image often as infrequently as one frame every 20 seconds. Rather than use special servers and file formats, webcams can use a regular Web server and software that saves the image in the Web server's file area. The Webcam software simply captures a frame from the camera, saves it as a JPEG (or other) format file, then sends the file to the Web server. The webcam software might be on the same computer as the Web server, or it may be another computer connected by the network.

Some webcam software will handle Web serving as well as the capture of images; both controlling the camera and serving the images and HTML files. On the face of it, this seems like the simplest solution, but the reality may be different. Having a single machine handle both serving and capture tasks can cause problems when traffic gets heavy. It also increases the likelihood of the server crashing.

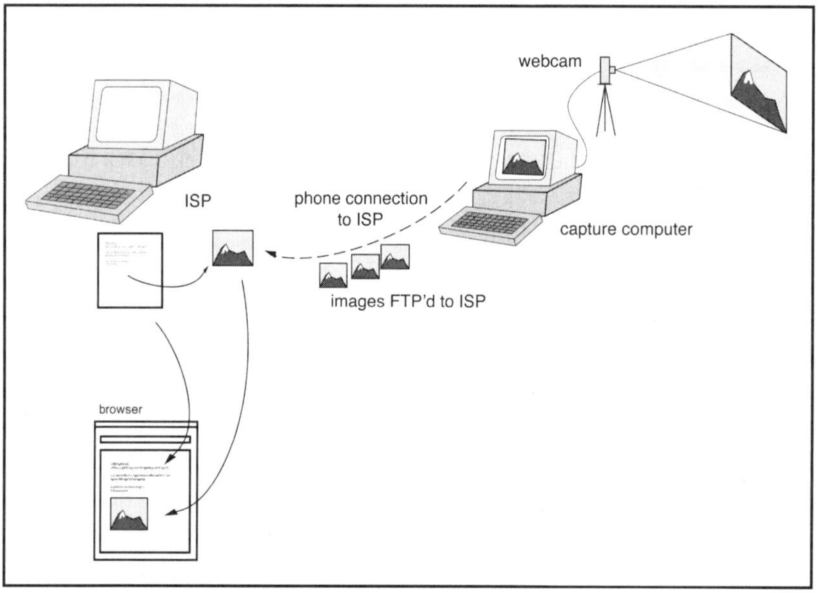


*Figure 18-1. Diagram of webcam solution.*

The rest of this chapter will focus on webcams, the low-bandwidth brother of video servers.

## WHAT DO YOU NEED TO SET UP A WEBCAM?

At its simplest, you need a camera to take the pictures, some software to capture those images, and a Web server application to serve the images. This assumes that the camera is connected to the Web server computer. In this scenario, the camera software simply has to save the file to a place where the server expects to find it. The capture of the image, and saving it to the correct directory for the Web server can often be done with existing software and a macro facility. For Macintosh users, AppleScript provides a scripting engine that can be used to update images with some camera software.

A better alternative is custom software that will work with a camera to handle the capture, naming and even archiving of captured images. This is often easier and more reliable than scripting an application.

Of course, attaching the webcam to the Web server computer isn't always possible or desirable. In those situations the camera can be connected to another computer. The software captures the image, then sends it—most probably through FTP—to the computer running the Web server.

This solution is ideal for the many webcam users who can't afford a dedicated connection to the Internet. For them, a semipermanent connection can provide an intermittent webcam. To set up the site, they place a Web page on their ISP that points to a JPEG image saved in the same file area. Then they dial in and send photographs from their webcam to the file area when connected. If their webcam isn't live, the HTML page still points to the last picture that was uploaded.

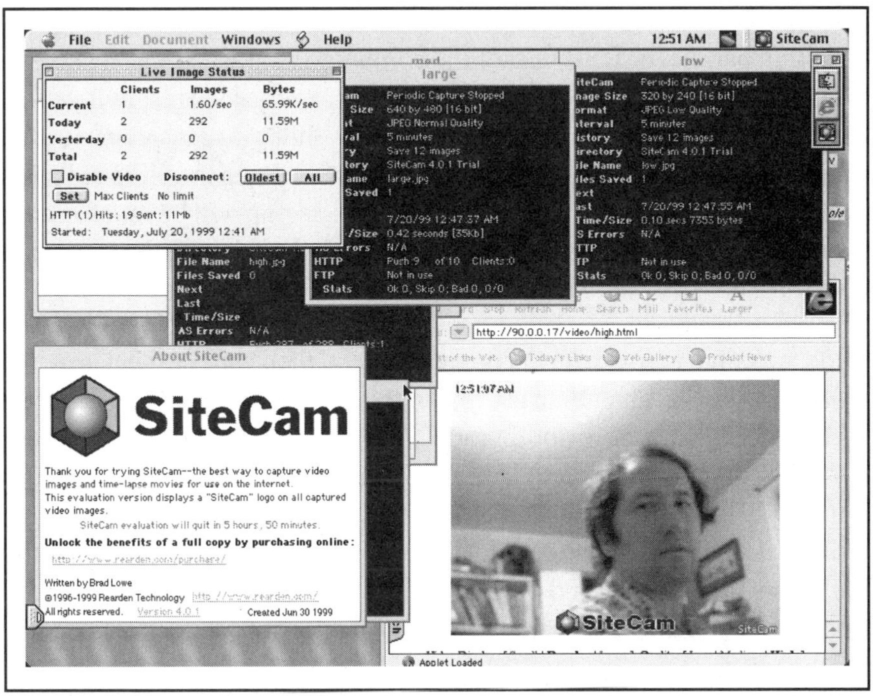


Figure 18-2. SiteCam server software.

A third option is to have the Web page on the ISP point to the IP address of the machine that has the webcam. This adds complexity because it still requires that the webcam machine include a Web server. It also means that if the webcam isn't working the visitor to the Web page will display an error message, not an old photograph. Another limitation of this method is that many ISPs don't provide fixed IP addresses. This means that every time the webcam computer is connected to the ISP it will be assigned a different IP address. The Web page on the server will have to be updated each time the camera reconnects.

Perhaps the best reason for having the webcam directly connected to the Web is for the new controllable cameras.

## WHAT KIND OF CAMERA?

The available cameras divide into two categories; Direct Connect and Video Cameras. Direct connect cameras are connected to the computer using either the serial, parallel or USB port of the computer. A direct connect camera is usually cheaper and easier to set up, and produces very good results if you only want to update the picture at low frame rates. While you can get inexpensive video cameras and capture cards, you're probably going to still spend at least twice as much as a direct connect camera. And setup will be much simpler, provided the webcam software you are using will work with the camera you have chosen.

If you are going to use a direct connect camera, then your best bet is to buy a dedicated camera rather than use one of the consumer digital cameras being sold for general-purpose photography. You can buy a Logitech QuickCam camera for about a hundred dollars, which is probably cheaper than most deals you can get on a still digital camera. These cameras are designed to be used while attached to a computer; they have no memory for storing images, and don't work unless connected to a computer. They usually include software that can control the camera, or third-party software is available for doing that.

Many of the webcam applications include support for the popular cameras.

The primary disadvantage of cameras such as the QuickCam is that the image quality is not nearly as good as is possible with one of the newer digital cameras. They also are often capable of only very low frame rates. This is not a problem for most webcams; many operators want to limit the traffic through their site and limit the updates to every 30 seconds or a minute—more than enough time for any of these cameras to update the image.

Another reason for not using one of the general-purpose cameras is that they are not really designed to be connected to the computer permanently, or even to be left on for long periods of time. They turn off after a few minutes of inactivity, often don't include an AC power supply, and software control can be limited.

## VIDEO CAPTURE

Rather than using a still digital camera, some webcams—and all video servers—use a video camera attached to a video capture card as the source for the images. Using a video capture card makes it possible to capture at much higher frame rates, but the quality is not always better. Also, the cost of a video capture card can be a significant additional cost if you don't already have a capture card. There are some low-priced options, such as Play's Snappy parallel port video capture card, but even these cost about $100, and the camera is extra. Remember that the higher the frame rate, the more processing power will be needed to process the frames.

## HOW THE VISITOR GETS THE IMAGE

Adding a webcam image to an HTML page can be very simple; if the image is saved as a JPEG file the HTML page can simply point to that image. If the image is updated every few seconds, the user will have to refresh the page to see the updated page. This technique of displaying images is called *Client Pull.*

It's possible to use an HTML Meta Tag that automatically refreshes the page after a certain number of seconds:

<META HTTP-EQUIV="Refresh" CONTENT=20>

Adding the above tag to a page before the body text will cause the page to automatically update every 20 seconds. The command will reload the entire page. To minimize the performance drag of reloading the entire page, you could place the image within a frame and place the meta tag on that frame.

Some browsers will stop the automatic update after a period of time—this appears to be a bug—but you can circumvent that problem by setting the Expires and nocache options as the example below shows.

<META HTTP-EQUIV="expires" CONTENT=" Mon, 1 Jan 1990 05:00:00 GMT ">

<META EQUIV="pragma" CONTENT="nocache">

There are other ways to force images to be updated. It's possible to write a DHTML script that uploads the image, and there are Java applets that can be used to do this. One example is JavaCam, which is a Java Applet from the author of the webcam software Webcam32. The Java applet and the webcam software can be found at:

*http://www.kolban.com/webcam32/*

An alternative method of sending images to the user is called *Server Push*. This is the method used by live streaming technologies such as RealNetworks RealVideo. With Client Pull, the image is downloaded and the connection between the browser and server closed after the data has been transferred. In Server Push, the connection between the browser and server remain open, and data is sent by the server, which is then displayed by the browser. The server has control over when and how much data it sends, and this method is more efficient because connections don't have to be opened and closed. But it also means that the number of connections must be carefully managed by the server; since the number of total connections that a server can handle is limited by server speed and network performance.

If you're using server push then you're probably trying to send a fairly high frame rate, and you'll probably want to use a video card capture device rather than a camera connected through the serial or parallel port.

## WEBCAM SOFTWARE

There's a range of software available that supports streaming video, from the expensive server solutions such as RealVideo, to much lower priced alternatives. TrueTech WebCam is a streaming video/audio application for the Windows OS. It supports plug-in free streaming, real-time encoding, a scripting language and remote control for Canon VCC cameras and a frame rate from 1 to 25 frames per second. You can find out more about it at:

*http://www.truetech.com/*

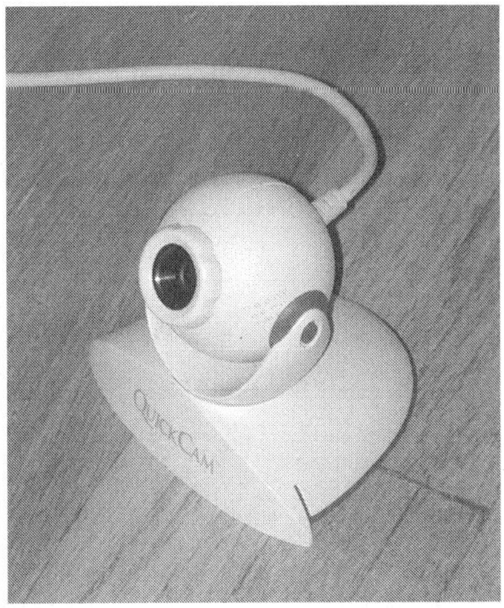

*Figure 18-3. Logitech QuickCam.*

Other software available for webcams include:

| | | |
|---|---|---|
| **ChillCam** | Windows | *http://www.thegeek.com/chillcam/* |
| **iCam** | Windows | *http://www.perceptualrobotics.com/* |
| **Oculus** | Macintosh | *http://www.poubelle.com/Oculus.html* |
| **SiteCam** | Macintosh | *http://www.rearden.com* |
| **WebCan32** | Windows | *http://www.kolban.com/webcam32* |

# CONTROLLABLE CAMERAS

One of the most sophisticated implementations of a webcam is the controllable webcam. In these systems the Web visitor is able to actually control the movement of the camera. This requires a special camera mount (or camera) as well as software and extra hardware to handle communication between the user and the camera mount. Logically, only one person can control the camera at one time; usually the camera control software allows one person to control the camera for a set amount of time, then another user gets control and can decide where to point the camera. Those not lucky enough to have control can only watch what the others are doing.

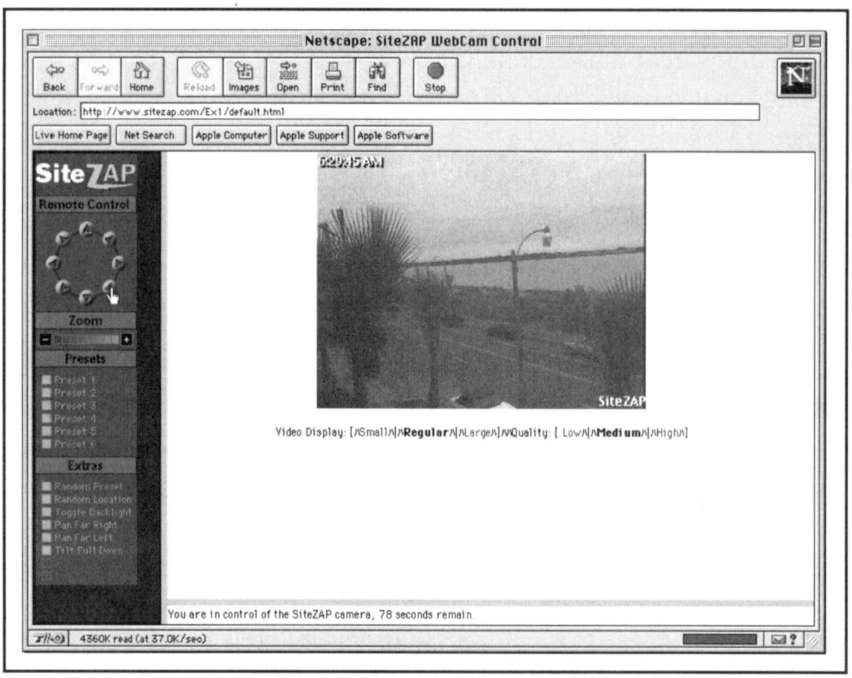


*Figure 18-4. Controllable webcam.*

There are several options available for implementing a controllable webcam. Reardon Technologies, which offers the webcam software SiteCam, also offers SiteZAP. The SiteZAP package includes a Sony EVI-D30 remote control zoom camera as well as the server and client software needed to operate the camera via a website. You can add controls to your Web page which point the camera at a preset location, or allow the user to zoom, pan and tilt the camera. The camera features autofocus, motion detection and autotracking.

SiteZAP runs on a Macintosh server and in addition to the package you'll need a video digitizer (if the computer doesn't already have one; the EVI-D30 is a video camera) and a WebSTAR API compatible server. Adding a remote control web camera isn't cheap; the SiteZAP package costs $2,000.

Reardon Technology

(510) 523-2267

*www.rearden.com*

For those who want to provide their own camera, there are motorized pan/tilt heads available that can be controlled using any computer. Both of the following pan heads include Windows software, but since they are controlled using a serial (RS-232, also referred to as COM) connection, it should be possible to write software that will control these heads that run on any computer system.

**CamMotion.** CamMotion is a motorized Pan/Tilt head that is controlled via a serial port connection and includes support for zoom and focus with the LANC connection found on most Sony consumer camcorders. Remote control software is included. More information is available at:

*www.magagem.com/commotion.html*

**Transit RCM.** Surveyor Corporation's Robot Control Mount is also controlled through a serial interface. Pricing starts at less than $300 for a unit that supports a 1.5-pound camera. A heavy-duty version supports cameras up to 4.5 pounds. Surveyor also provides control software and a developer SDK for writing your own software to control the pan head. You can "survey" their site at:

*www.surveyorcorp.com*

# VIDEO CONFERENCING

Videophones have been a fixture of science fiction for many years, appearing in movies ranging from *2001: A Space Odyssey* to *Star Trek*. Some time back in the '60s, Bell Telephone demonstrated a videophone—a telephone with television and video camera—which enabled callers to see one another. It was a very exciting technology demonstration that promptly was never heard from again.

In the early to mid '80s a few manufacturers developed the first consumer videophones that worked over standard phone lines and transmitted at very low frame rates. The units cost less than $1,000 (each), but perhaps the biggest problem was that they only spoke to the same company's phones; you couldn't call someone with a different make of phone.

More time passed.

Finally, the arrival of the Internet brought with it new communication software, which started with audio only capabilities, then proceeded to add video capabilities. The size of the frame, and the frame rate of the software can be very low, but is often no worse than the dedicated systems of the '80s. Using these programs doesn't always result in the productivity or cost savings you might imagine. The first time I tried an Internet audio conferencing system it took us quite some time to get the software and install it. Then we had to phone the person we were calling to tell them that we were going to connect. The actual connection was simple enough, though communication was rather stilted and of poor quality. At the end of the "call" we used the telephone to call them up again and compare notes!

Video conferencing is really just a videophone call, but was dressed up as a separate category because until recently a good video phone call could only be made with very expensive equipment that was specially installed in a location.

Some important things to think about when using videophones:

It's true; you really do find yourself tidying up before using it!

It's really useful to be able to see what your camera is sending; you never know when you might move out of the frame.

If you don't have 100% access to the Web (and even when you do), setting up a phone call takes a lot more doing then just picking up the phone.

There are now at least three ways to make a video phone call.

Internet Connection

Computer-to-Computer

Standalone Video Phone

## INTERNET CONNECTION

This requires a computer with "video" camera (either a digital camera or a video camera) and software to handle the connection between the two callers. The software may use proprietary transmission protocols to transfer the data, in which case both callers must be running the same software.

Several companies and organizations have attempted to create standard transmission protocols for videophones. By making a standard, all callers would be able to speak to each other. Unfortunately, at the moment there are several such standards.

H.323 was created by the International Telecommunications Union (ITU) specifically for transmitting video and audio streams over the Internet. It is similar to the H.324 standard used for direct video calls (see below). Software that supports H.323 include Intel's Video Phone software.

Other examples of software that can be used for Internet-based phone calls include CUSeeMe. The application, which arguably started the videophone trend on the Internet, is available for **both Macintosh and Windows** systems, and uses its own format for transferring data. It can be used over the Internet or any TCP/IP network. It supports full color and audio, and also has a Caller ID feature. A Who's on-line feature will find who is connected to a server. Find out more at:

*http://www.wpine.com*

## COMPUTER-TO-COMPUTER

With computer-to-computer calling, the connection is made by dialing on a telephone line, and then if both callers have a computer that supports the feature (and they're connected to the phone line), then they can switch to a video call. The connection takes place through the phone line, rather than via the Internet. This will in many cases make for better performance because the data being sent between the two machines is not going through any other servers or network connections.

The H.324 video phone standard was developed for sending audio, video and data over an analog phone line. It is supposed to be a standard for such calls. As has been said, standards are wonderful things; there are so many of them, if you don't like one, then you can pick another.

The H.324 standard assumes at least a 28.8 (that's 28,800 bits-per-second or bps) modem. The audio (voice) signal is compressed to about 6,000 bps, leaving the rest of the 28,800 bps bandwidth for video. Modems must support the industry protocol V.80 to use H.324.

## STANDALONE VIDEO PHONE

These are units which are designed strictly for making a video phone call. A few manufacturers have started releasing such devices which range from a completely self-contained unit that has an LCD display, to TV-based devices which display the video image on a television. Most of these are using the H.324 protocol, so they should work with the PC-based systems that use this protocol.

The quality of the images depends on a number of variables, but the biggest limitation is the low bandwidth available over an analog phone line. Frame rates between 4 and 12 frames per second, with an image size of 176 x 132, are standard. This is much smaller than television or even MPEG-1.

8x8, Inc. is the maker of the ViaTV range of videophones, which support the H.324 protocol. You can find them at:

*www.viatv.com*

# CHAPTER 19
# ANIMATION

While this book is about digital cameras for taking still photographs, sometimes you want to animate images or sequences of images, and that's what this chapter is all about. First it describes the different software, hardware and techniques used to create animated sequences.

The section on authoring software looks at some of the choices for creating animation. The last part of the chapter looks at DV cameras and the features and functionality they offer. For creating sequences of images, and for use in place of a digital camera, these units offer some compelling capabilities. Should you buy one of those instead of a digital camera? Find out below.

## WHY CREATE ANIMATED SEQUENCES?

If you operate a webcam one day, have a time-lapse sequence, or are dabbling in claymation, then you'll find yourself with a sequence of images that you want to turn into a single file that you can play and send to others. This means moving from the JPEG, GIF and TIFF single-image file formats and using another file format. It may also mean getting editing software to create the file, and playback software to enable viewing the software. And if you want to place the sequence on your Web page, then you need to consider what file format will work best for the people viewing the file.

There are several file formats you can use to distribute a sequence of images:

### GIF 89A: ANIMATED GIF

Most browsers will play an animated GIF file. There are several tools available for assembling a GIF sequence, many of them shareware. Also, some of the

Web graphics tools (Macromedia Flash and Adobe ImageReady) can be used to create an animated sequence.

Unfortunately, the files are as limited to an 8-bit palette, which can seriously degrade the quality of the sequence. Also, even though compression is used, the file sequence may be no smaller than the same sequence saved in other formats.

Building GIF animations, and the applications that can be used to do this, are described below.

**Pro**: Only format supported by most browsers without additional plug-ins.

**Con**: Limited to single 8-bit palette. Limited compression.

## MPEG

A file format and compression technology developed for compressing video sequences. It is a standard supported by many applications. Most people will be able to play these files, as they are supported by QuickTime (Macintosh) and Windows MediaPlayer (for Windows). Some digital cameras create MPEG files.

MPEG is really several standards. MPEG-1 was developed for standard television images, and is based on compressing 352 x 240 frames at 30 frames per second (the frames are doubled to create a full screen that's 704 x 480), though it can be used for other sizes.

MPEG-2 was developed as a high-quality television compressor, and is used in DVD video discs. It requires a fair amount of processor power to compress sequences, but it is already being adopted for some editing systems. Systems for compressing MPEG-2 have been fairly expensive, though prices are falling.

MPEG-3 was going to be used for video, but it was found that MPEG-2 could be adapted for HDTV, and MPEG-3 was abandoned. MPEG-3 should not be confused with the MP3 audio compression format, which is actually part of the MPEG-1 format.

MPEG-4 adds interactive capabilities to existing MPEG formats, but is not yet finalized.

**Pro**: Supported by variety of software.

**Con**: Good general-purpose tool, but limited range of capabilities and not good for editing.

## QUICKTIME

A technology developed by Apple, QuickTime is both a file format and system software that is used to edit and play the files. As such, QuickTime differs from MPEG (which is a combined file format and compression technology). QuickTime can support MPEG compression within its file format. It also includes several other compressors, providing a range of choices for different jobs. QuickTime is widely used on the Macintosh, and is also available for Windows. It really shines as an editing technology, and because it is cross-platform, it can be used on the Web.

QuickTime can be downloaded from Apple's website (www.apple.com). It's a large download, but you get two applications; MoviePlayer and PictureViewer, that can be used to view movies and images. If you buy the Professional version (it costs $30, and is an unlock key for the software you downloaded), then MoviePlayer can be used for simple editing, including assembling sequences of images. It's not nearly as sophisticated as a program like Adobe Premiere, but it might be all you need.

NOTE: Some applications (including MediaCleaner Pro from Terran, and Electrifier Pro from Mari Software) include an unlock key for QuickTime.

**Pro**: Lots of features, easy to use, many different compressors.

**Con**: Requires the user install QuickTime (which can be a fairly large download).

### Assembling a Sequence using MoviePlayer

The following describes the steps to assemble a sequence using MoviePlayer, the viewer/editing application included with QuickTime. You must purchase the Pro version to be able to do this.

**1.** Launch the program and choose "Open Image Sequence" from the File menu.

**2.** A dialog appears that lets you choose multiple files in the same directory (folder.)

**3.** A second dialog appears asking you to choose the frame rate. The program then opens the sequence as a new movie.

To save the movie in a compressed format, choose *Export* from the Edit menu. A dialog appears, letting you choose the compression settings (by pressing the *Options* button) and save the compressed file.

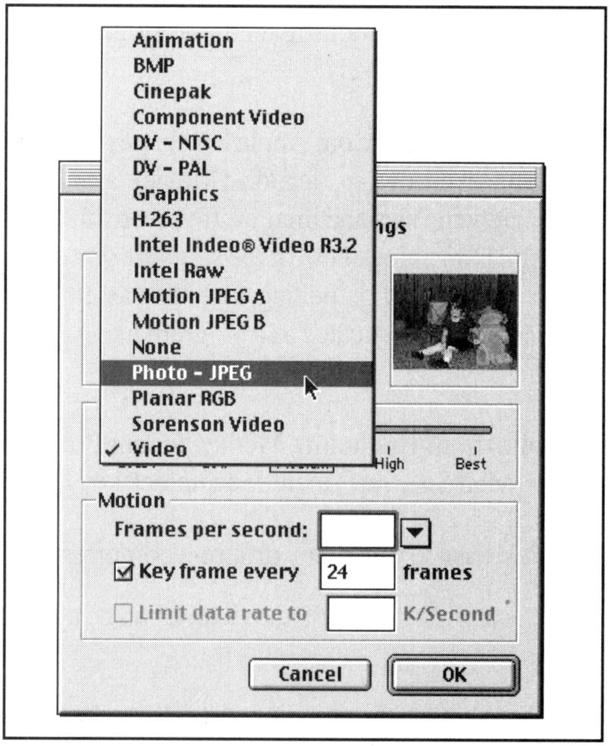


*Figure 19-1. Choosing the compression option.*

## AVI

The original motion file format for Windows, AVI is widely used under Windows, but is not as powerful as QuickTime or the newer Windows Media technologies. Not recommended for use on the Web.

**Pro**: Widely used on the PC.

**Con**: Performance is poor; not supported on other platforms.

## WINDOWS MEDIA FORMAT

Microsoft has developed a number of motion file formats, and this one is the latest. Lots of power, will be widely adopted. May be problematic for Macintosh users.

**Pro**: Microsoft's latest software.

**Con**: May not work well for other platforms.

## REAL VIDEO

A popular streaming technology for the Web. This format makes most sense if you want to stream video.

**Pro**: Excellent for streaming video.

**Con**: Not so good for animation.

## FLASH AND SHOCKWAVE

Formats used by Macromedia Flash and Macromedia Director for playing content on the Web. Works best for animation and interactive presentations, but could also be used for animating a sequence of images. These tools are described in more detail below.

**Pro**: Capabilities of the authoring tools are many and varied.

**Con**: Requires the authoring tool to create the files, and user must download and install plug-in to view content.

## BEFORE YOU GET STARTED

Before you start assembling your sequence—perhaps even before you start taking the pictures—you need to consider a number of things:

1. What resolution for the capture file and the output?
2. How many frames per second?
3. What naming scheme will you use?

## PRINCIPLES OF MOTION COMPRESSION

All of the options described above use some kind of compression to reduce the size of the files created when you assemble a sequence of images. MPEG is both a file format and a compression algorithm, but some of the other formats offer a choice of compression algorithms. QuickTime includes MPEG, CinePac, Sorenson, and several which are compressors useful for specific applications. Windows Media also supports multiple compressors.

Motion compressors attempt to do for sequences what the JPEG compression algorithm does for a single image. But motion compressors also take advantage of the fact that the current frame is often very similar to the previous frame. Rather than compress the frames independently, it can take the information from the previous frame and work out the difference between it and the current frame, and then save just those changes. The first frame in a sequence has to be saved entirely. That's called *intraframe compression*. The subsequent frame can be saved as changes from the previous frame. This is called *interframe compression*.

In addition to saving differences between the frames, the compressor can use motion analysis to find parts of the current frame that have moved relative to the previous frame. It can then save information about the movement of the region. Saving the measurement of the movement is more efficient than saving the changes in all of the pixels that change as the region moves.

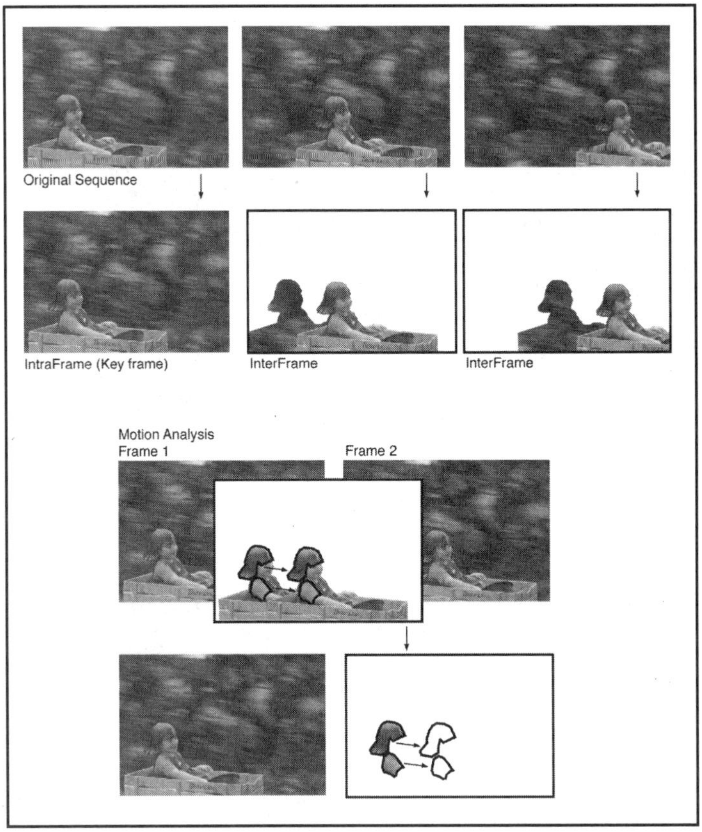


*Figure 19-2. Motion compression techniques.*

When you have a long sequence of images, and if each is based on the previous frame, tremendous file savings can be achieved using interframe compression. However, this savings comes at the expense of random playback and the ability to move back and forth within a sequence. If you have a ten-minute sequence, and want to start at the ninth minute, the playback engine would have to read the entire sequence before it could start at that point. For this reason, *key frames*—frames that have been compressed using intraframe compression—are placed within the movie to facilitate jumping about within the movie. This increases the file size, but is usually considered worth doing. Typically the key frame rate is set as the frame rate of the movie; a 15 frames-per-second movie would have key frames set to 15, a 30 fps movie would have key frames every 30 frames. However, you can increase this as you see fit.

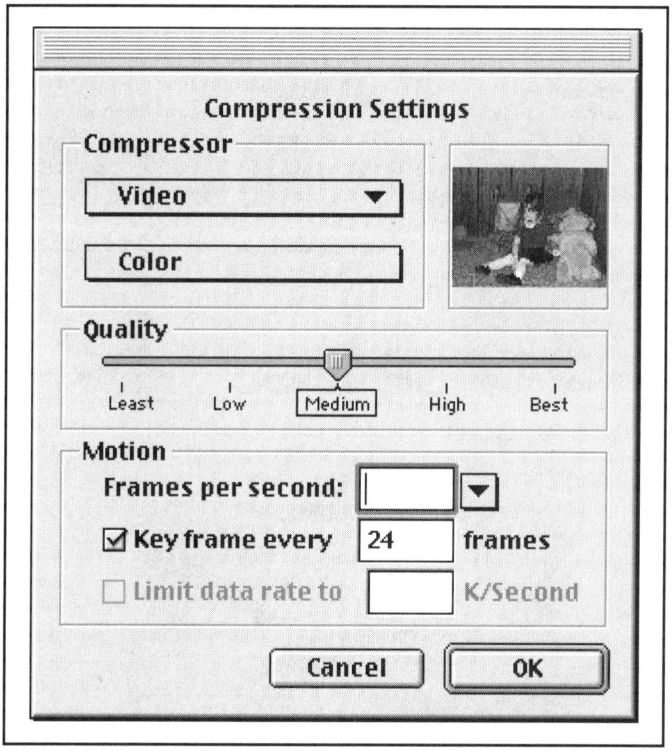


*Figure 19-3. Compression options in QuickTime.*

# CREATING GIF ANIMATIONS

One feature that sets GIF apart from JPEG is that it can be used to create animated sequences, which is great for animating logos and other simple artwork. For video, the results can be less than spectacular (and the file sizes can be as big as files create using other video formats). But, most browsers can play these sequences without the user having to download extra software, so turning a video into an animated GIF file can have advantages.

If you want to create an animated GIF you'll need to find an application that supports this option. Many graphic programs that can create single GIF files, do not support Animated GIF. Here's a list of some programs that do, along with a short description of their functionality.

## EQUILIBRIUM DEBABELIZER

DeBabelizer is a graphics optimization tool that can be used to change several files at once. Need to change all of your graphics to a single palette? DeBabelizer is the tool to have. This is purely a batch tool; it's not designed for pixel editing individual files, so it doesn't replace a tool such as PhotoDraw or PhotoShop. DeBabelizer can take a sequence of still frames and convert them to an animated GIF file. In the process, it will create a single optimized palette for the file (a necessary requirement for GIF files).

## MACROMEDIA FIREWORKS

This general-purpose Web graphics tools is great for creating Web graphics. It has an Export Wizard that will show you what the image looks like saved with different compression settings, and you can slice up a large graphic and assign different compression settings to each slice. The program will output the different slices and an HTML file that puts the whole thing together. If you're working with a lot of Web graphics it's the program to get.

It also supports GIF animation; different timings per frame can be added, the animation previewed, and onion skinning (displaying a faint image of the preceding or following frames) is available. You can also clone an object between frames so that if you make changes to the master object all of the other copies are automatically changed. However, the program does not have a timeline interface, and while it does support tweening, it's not as easy to use as some dedicated programs. If you want to create complicated animation, then Macromedia's other tool, Flash, might be a better choice.

### TWEEENING

Short for in-betweening, tweening is a function that calculates the position of an object based on its starting and ending location. For example, in frame 1 of an animation you want the object to be on the left side of the screen, and in frame 10 you want it to be on the right side. Tweening will calculate the position of the object in frames 2 through 9 to create the movement of the object from one location to the other. Tweening may include options for adjusting the apparent motion of the object by adding acceleration (sometimes called ease-in and ease-out) and specifying the smoothness of any changes in direction of the object.

### GIF CONSTRUCTION SET PROFESSIONAL

A shareware ($20 US) application for Windows that creates GIF animations, it includes an Animation Wizard for quickly assembling animations, as well as a Banner generator for creating animated text banners. Other features include animated transitions, rotation, crop and color-adjustment tools. Animated GIF files can include a transparency color. You can download it from:

*http://www.mindworkshop.com/alchemy/gifcon.html*

### OTHERS

Other GIF animation tools include Ulead Gif Animator (*http://www.ulead.com*) and Extensis PhotoAnimator (*www.extensis.com*), both of which offer extensive animation features.

## SPECIFIC CAMERA FEATURES

Some cameras have features that are tuned towards creating animated sequences; some even capture short sequences saved in MPEG format.

Both Sony and Hitachi offer cameras with this ability. Hitachi's M2 camera is billed as a multimedia recorder and can capture an MPEG sequence of up to 27 minutes in length. It uses a 260MB PCMCIA Type III card for storage. The MPEG-1 sequences are captured at 30 fps with a frame size of 352 x 240 pixels. Audio is recorded at the same time at 32kHz, and a one-minute video is about 9MB. The camera can also capture still images and save them in JPEG format. The still images have a resolution of 704 x 480.

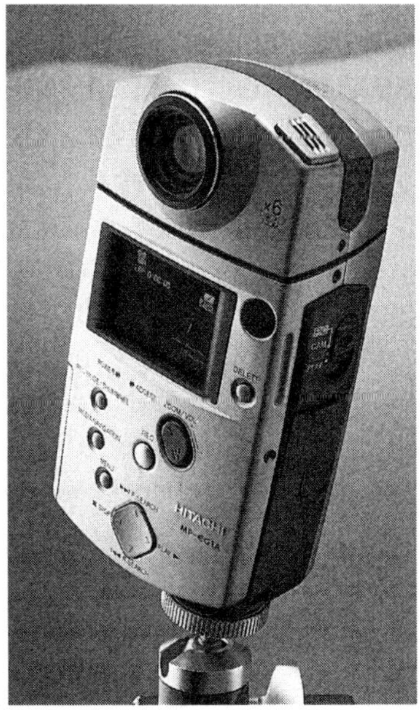

*Figure 19-4. Hiatchi M2 (photo courtesy Hitachi America, Ltd).*

Several of Sony's cameras will capture in MPEG format too, but they are limited by the size of the media they use to store the images. The MVC-FD91 stores images on a floppy disk, and can capture 15 seconds at 320 x 240, or 60 seconds at 160 x 112. In both cases, you cannot use the flash while capturing in movie mode.

## MULTI-PICTURE CAPTURE

A few cameras feature a multi-picture capture mode. This is different from the burst mode, available with some cameras, that captures several images at any resolution in a buffer then saves them to storage. The multi-picture mode captures several images but saves them as one frame. In the case of the Sony MVC-FD91, nine images are captured in a couple of seconds, and then saved as nine images of 320 x 240, arranged as one frame that is 960 x 720. This image can then be broken up into a sequence and played back

## ANIMATOR-9

Animator-9 is a freeware application, written by Paul Law, which takes a multi-picture frame and will play the animation and save it as a GIF sequence. It will also save the individual images as a sequence of JPEG files. A Java applet is also included, which will animate a sequence of JPEG images. This provides much better color since the JPEG images are 24-bit, not 8-bit, and the files are often smaller than the same GIF file. Of course, if the user's browser does not have Java support they won't see the JPEG sequence.

To create a JPEG sequence you copy the *animator9.class* file to a directory and use Animator-9 to split the image into individual JPEG files and save them in the same directory. Then use the HTML button to create an HTML file containing the code that calls the Java applet. As long as your browser supports Java (you may need to upgrade), when you open the HTML file you should see the sequence.

When uploading to a website you must upload the HTML file, individual images and the *animator9.class* file.

Cameras supported include the Sony DSC-F1, MVC-FD51, MVC-FD71, Epson PhotoPC 600 and Nikon CoolPix 900 and 600.

Website: *http://members.xoom.com/animator9/*

**Framer** is another simple tool that will play animated sequences from a multi-picture images. Its website is at:

*http://members.tripod.com/~tawba/framer.htm*

## CAPTURING SEQUENCES OF IMAGES

If you want to capture a sequence of images over a period of time, there are at least two ways of doing this. Some cameras feature a time-lapse function where you can specify the frequency with which photographs are to be taken, as well as the number. Kodak's Digita-enabled cameras (the DCS260, etc.) and Minolta's Dimage 1500 camera have such a time-lapse option.

Alternatively, if the controlling software that comes with the camera can be used to trigger the shutter, then you can use the software to take time-lapse sequences. Even if the software includes no options for taking images over a period of time, you might be able to use a macro language to trigger the software. For Macintosh users, AppleScript may be usable with the application. Find out more about AppleScript at:

*http://www.apple.com/*

For Windows users, the tool Macro Scheduler may work. Macro Scheduler is a utility for Windows 95/98 and NT that records macros and scripts to automate most Windows tasks. It essentially records mouse clicks and keystrokes and then plays them back one or many times. See more at:

*http://www.mjtnet.com/*

## ANIMATING A SEQUENCE IN A WEB BROWSER

To animate a sequence in a browser, the easiest option is to use a GIF file. For improved quality, the other options are to use a custom file format (MPEG, QuickTime, etc.). In some cases the visitor to your website may need to install a plug-in to be able to see the file.

Another alternative is to use a Java applet to load and play a sequence of JPEG files. This is what the applet included with Animator-9 does. Another Java applet that does the same thing, but with more options, is The Animator applet. For those wanting to animate a sequence of JPEG files, The Animator provides a general-purpose tool with many optional parameters, including order, repetition, sound, pause between frames, and a startup image to display while loading.

Developed by *jellinek@eng.sun.com* at Sun, the applet source code is available at:

*http://java.sun.com/applets/jdk/1.1/demo/Animator/*

## AUTHORING SOFTWARE

There are also several multimedia authoring and animation tools that can be used to assemble sequences of frames into animations. Two very popular such tools are Macromedia's Flash and Director. These are different from the tools already covered in that they let you create interactive presentations. Both tools are available for Windows and Macintosh, and create files that can be played in Web browsers, provided the website visitor has downloaded the plug-in that supports the files.

Flash is the simpler and cheaper of the two programs. It provides a timeline interface in which you can arrange and animate objects. It can also be used to assemble a sequence of images into a movie or animation—though if that's all you want to do you would be probably better off with a video editing tool or something that can save in formats such as RealVideo and QuickTime.

Director is much more expensive, and more sophisticated. It includes a scripting language which lets you create almost any kind of interactive presentation. Again, if you don't need that kind of functionality, then another tool may be a better choice. Director files are saved in ShockWave format; it's a protected version of the Director file format designed to be streamed over the network.

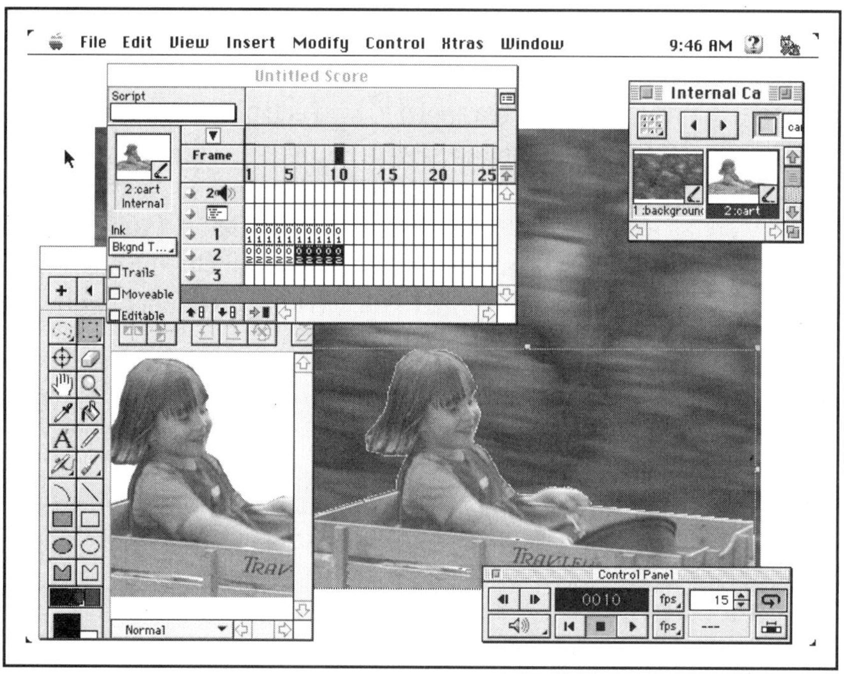


Figure 19-5. Director.

There are several tools available for creating animations in the QuickTime file format. Unfortunately, all of these programs are currently only available on the Macintosh.

Two programs, LiveStage and Electrifier, provide an authoring environment for creating complex QuickTime animations that include transitions and interactive capabilities. These programs are easy to use and include tools such as tweening. Use either to create all kinds of animated materials:

*http://www.electrifier.com*

*http://www.totallyhip.com*

## DV (DIGITAL VIDEO)

One of the most exciting recent developments in the consumer and low-end professional video market has been the arrival of the DV digital video format. These new video cameras offer improved video quality compared to Hi8 and S-VHS camcorders. But this isn't enough to warrant changing to yet another tape format (DV uses small tape cassettes that are half the size of a Hi8 cassette)—the real selling point is for those who are using computers to edit their video. For while DV camcorders use CCDs to capture the video signal (just as their Hi8 analog cousins), the image is then converted to a digital image and then compressed using the DV codec and saved to tape as a digital signal rather than an analog one. A new interface called IEEE-1394 (also called FireWire by Apple, and i.Link by Sony) can be used to connect the camera to another DV recording device, or to a computer. Computers equipped with the IEEE-1394 port (and appropriate software) can be used to transfer the DV file from the camera to the computer without any change in the data file. It's possible to perform simple edits on the DV file and send it back without having to recompress the images and without any loss in picture quality. This compares to editing Hi8 or S-VHS, where a digitizing card must be used to convert the analog video signal to a digital format and then convert it again when exporting back to video tape. Each time the file is transferred between the analog video and the computer there must be an analog-to-digital conversion, which causes some change in the image from the original.

Of course, it's not all a free ride. While it's true that simply transferring files from the camera to the computer and back doesn't result in any picture loss, when the video is edited on the computer (and that's why you transferred it to the computer in the first place, right?), then there must be a decompression and recompression of the image. The recompression can introduce slight changes to the image. Some DV codecs may produce better results than others. Taking the same image and recompressing it multiple times without performing any changes to the image itself can, over a series of recompressions, result in significant loss of image quality. So while it's true that DV offers lossless image transfers, when editing and adding effects to DV video you still should try to minimize the number of recompressions that you perform on your DV files.

DV cameras have been available for a couple of years now, and the price of these cameras, like the prices of other electronic devices, have been falling. Unlike still digital cameras, which have seen falling prices and rising image quality, the DV standard is a fixed resolution size (i.e., it's designed for producing standard television), so to a certain extent the DV camcorders haven't been through the many changes that have occurred in still digital cameras.

## WHEN IS DV NOT DV?

Some three years after the introduction of DV camcorders, Sony—one of the pioneers of the DV standard—introduced Video8 cameras, which offer many of the features of DV camcorders, but use Hi8 tapes instead of the miniDV tapes. Sometimes it seems electronics manufacturers just like to shoot themselves in the foot by confusing consumers, but Sony's justification is that these units are designed for people who already have a Hi8 camcorder; they can use them to play existing tapes. And the price of Hi8 tapes is actually much lower than DV tapes (though note that Digital8 records only one hour of program on a two-hour Hi8 tape).

So should you buy a Digital8? There certainly are advantages to the Digital8 format, not the least of them being that Sony is pricing the Digital8 camcorders much lower than their DV camcorders. But then again, their DV camcorders offer more sophisticated features. It seems that Digital8 is designed for the consumer, while DV is aimed at the prosumer.

How successful will Digital8 be? One problem is that it surrently isn't offered by any other manufacturer. In the past this has been a big negative for new consumer formats. However, the fact that the cameras use Hi8 tape makes the calculation much more complex. We'll have to wait and see.

## DOES A DV CAMCORDER MAKE A GOOD ALTERNATIVE TO A STILL DIGITAL CAMERA?

One of the features available on most DV camcorders is a Photo Mode. Part of the DV specification, Photo Mode saves a single frame to tape. It actually records five seconds of tape (simply recording the same frame repeatedly). This means that you have to wait while the camera writes that video to tape. Some cameras offer features that accentuate this option; Canon's Optura camera is shaped more like an SLR camera than a video camera, and can be used with a standard flash unit. In its Motor Drive mode the camera captures frames in progressive mode rather than in interlaced mode. This means that no feathering is visible in individual frames. (Traditional video is displayed as fields that make up an entire frame, see Chapter 4.)

Once you've captured your images you need to transfer them to your computer. That means that you'll need an IEEE-1394 port for your computer (it's included on newer Macintosh models, or you can buy a PCI card for your existing computer). You'll also need software that lets you handle the downloading of the DV data. This software may be included with the card; you should check. Just note that some software assumes that you are downloading video segments rather than individual frames. The process of downloading images may not be as quick and convenient as you might hope for.

Sony's TRV-900 camera goes one step further by offering a floppy disk that can be used to store the captured images, or the camera can be connected to a computer using a PCMCIA card. This makes it much faster for transferring images to your computer.

Comparing DV camcorders to still digital camcorders, you'll find that the comparative quality of the DV camcorder images is as good as any of the 640 x 480 resolution digital cameras available. It's only when comparing the images with those from the newer, high-resolution cameras (such as the Nikon CoolPix 950 or the Kodak DC265) that you'll see a clear difference between the images.

And while a 640 x 480 image may be acceptable for printing at sizes up to about 3 x 4 inches, if you want to print an 8 x 10, or fill a high-resolution screen, then a digital camera will serve you better. Also, if you're primarily taking still images, then the digital camera generally is more convenient to use, and is more convenient for transferring images to your computer.

But, if you're primarily interested in editing video on your computer and want to take occasional still images, get a DV camcorder.

If you want to shoot video (but don't want to edit it) and also want to take still images, buy a cheap Hi8 camera and a digital camera.

If you want to shoot a lot of images, buy a digital camera (and if you have to buy a DV camera as well, then that's the way it is).

**Great Tricks with Digital Video**

There are a few things that a video camera can be great for; time-lapse photography is one example. Assuming that the period is not much longer than the length of your tape, why not shoot the entire process, then go back and take out the frames that you want?

An interesting trick is capturing several frames of the same scene, and then combining those images to create a new image that reduces the noise in the image.

Here's how it works. A friend of mine was at a bar and he saw a plate on the wall. It was dark, but he wanted to capture the plate because it had the name "Borg" on it. So he took out his DV camcorder and shot a few seconds of the plate.

*Figure 19-6. Original frame taken in low light. (Courtesy of Kent Borg.)*

Since there was so little light the resulting video was very dark and noisy. You could barely read the words on the plate. But using a piece of software called Comotion he was able to take this video and produce a final frame that had most of the noise removed.

First he had to use motion tracking; because the camera was moving he had to tell the software how the plate was moving within the frame. The software then produced a new segment in which the plate remains in the same position. Then using FrameBuilder QT (from Dragonfly Digital Design, www.dragonflydigital.com) he "blended" together the frames to create an enhanced image. The software is primarily useful for removing random noise. Since the noise is different in every frame, blending many frames together acts to cancel out the noise. Simply put, if there were three frames and noise was randomly appearing throughout these frames, then it's unlikely that noise will appear in the same place in all three frames at the same pixel. So if in two of the frames a pixel has one value, but in the third frame it has a different value, it can be assumed that the value that appears in the two frames is the accurate one.

This technique could be used with a sequence of images taken using a still digital camera.

Nikon's CoolPix 950 has a "best shot" selection program that will take a sequence of pictures and then choose the "best" image. This is a very different process; the algorithm tries to determine blurriness in an image by comparing contrast between pixels (the more contrast from pixel to pixel, the "sharper" the image).

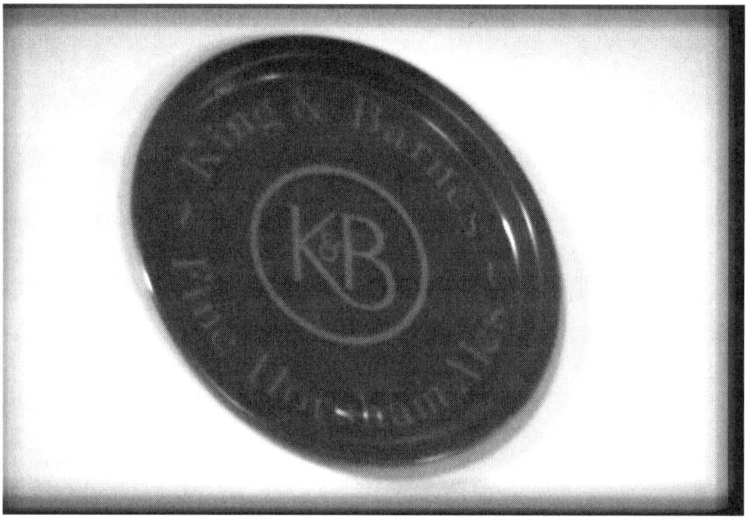


*Figure 19-7. Interpolated image (courtesy Kent Borg).*

# CHAPTER 20
# DIGITA OPERATING ENVIRONMENT

While a digital camera may seem like it's a regular camera with a CCD instead of film, in reality it is a very different beast. What you may not realize is that it is actually a computer; all digital cameras include a chip to manage the menus or interface and to handle compression and the transfer of files. This chip could be made to do much more. Most cameras already record capture information, and manufacturers are adding special processing features and effects as well. Turning cameras into computers has other advantages; they can be updated and improved with new software. Epson has released an image marking system that marks an image in the camera as it is captured (see *Copyright*, Chapter 21).

Enter Digita Operating Environment. Written by the company FlashPoint, Digita is a digital camera operating environment. It's already available in camera models from Kodak and Minolta. And FlashPoint has also licensed the operating system for use in printers. The theory of adding an operating environment to a digital camera or digital imaging device is that it provides better access to the camera features and also lets users perform functions that would previously have required images to first be transferred to a computer. In the case of Digita printers, it also allows users to skip the computer altogether.

This chapter covers Digita, not because it is the best or only way to add functionality to digital cameras, but because it suggests the future of intelligent cameras.

## DIGITA

FlashPoint grew out of imaging technology development at Apple, and focuses on the capture, output and storage of images. Their goal is to add intelligence to the cameras and printers to make life easier for the average consumer, while making the devices powerful enough for professional use. The Digita operating system, which includes camera control and other features, can be customized using scripts or by adding additional applications.

## THE PARTS OF DIGITA

The Digita operating system is very similar to a computer operating system. It controls the camera and provides a basic interface for interacting with the camera. The Digita operating system itself can be updated as well.

Digita uses an LCD panel and control buttons to provide interaction with the operating system. Users can select and change camera parameters, as well as choose scripts that they want to execute.

Digita scripts, which can be written using any text editor, can be placed in the ROM of the camera by the manufacturer, as well as located in the RAM or removable memory. Scripts can be executed at startup, or from a user menu.

Programs that run under Digita can also be placed in memory (see Digita FX).

## DIGITA IN USE: THE KODAK DC260

The Kodak DC260 really has two interfaces, not one. The first interface, which is common to many digital cameras, is a small LCD display at the top of the camera with some small pushbuttons arranged at the edges of the display. The LCD displays icons for flash mode, image quality and other options, as well as displaying the number of images available. One button is used to move through the choices; clicking the button moves to the next selection. A second button is used to toggle the various options for that selection. In the case of flash, pushing that button will toggle through AUTOMATIC, ALWAYS FIRE, FILL, and OFF. Separate buttons are available to zoom the lens and a shutter button for taking the picture. Using these controls, you can change basic parameters and take pictures.

The Digita system is accessed through the large display LCD.

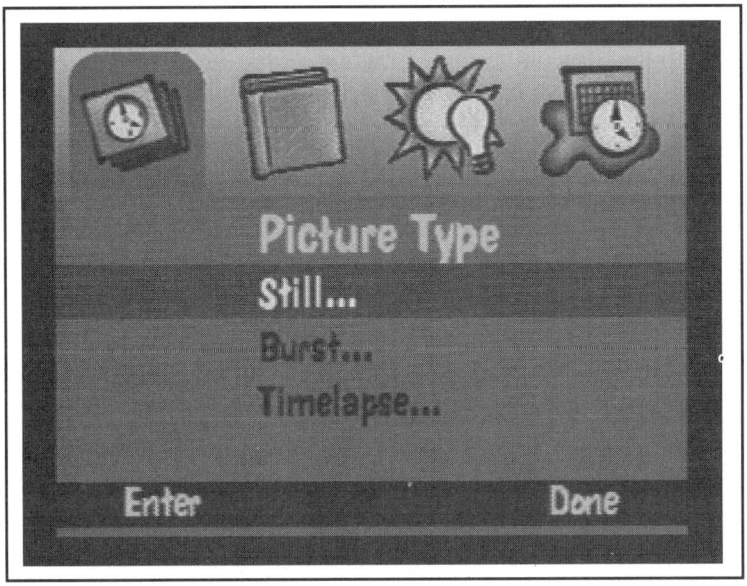


*Figure 20-1. Digita Interface.*

The Digita interface consists of icons that represent major functions (image type, exposure type, scripts, etc.). In each section there are either items you can select or a list of parameters that you can change. In this case, choosing TIME-LAPSE brings up the next screen, *Figure 20-2*.

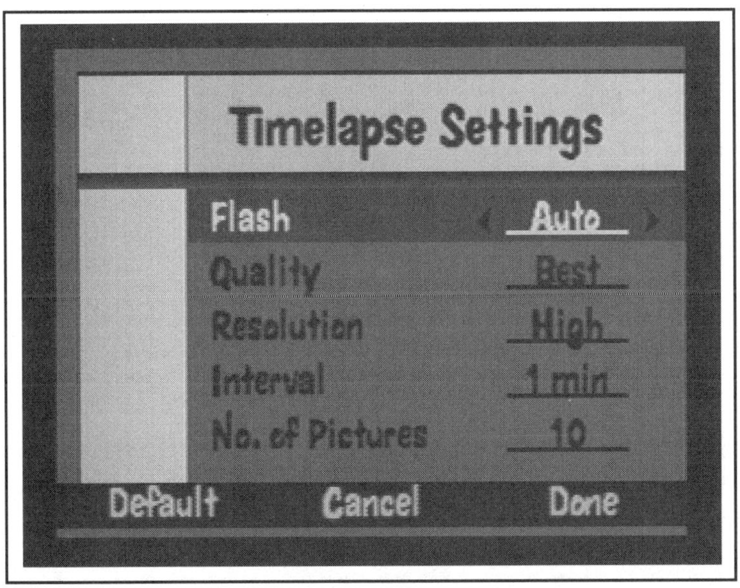


*Figure 20-2. Time-lapse Settings for DC260.*

Both of the screens appearing here were captured from a Kodak DC260 using the video output option of the camera, which sends the same image as appears on the LCD, as an NTSC video signal that can be viewed on a television.

## CHOOSING AND SETTING THE TIME-LAPSE SETTINGS FOR THE KODAK DC260

Navigation through the Digita menus, on the Kodak cameras, is accomplished with a single "four-way controller button." This buttons has four arrows on it, much like a compass rose, and you can rock the button left, right, up or down. This enables navigation across and down menus.

This method of camera control has both advantages and disadvantages. While it provides access to lots of options, it's more tedious than an SLR camera that provides individual buttons for switching between exposure modes and manually selecting shutter and f-stops. Unfortunately, most cameras seem to be moving away from that model of camera interface.

The interface includes options for choosing images from the memory card and displaying them on the LCD panel. You can view them shrunk to fit the screen, and you can then zoom in to view a portion of the image.

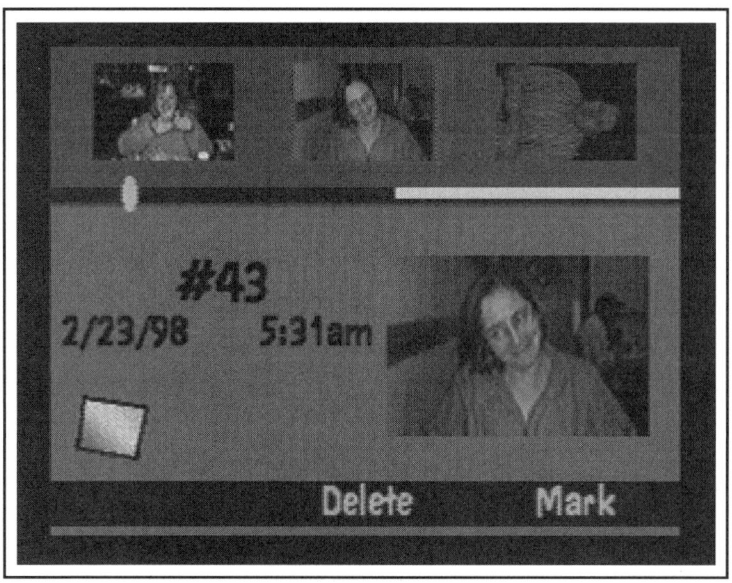


*Figure 20-3. Reviewing images.*

The biggest advantage of Digita is that you can create scripts that perform several functions, either before or after the image is taken. Or a script can be written that captures multiple pictures with one click; for example, taking a picture every five seconds, or zooming the lens slightly as it takes each picture.

Scripts are stored on the CompactFlash memory card of the camera. Simply placing a script in the card (either using a card reader, or the connection from the computer to the camera) adds the script.

## DIGITA SCRIPT

Digita script files are simply text files that contain Digita programs. The files must be named using the DOS naming convention, which is an eight-character name followed by the extension ".CSM". Examples of valid file names include:

BACKLGHT.CSM

MULTIPLE.CSM

Scripts can be edited by any program that can edit text files. Windows users can use the NotePad application, while Macintosh users can use SimpleText as the editor. You can also use a word processor; just make sure when saving the file that the file isn't changed to the word processor's file format or Digita won't be able to understand the file.

Scripts are placed in the SYSTEM directory on the memory card of the camera. In the case of the Kodak DC260, the SYSTEM directory is contained on the CompactFlash card. Interestingly, this could allow you to create different camera configurations using multiple CompactFlash cards. To switch to a different configuration you could simply switch Compact Flash cards. Once a script has been added you must eject the CompactFlash and then reinsert it for the script to appear in the script menu. While the interface includes a section for user scripts, you can actually add a script to any of the menu tabs in Digita.

Scripts are selected using the Digita interface. A tab for user scripts lists the scripts added to the System folder. The user selects the script and uses the Start key to run the script.

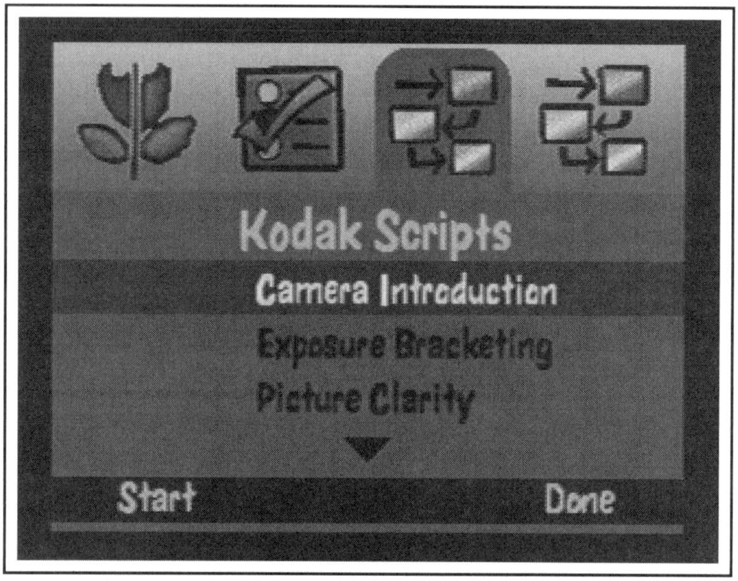


*Figure 20-4. Selecting a script from the Kodak menu of the camera. The interface displays a tab for Kodak scripts, and one or more for user scripts.*

While the Digita system is designed and developed by FlashPoint, it may be implemented slightly differently on the cameras of different manufacturers. Certainly some of the options may change depending upon the functionality of a particular model; for example, if the camera did not support a particular exposure mode.

The Digita script language resembles other programming languages, so if you are familiar with programming you should have little problem writing scripts. If you have no prior programming experience, then you might find it difficult to simply knock out a script that does what you want. However, Digita does offer a large amount of documentation at their site, so you might be able to learn from the examples. They also make available scripts that others have written, which for many of us, may be all we need or want.

One thing to note about writing programs for Digita. It's possible to write a program that loops endlessly and can't be stopped. If you find that the camera freezes up you'll have to disconnect all power (remove the batteries) to halt the program.

For those interested in the Digita operating system and scripting language, both Kodak and Digita have a variety of information available at their websites. Both companies are offering Scripting SDKs (Software Developer Kits) for those who want to develop scripts for the camera. The SDK includes a Reference Manual, User Interface Guide and Overview, and sample scripts.

Kodak also offers an ActiveX control that can be used with Visual Basic and other ActiveX interfaces to control their cameras. FlashPoint plans to release an Application SDK for programmers who want to write applications that run on the cameras, rather than scripts. Programs usually run faster and have greater access to the features of a camera than a scripting language does. Visit:

*www.digitacamera.com*

An example of a Digita script appears at the end of Chapter 21.

## EXTENDING DIGITA; THE DIGITA FX APPLICATION

FlashPoint has also released an in-camera application, Digita FX. It works with cameras that are Digita-enabled, including Kodak's DC220, 260 and 265,

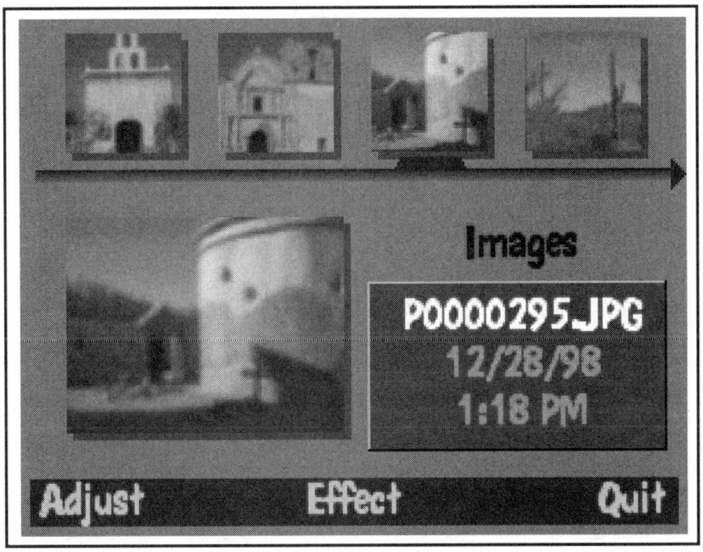


*Figure 20-5. Choosing an image to edit in Digita FX. This screen closely parallels the image selection screen in the Digita operating system. (Courtesy of FlashPoint.)*

and the Minolta Dimage 1500 camera. It lets you edit images in the camera, adjusting the color balance, contrast and brightness and applying filter effects. Filters include stucco, emboss, water color, mosaic and pixellate. Sharpen, rotate, blur and hue/saturation tools are also provided.

Images are chosen and then effects applied using the LCD screen of the camera as the interface. Some of the controls, such as brightness and contrast, provide a bar with an UP and DOWN selection for each property. Choose the property and then use the toggle button to increase or decrease the contrast.

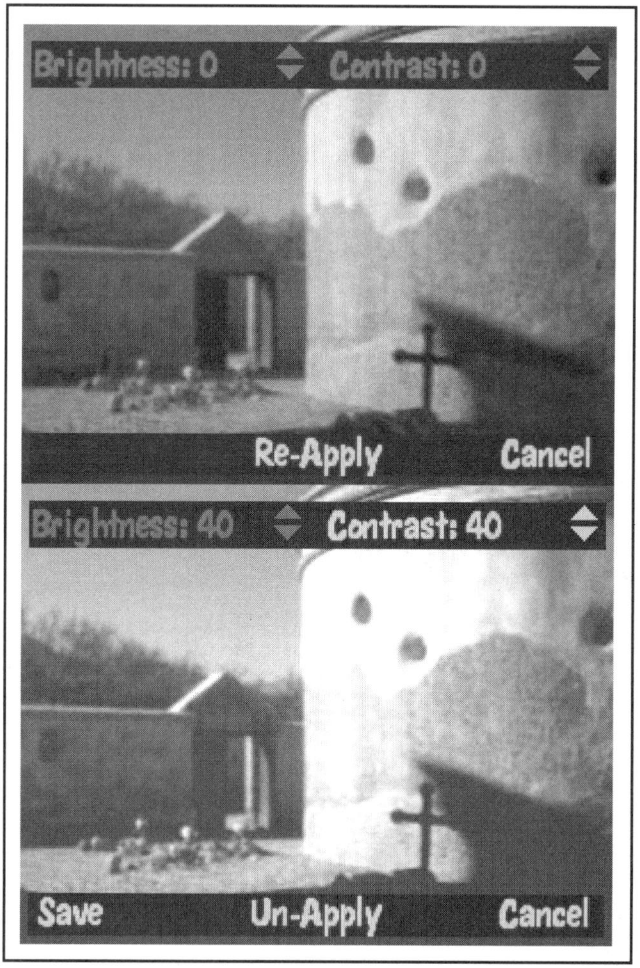


*Figure 20-6. Adjusting brightness and contrast in Digita FX. Each of the parameters is selected and then the value changed by clicking the camera's controller button. (Courtesy of FlashPoint.)*

Once you have finished editing an image you can save the file with a new name; the original file remains untouched.

While the program functionality may be too limited for some, the price is right: $19.95.

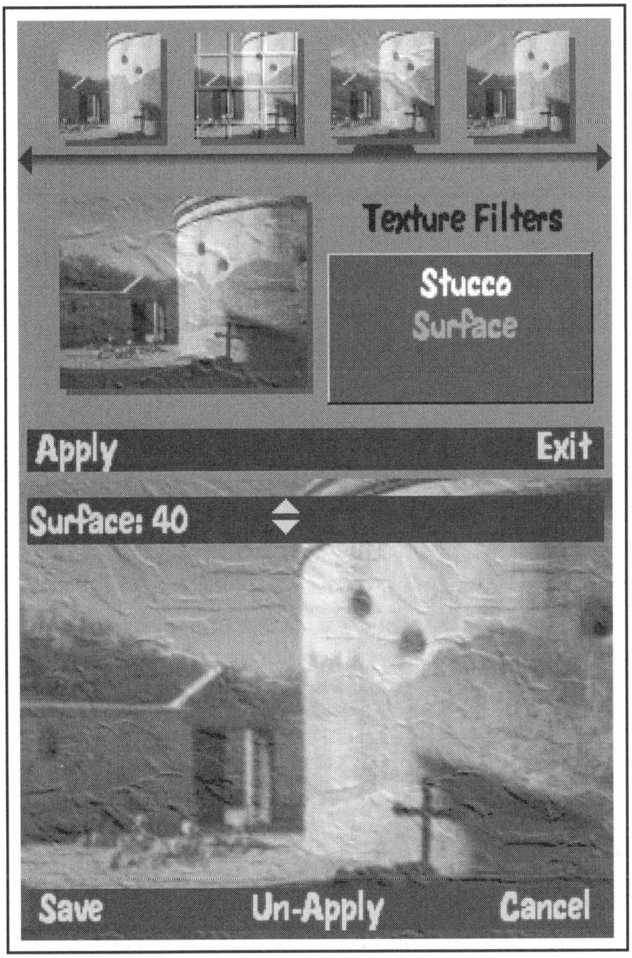


*Figure 20-7. Choosing the stucco filter to apply to an image (top), and viewing the effect (bottom). The strength of the filter can be changed and is represented by the surface field at the top of the screen. (Courtesy of FlashPoint.)*

### EPSON'S PRINTON PRINTER

Epson's PrintOn photo printer includes the Digita system and is currently available in Japan. It may soon be released in the United States. The printer features an LCD panel, much like those found on digital cameras that are used to interact with the Digita operating system. It also has a PCMCIA card slot which can be used to load images; a SmartMedia or CompactFlash card requires an adapter to work with the PCMCIA slot.

The camera will operate without being attached to a computer; you simply turn it on and insert a memory card. The Digita interface lets you choose the image you want to print, as well as the type of paper you are using. It will even print an index page containing small thumbnails of the images on the memory card. The printer can also be used as a regular printer, connected to a camera. In most respects it is the popular Stylus Photo 700 with the Digita operating system added.

Other companies are also developing add-on products for Digita cameras. Digitella Technologies is developing IntelliPix, which is designed to help add descriptive information (meta information) to an image when it is captured. The user can choose from predefined descriptions, or enter information using the camera's interface buttons. Computer software can then use this information to file the images when they are downloaded to a computer. They also offer a ScriptGenerator tool.

## FUTURE

Will all cameras become superintelligent devices that talk to you? Probably not. But moving intelligence to the cameras and printers makes sense for those who don't want to bother with computer-based editing software. And if you're not going to do much editing, why should you have to put the images onto a computer just to print them?

# CHAPTER 21
# COPYRIGHT

With the arrival of the wonderful world of digital media and instantaneous communications, the principles of copyright have sometimes been forgotten. But since you have a digital camera, and you might be the person who takes the picture proving Elvis is alive, you need to know how to protect that image from all the tabloids that will want to print it; unretouched or not!

This chapter explores copyright law. It explains the basic principles and what you should do to protect your work. Other legal issues such as model releases and work for hire are also explored. Software that you can use to protect your work is also examined. The chapter concludes with some useful information and procedures.

## COPYRIGHT

Copyright is a form of legal protection granted to the authors of "original works." You can copyright just about any specific expression, including photographs, movies and written works. Though there are some exceptions, copyright lasts for the life of the author plus 90 years (this was changed from 50 years in 1998).

Copyright, however, *does not* protect an idea, just the expression; so while you can't prevent other people from taking pictures of puppy dogs, you can protect the picture you took.

Copyright *does* provide you with protection across mediums. As just one example, a photographer took a picture of a group of puppies and a sculptor then made a sculpture of puppies that was based on that photograph. The sculptor was found to have infringed upon the photographer's copyright. Only the work's author has the right to create derivative works; you cannot use another's protected work as the basis for something of your own without permission.

Copyright, like other law, is constantly evolving. Because of this, the rules and parameters have changed over time, so what you've heard about copyright before may or may not be true. The following outlines the major points of copyright as they are understood in the United States today. (Our lawyers insist we add that this section is provided for illustration purposes only. While theoretically simple, copyright law can get complicated "at the edges." You should always get legal advice.)

Unfortunately, there's no international copyright, though many countries are part of the Berne Union, which shares copyright across countries.

**Popular Copyright myths:**

**1.** You must add the copyright symbol, date and name to a work to protect it.

*Not anymore. While this used to be a requirement, it is no longer required. Copyright exists as soon as the work is created. Adding a Copyright notice does have some benefits (see below).*

**2.** Copyright expires after 28 years. Copyright expires after 50 years.

*Copyright now extends for 90 years after the death of the author, or 75 years after date of publication if the author is unknown. (The law has changed several times in the last 25 years.)*

**3.** "It's a Wonderful Life" is no longer protected by Copyright

*True. The movie isn't. Though the colorized version is(!) and so is the story the movie was based upon.*

**4.** Anything on the Internet is public domain unless it says otherwise.

*Not true. In reality, the reverse is true. Anything on the Internet is protected unless it says that it is in the public domain.*

There is one major limitation to copyright; the doctrine of "fair use." Fair use allows others to reproduce a work for criticism, comment, news reporting, teaching (including multiple copies for classroom use), scholarship, or research. Whether a use is considered "fair" depends upon several factors, chief amongst them being whether the use is "commercial." Other factors include the amount of the portion used in relation to the copyrighted work, and the effect of the use on the potential market for the work. In other words, if you are limiting the ability of the author to make money from the work, reproducing the vast majority of the work, or making money from it yourself, then it is likely *not* to be considered "fair use."

Unfortunately, there are no hard and fast rules about "fair use;" no rules about the number of pages, percentage of the image, or anything else which defines fair use. If you attract the attention of the copyright holder and they take you to court, it's up to the court to determine whether it's fair use. Even if you win, you're exposed to court costs, so it's best to be careful.

Additionally, parody is also an allowed exception, however the same limitations apply, and you have to be particularly careful when dealing with company logos, etc., which are often protected by trademark law. The courts often take a dimmer view of parodies that damage a company's reputation.

Interestingly, architecture is now protected by copyright law. Any building built since December, 1990 is covered by this protection. However, the law has an exception, which says that the copyright owner cannot prevent the making, distribution or display of photos or other pictorial representations of the work. You can still take pictures of the building.

## REGISTRATION

The Copyright Office of the Library of Congress acts as a register of copyrights; however, as has already been mentioned, you do not have to register a work to have it protected by copyright. Registration requires a fee (check the Copyright Office for latest fees), as well as lodging a copy of the work. While usually a separate registration (and fee) is required for each work, it is possible to lodge a collection with a single fee. A collection must be assembled in an "orderly form" and should have a single or common author.

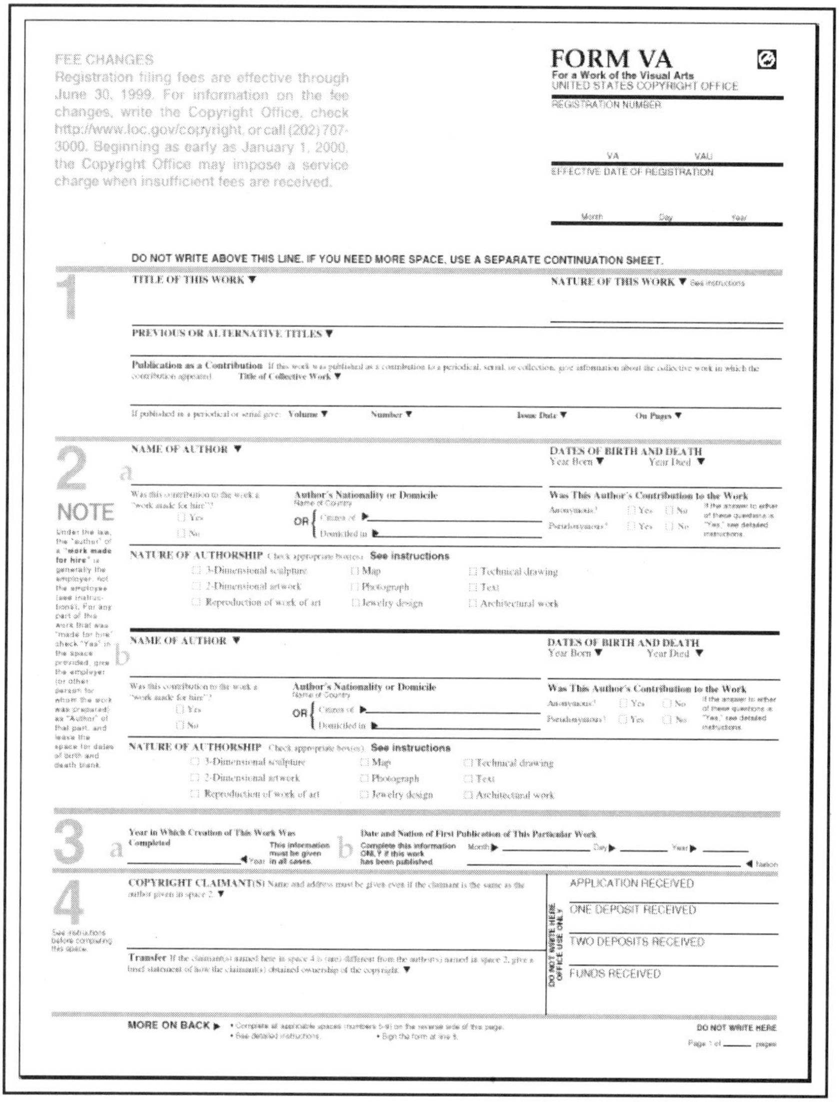

FEE CHANGES
Registration filing fees are effective through June 30, 1999. For information on the fee changes, write the Copyright Office, check http://www.loc.gov/copyright, or call (202) 707-3000. Beginning as early as January 1, 2000, the Copyright Office may impose a service charge when insufficient fees are received.

FORM VA
For a Work of the Visual Arts
UNITED STATES COPYRIGHT OFFICE
REGISTRATION NUMBER
VA VAU
EFFECTIVE DATE OF REGISTRATION
Month Day Year

DO NOT WRITE ABOVE THIS LINE. IF YOU NEED MORE SPACE, USE A SEPARATE CONTINUATION SHEET.

1 TITLE OF THIS WORK ▼ NATURE OF THIS WORK ▼ See instructions
PREVIOUS OR ALTERNATIVE TITLES ▼
Publication as a Contribution If this work was published as a contribution to a periodical, serial, or collection, give information about the collective work in which the contribution appeared. Title of Collective Work ▼
If published in a periodical or serial give: Volume ▼ Number ▼ Issue Date ▼ On Pages ▼

2 a NAME OF AUTHOR ▼ DATES OF BIRTH AND DEATH Year Born ▼ Year Died ▼
Was this contribution to the work a "work made for hire"? ☐ Yes ☐ No
Author's Nationality or Domicile Name of Country OR { Citizen of ▶ Domiciled in ▶
Was This Author's Contribution to the Work Anonymous? ☐ Yes ☐ No Pseudonymous? ☐ Yes ☐ No If the answer to either of these questions is "Yes," see detailed instructions.
NATURE OF AUTHORSHIP Check appropriate box(es). See instructions
☐ 3-Dimensional sculpture ☐ Map ☐ Technical drawing
☐ 2-Dimensional artwork ☐ Photograph ☐ Text
☐ Reproduction of work of art ☐ Jewelry design ☐ Architectural work

NOTE
Under the law, the "author" of a "work made for hire" is generally the employer, not the employee (see instructions). For any part of this work that was "made for hire" check "Yes" in the space provided, give the employer (or other person for whom the work was prepared) as "Author" of that part, and leave the space for dates of birth and death blank.

b NAME OF AUTHOR ▼ DATES OF BIRTH AND DEATH Year Born ▼ Year Died ▼
Was this contribution to the work a "work made for hire"? ☐ Yes ☐ No
Author's Nationality or Domicile Name of Country OR { Citizen of ▶ Domiciled in ▶
Was This Author's Contribution to the Work Anonymous? ☐ Yes ☐ No Pseudonymous? ☐ Yes ☐ No If the answer to either of these questions is "Yes," see detailed instructions.
NATURE OF AUTHORSHIP Check appropriate box(es). See instructions
☐ 3-Dimensional sculpture ☐ Map ☐ Technical drawing
☐ 2-Dimensional artwork ☐ Photograph ☐ Text
☐ Reproduction of work of art ☐ Jewelry design ☐ Architectural work

3 a Year in Which Creation of This Work Was Completed This information must be given ◀ Year in all cases.
b Date and Nation of First Publication of This Particular Work Complete this information ONLY if this work has been published. Month ▶ Day ▶ Year ▶ ◀ Nation

4 COPYRIGHT CLAIMANT(S) Name and address must be given even if the claimant is the same as the author given in space 2. ▼
See instructions before completing this space.
Transfer If the claimant(s) named here in space 4 is (are) different from the author(s) named in space 2, give a brief statement of how the claimant(s) obtained ownership of the copyright. ▼
DO NOT WRITE HERE OFFICE USE ONLY
APPLICATION RECEIVED
ONE DEPOSIT RECEIVED
TWO DEPOSITS RECEIVED
FUNDS RECEIVED

MORE ON BACK ▶ • Complete all applicable spaces (numbers 5-9) on the reverse side of this page. • See detailed instructions. • Sign the form at line 8.
DO NOT WRITE HERE
Page 1 of ____ pages

*Figure 21-1. VA Application form. The VA Copyright registration form for graphic arts. This form is included on the CD-ROM in Adobe Acrobat format.*

There are some advantages of copyright registration; if the work is registered within 3 months after publication or prior to an infringement, then statutory damages and attorney's fees can be recovered in court action. If a work is not registered you can only receive actual damages and profits. Therefore, if you have a valuable piece of work that you think is likely to be copied you should seriously consider lodging a copyright registration.

While registration of a work is optional, the law requires that you must deposit copies of a work if it is "published." You can be fined if you fail to comply with this obligation.

A published work is one that has been offered for sale to the public; in most cases the display of an item does not constitute publishing. The Copyright Office also says that if the "material object does not change hands, for example... displays on television," it does not qualify as publication, It's unclear, but it's likely that "displays on a Web page" is not considered publication.

If you register a work, whether it is published or not, you must lodge a copy with the Copyright Office, though the number of copies you have to lodge varies depending upon whether the work is published or not. If unpublished, one complete copy of identifying material is required. If published, two complete copies of the best edition must be filed (with some exceptions, notably for motion pictures and CD-ROMs).

To register a visual art you must use the VA form (included on the CD in Adobe Acrobat format). If you print the form make sure that you print it head to head (top of page 2 is directly behind the top of page 1) on a single piece of 8-1/2 x 11-inch white paper.

## NOTICE OF COPYRIGHT

After March 1989 it become optional whether you included a notice of copyright on a work. Use of the notice is still recommended as a reminder to others that the work is protected. Another advantage of using the notice is that if the work carries the notice, then courts will not allow a defendant to claim "innocent infringement" (that they had no idea the work was protected). There is no specific requirement where you place the notice.

The notice should consist of the following:

**1.** The symbol © (the letter C in a circle), or the word "Copyright," or the abbreviation "Copr."

**2.** The year of first publication of the work.

**3.** The name of the copyright owner.

Example: © 1999 John Doe

The year can be omitted on greeting cards, postcards, stationery, jewelry, dolls, toys, or other useful articles. For a compilation or derivative work the original date of publication is used.

You should place the notice in a prominent—or at least noticeable—location; otherwise, you negate the purpose of the notice. Some files formats (such as QuickTime and JFIF) have a copyright field. Unfortunately, you have to specifically look for this field; few applications display this by default and many users aren't even aware of its existence, so it's unclear whether placing a notice in this field would be considered sufficient notice in its own right.

In most cases copyright belongs to the author of the work; however, this is not true for a "work for hire." A work for hire is anything created by an employee of a company. If you are hired by someone to take a photograph for them, generally the copyright would remain with you, unless the other person could claim that you took the picture under their direction. So if they told you where to put the camera, what to point it at, and when to take the picture, then things get kind of murky. It's always a good idea, therefore, to get things in writing (it saves legal bills). Most companies, when they hire consultants, will stipulate that any work done is a work for hire. Whether you agree to that, as a photographer, is up to you, and a subject for negotiation. But if you sign a work-for-hire agreement, then copyright does go to that person.

It is also possible to transfer copyright to another. It's also possible to license a copyrighted work (which is different to transferring copyright).

It's important to note that copyright is different from ownership of the work. If a photographer sells a copy of a photograph to another (even if it's the negative or slide), this does not transfer the copyright. Copyright remains with the author unless transferred through legal agreement. That's something to remember if you hire someone to take a photograph for you; unless you stipulate the agreement prior to the taking of the picture, the photographer maintains the copyright.

Finally (you never thought it would be this complicated, right?), a "moral right for visual artists" has been added to law. For certain, limited one-of-a-kind visual art and limited editions, authors are accorded rights of attribution and integrity. This is designed to protect an artist's work against modifications and destruction (for example, if you paint a mural and someone repaints it).

## REGISTRATION PROCEDURES

For information on registering (you should always check, because rules and regulations have a habit of changing every few years) contact:

Library of Congress
Copyright Office
Register of Copyrights
101 Independence Avenue, S.E.
Washington, D.C. 20559-6000
*http:// www.loc.gov/copyright*
or call (202) 707-3000

The Copyright Offices records since 1978 are available on the Internet via Telenet through LOCIS (Library of Congress Information System). Connect to LOCIS through the World Wide Web or directly by using Telnet.

World Wide Web: *http://www.loc.gov/copyright*
Telnet: *locis.loc.gov*
Gopher: *marvel.loc.gov* (port 70)

Application Form VA is the appropriate form for registration of a work of the visual arts.

Dealing with the Copyright Office can be a bit tedious, and it can take a long time for them to finally issue a copyright, but they can't be all bad. The following answer appears in their FAQ on the Copyright Office website:

**How do I protect my sighting of Elvis?**

> *Copyright law does not protect sightings. However, copyright law will protect your photo (or other depiction) of your sighting of Elvis. Just send it to us with a form VA application and the $20 filing fee. No one can lawfully use your photo of your sighting, although someone else may file his own photo of his sighting. Copyright law protects the original photograph, not the subject of the photograph.*
>
> (From: Copyright Office Library of Congress FAQ)

## OTHER LAWS TO CONSIDER

While copyright law protects your images from people stealing them, there are other laws you must be aware of when taking pictures, specifically, rights of privacy and publicity restrict the use of a person's image. If you take a picture of a person and publish the picture, the person pictured can sue you for invasion of privacy. There are some exceptions to this, such as famous people, public figures, and certain events that occur "in public." Also, the use of the picture can have a bearing. Famous people can control their image; if you take a picture of a famous person on the street you could sell that picture to a newspaper; however, you could not use that picture in an advertisement.

As with many things, it's always best to be careful. Therefore, a model release should be used whenever possible. Most people, if approached in the right way, will be only too happy to sign a model release. So have them with you, and where possible, get a model release. Use your best judgment here. If you're planning on using a photograph for commercial purposes (advertising, company annual reports, etc.), definitely get a release.

An example release is included here and on the CD.

**MODEL RELEASE**

In consideration of ________ and/or other valuable consideration, I hereby grant to ______________________________, ("Photographer") his/her heirs, representatives and assigns, full rights and exclusive permission in perpetuity to copyright and/or use, re-use, publish, republish or otherwise transmit any photographs, portraits, video images or electronically generated images of myself (with or without my name) without restriction as to changes or alterations.

I release and save harmless Photographer, his/her representatives and all persons acting under his/her authority from any liability by virtue of any distortion, editing or composition, whether intentional or otherwise, that may occur or be produced, as well as any publication thereof, including without limitations any claims for libel or invasion of privacy. I waive the right to inspect or approve the finished product or products or the use to which it may be applied.

I certify that I am of legal age and am possessed of full legal capacity to execute this authorization. This release shall be binding upon me and my heirs, legal representatives, and assigns.

NAME ______________________

ADDRESS ______________________

______________________

SIGNATURE ______________________

DATE ______________

WITNESS ______________________

DATE ______________

*Figure 21-2. Model release form. This form is included on the CD-ROM*

Similarly, taking photographs of trademarked property (signage, artwork etc.) can also be a problem, and you should consider getting a release on those occasions too.

# SOFTWARE PROTECTION

We've explored copyright law, and copyright law should be your first defense against unlawful copying of your work. If you place your images on the Web, make sure that you include a copyright notice in a prominent place (either on the image, or just below the image). This at least informs people that you are claiming copyright on the work.

Can you go further? Unfortunately, the Web is a very open system. That's its advantage and disadvantage. Unlike television, where it requires a little work to make a copy of something (using a VCR), with the Web the browser does the hard work of downloading the images to display them. All the user has to do is click a graphic and choose SAVE to have an exact copy of the original (something very hard to do even with a VCR).

Some companies have developed technologies to help track down those who are using others' works illegally. Digimarc and Signafy are two companies offering such products.

Digimarc's software encodes a key in the visual data of the graphic itself. The graphic can be saved in JPEG format and displayed by any browser. Using the Digimarc software, you can go back and check a graphic to see if the key is there. Because the key is encoded in the visual information, and not in the file, someone copying part of the graphic from one file to another doesn't remove the key. However, depending upon the level at which the key is encoded, the key can be removed by heavy filtering or other image editing.

The first step with Digimarc's system is to get a Digimarc code and download their software. They also ship a plug-in that works with Adobe Photoshop. Digimarc's pricing has changed (and may change again), but currently they offer a basic ID for free, and a Smart ID which costs $500 annually. The basic ID points back to a Web page on their website that has information about you. The Smart ID can point to any website anywhere. They also sell a batch embedding tool for less than $100.

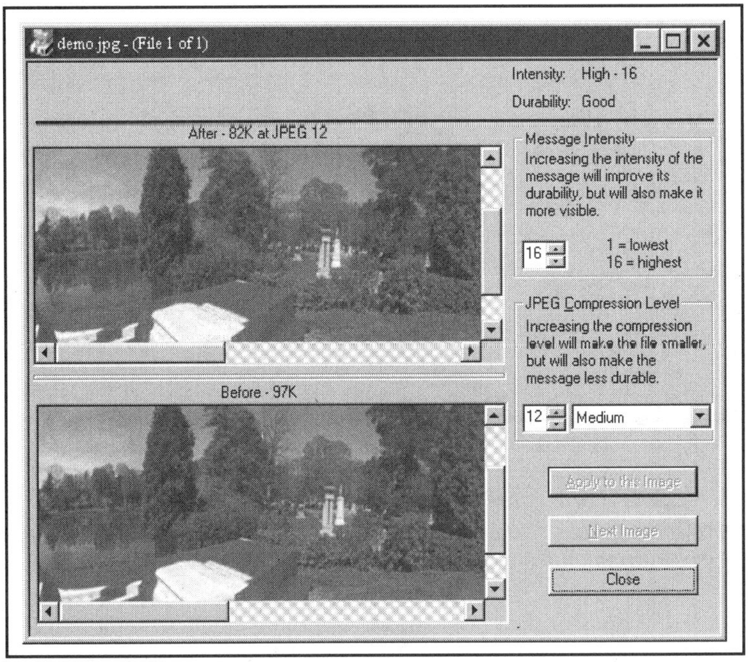


*Figure 21-3. Encoding Digimarc code. The software displays the image before and after adding the Digimarc code.*

Digimarc also has a MarcSpider search engine which searches the Web and reports back the location of copies of graphics encoded with your ID. This service costs extra.

Once you have a key, you run the embedding application or plug-in, choose the graphic and enter the date of copyright. The application lets you choose the Message Intensity and the level of JPEG compression, and displays a before and after view of the image.

The level of image intensity is an important consideration because, like JPEG compression, adding the key can impact the appearance of the image. The message intensity parameter adjusts this level of change. Increasing the message intensity increases the amount of image change. Unfortunately, if you don't use a very strong message intensity, then it can be relatively easy to adjust the image using an image editor so that the key is removed. Some filter effects will "remove" the key (make it unreadable by the viewer software) very easily if the message intensity is not strong. Increasing the message intensity, though, degrades the image.

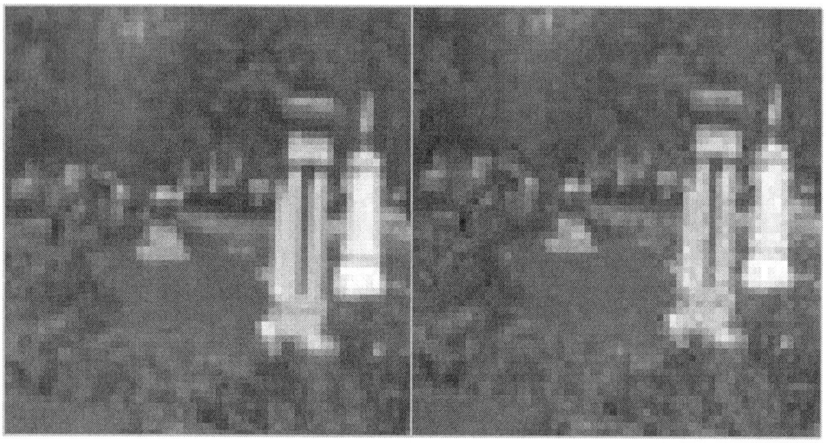

*Figure 21-4. Before and After image. A close-up of a section of the image before (left) and after adding the Digimarc key at high intensity. Random pixels are particularly visible in the light colored columns on the right hand side of the picture.*

The Digimarc software reviews an image and looks for the Digimarc key encoded within the graphic information. If it finds the key, it displays information about that key (creator code and Web address).

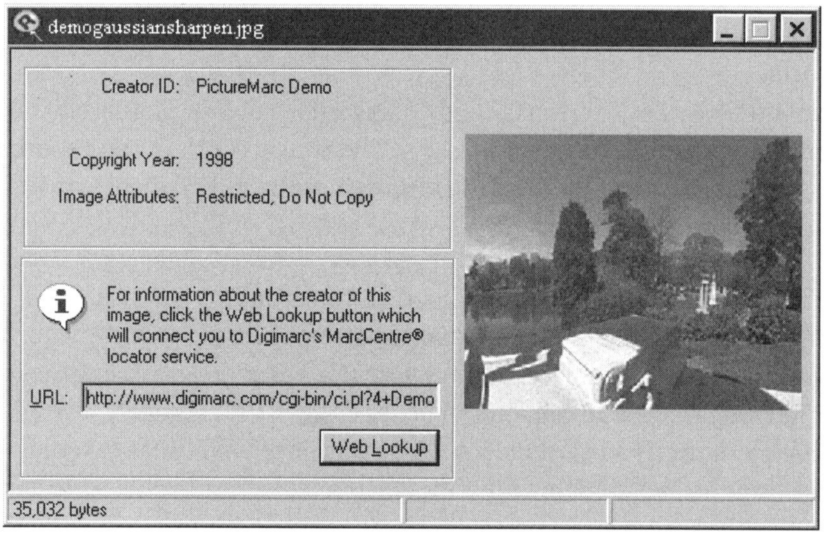


*Figure 21-5. Digimarc Reader software. The reader displays the Digimarc key information.*

Digimarc's spider application is perhaps the most appealing part of their technology, since it checks the Web for copies of your graphic. However, it's a relatively expensive service. For most people, simply placing a copyright notice on the graphic itself is just as efficient. And remember that Digimarc is not a form of copyright registration, and doesn't really help you in any legal proceedings, except that it provides a stronger evidentiary link if you did not register your image with the Copyright Office.

DigiMarc's software is available for both Macs and PCs. Get more information on DigiMarc at:

*<www.digimarc.com>*

Signafy is another company which offers a similar watermarking technology; however, their system is a little different. Their tool, OwnerMark, uses "spread spectrum" image processing techniques to embed a digital watermark in an image's "perceptually most significant areas" which, though it may not be the same algorithm as Digimarc, sounds kind of similar in principle.

The OwnerMark application works differently from DigiMarc's. Where Digimarc requires that you download a user key that is embedded in the file, Ownermark simply requires that you open the original file and a copy that you want to watermark. The software then adds the watermark and you can distribute the copy. You must keep the original in a safe place, as it is the "key" that is used to check if a file contains part of the protected file. So when you get a file you want to check, you open the original and the "copy" and have the software check for the key. The software isn't comparing the original to the copy, it's using the original to reconstruct the key that it then looks for within the other file.

*Figure 21-6* shows the software comparing the original file with a file that has been sharpened and brightened. The application reports that the ownermark is "strongly detected."

As an experiment I then painted over a large part of the image (but about half the image remained untouched). When I ran the software on the altered image it still found the ownermark, though it wasn't as strongly detected.

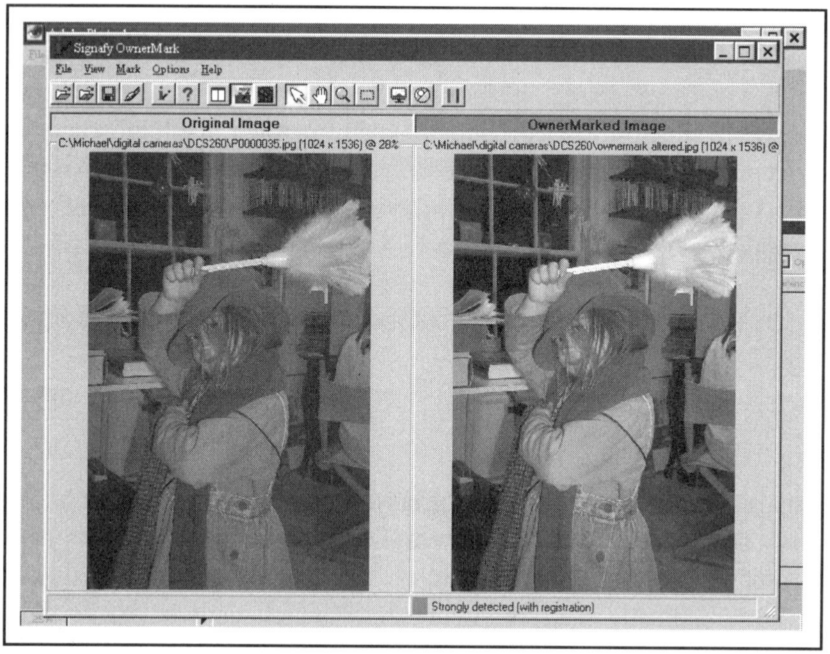


*Figure 21-6. OwnerMark detecting a copy based on the original file.*

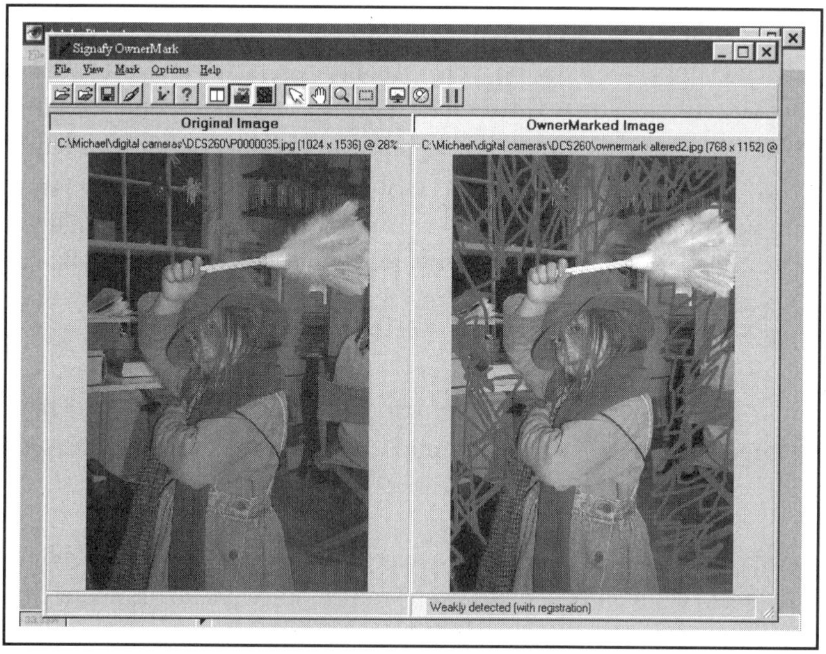


*Figure 21-7. OwnerMark copy is painted over, degrading the identification.*

I then applied a blur to the image and the ownermark was no longer detected. This testing was all done with their standard settings. Like Digimarc, you can adjust the strength of the signal applied to the image, but this can degrade the image.

Finally, JPEG Wizard software will batch add a comment or a watermark to images without uncompressing and recompressing the JPEG file. Its website is found at:

*http://www.jpg.com/product.html*

## CONCLUSION

The watermarking software does very well if the image is simply trimmed or rescaled, but starts to fail after some editing. While you can strengthen the watermarking, this starts to impact the image. Even at their strongest setting, if a heavy filter is used on an image, the applications will not detect that the manipulated image was based on a protected original.

While I think watermarking makes sense for some uses (the Digimarc spider is a particularly interesting concept), for most people simply applying a copyright notice will provide all the protection they need.

## OTHER RESOURCES

### ADDING THE COPYRIGHT SYMBOL TO THE DC260

The Kodak DC260 provides a feature for adding watermarks to images. Unfortunately, the text system only supports 7-bit ASCII characters, which does not include the ASCII character. It is possible to add the symbol using a Digita script.

```
name "CPYRGHT.CSM"
mode 1
menu "Scripts"
label "Copyright"
#set the watermark string — the \xA9 is
# the character value for the copyright symbol.
SetCameraState("wstr", "\xA9 1999 John Doe")
```

(script courtesy FlashPoint Technologies)

## HOW TO TYPE THE © SYMBOL

It seems like a fairly simple thing to do, but how do you type the © symbol? Most computer typefaces contain many more character than the ones represented on the standard keyboard. The © symbol is one such character.

For the Macintosh, the © symbol can be typed by holding down the option key and pressing the lowercase g.

Option - g = ©

Some typefaces do not include the © symbol. If you don't see it in the font, try using a font like Palatino or Geneva.

For the PC, you'll need to use a special character option in your word processor.

# CHAPTER 22
# HIGH-END CAMERAS

This book has primarily focused on consumer cameras priced under $1,000, but much of what is true about the sub-$1,000 cameras is true for the higher priced cameras. This chapter is designed as a quick introduction to the cameras in the professional market. At the moment, these cameras are priced from $4,000 up to $20,000 or more. Much of this price is due to the high cost of production for the units, which is compounded by the limited market for the units due to the cost. It's a chicken-and-egg situation. However, as the market for digital cameras increases, and technology improves and comes down in cost, the cameras available to the consumer will rival these professional units.

Professional units differ from consumer units in a number of ways; chief amongst them is the higher resolution that can be captured, and the improved color accuracy. Professional cameras can be divided into two broad categories; studio and portable cameras.

## STUDIO CAMERAS

Studio cameras, due to design and functionality, must be connected to a computer to immediately download the image, and/or use multiple shots to take an image, requiring that the subject be stationary while the picture is taken.

Such cameras include The Leaf Volaire from Scitex, which is described in more detail below. Several of these digital cameras aren't cameras at all; they are digital backs that work with one or more professional cameras such as the Hasselblad, Sinarcam or Mamiya. Often the price of the camera and lens is not included in the price of the unit.

# PORTABLE CAMERAS

For those professional photographers looking for a portable camera, several companies, including Nikon and Kodak, offer cameras that are based on film SLR camera bodies with digital backs. Kodak offers a whole line of such cameras with a range of features and price points. These cameras are built on top of an existing SLR body, use a single chip and built-in memory much like the low-priced cameras. But they offer much higher resolution. Since they're built upon existing SLR bodies, they look and operate very much like their SLR brothers, though the added electronics can increase the weight and size of the unit. They can also use existing lenses. These cameras feature much larger chips than the consumer models; but they still aren't quite as large as a 35mm film frame, so the apparent focal length of a given lens is different from what it would be if the lens was used on a 35mm camera. The other major difference between these cameras and their film counterparts is the burst rate—the ability to quickly capture multiple images. Unlike film, where you are limited only by shutter speed and the number of images on a roll of film (for argument's sake, a film camera has a burst rate of 36 frames), digital cameras are limited by the amount of RAM memory in the camera available to store images while they are being processed and saved to storage, as well as the time it takes to store an image. The burst rate for these cameras varies from 12 frames down to as low as three. This is determined by the resolution of the images the camera captures; the higher the resolution of the images, the lower the burst rate.

Because these cameras are portable, they can be used for a wide variety of purposes outside the studio. Of course, they can also be used as studio cameras.

Is now a good time to buy such a camera? If the camera produces output that you can use, and can pay for itself in a couple of years, then buying one of these cameras makes sense. Otherwise, you might want to wait a little longer.

## KODAK DCS 560

The DCS 560, like all of Kodak's professional cameras, is based upon an SLR body built by another company. Currently Kodak is offering cameras based on bodies from both Canon and Nikon—the DCS 560 is based upon the Canon EOS-DCS. While the camera is built by Canon, it is modified for the DCS 560, and you can't simply buy a digital back for an existing EOS camera.

The DCS 560 is a single-chip camera with a six megapixel CCD which has dimensions of 3072 x 2048 pixels. The final images are 3040 x 2008 pixels. The ISO of the chip is 80, though ISO can be adjusted up to 200. The images are captured at 12 bits. The camera supports standard EOS lenses, but the focal length multiplier is 1.3 (a 50mm lens is actually the equivalent of 65mm). This is as close to a 35mm frame as Kodak has yet achieved; other cameras range from the DCS 520 (1.6) to the DCS 420 (2.6). At 2.6, a 50mm lens is the equivalent of 130mm.

Images are stored on a PCMCIA card. The number of images that can be stored depends upon the size of the card. Images are stored in a proprietary Kodak TIFF format, and are about 3.5 megabytes on disk (18 megabytes when uncompressed). The burst rate is 3 frames; after three frames you have to wait while it saves the previous images; however, it can save an image in a second, so it can take an image every second for as long as there is memory in the card. For those who want a higher burst rate, the DCS 620 and 520 models offer burst rates of 12 frames; but the resolution of the images they capture is 1/3 the size of the images captured by the DCS 560.

A 1.8-inch LCD on the back of the camera can be used to review images, and also features a histogram display to show the exposure of an image. The camera stores camera settings with the image, and you can even voice-annotate the images using a built-in microphone.

All this sophistication doesn't come cheap. When released, the camera was priced at $28,500. You can find out more at:

*www.kodak.com*

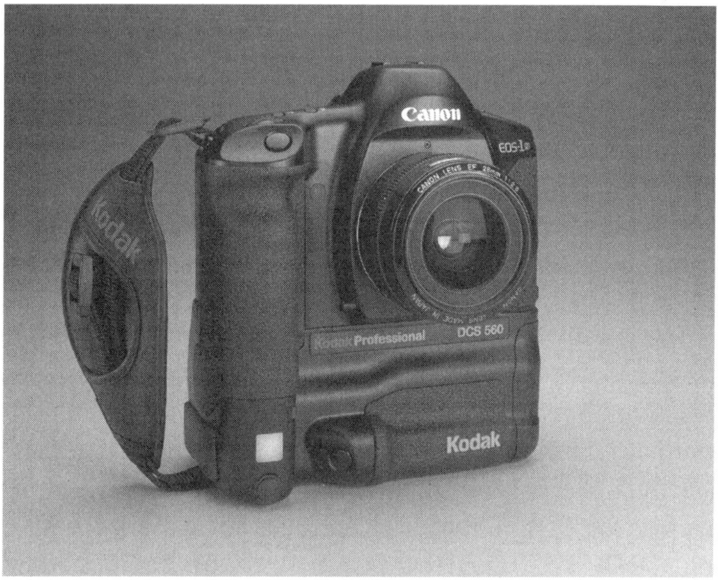


*Figure 22-1. Kodak Professional DCS 560 camera (photo courtesy Eastman Kodak).*

## SCITEX LEAF VOLARE DIGITAL CAMERA BACK

The Leaf Volare is an excellent representative of the digital studio camera, not just because it's a three-shot camera, but because it is so different from the other digital cameras covered in this book. The Volare takes three images—one for each color—then combines the three images to create the final color scene. It offers very high resolution and very accurate color reproduction. The company claims a dynamic range of 11 f-stops. The camera back can be used with a range of different medium-format cameras, and can be rotated to shoot horizontally or vertically. The CCD has a resolution of 3072 x 2048 pixels, and the data is captured in 14 bits, resulting in files of 36MB. The camera includes active cooling to reduce noise in the CCD.

The Volare back must be used while attached to a computer. While this is a negative for some uses, it's also a positive because the software can be used to provide live feedback of what is being taken. For example, parts of an image can be enlarged to check for sharpness using the live video preview. The software also includes the "Leaf Digital Layout" which lets the photographer superimpose the live preview image onto a layout to see how the image fits with other design elements. The software is available for both Macintosh and Windows NT.

This camera costs almost as much as the Kodak. It's $25,000 for the back, plus $900 for the interface plate to the camera. Oh, and you have to provide your own camera. Get more details on this gem at:

*www.scitex.com*

## OTHER CAMERAS

There are a range of other cameras and camera backs available, including several digital backs from Phase One (*www.phaseone.com*).

In late 1999 Nikon will release a new professional SLR digital camera, the D1. It's 23.7 c 15.6mm CCD contains 2.74 megapixels and produces an image of 2,012 x 1,324. It includes three exposure modes; Programmed, Shutter-Priority and Aperture-Priority. ISO can be set to 200, 400, 800, and 1,600. The camera can shoot 21 shots at 4.5 fps and has a maximum shutter speed of 1/16,000. It supports Nikon lenses and has a focal length multiplier of 1.5. All this sophistication should cost about $5,000.

Nikon already offers a high-end SLR camera, the E3/E3s, which offers a resolution of 1.4 million pixels (now excelled by the Nikon CoolPix 950). The camera resembles a SLR camera and supports Nikon 35mm lenses. It includes an optical system that adjusts the image coming from the lens so that the field of view and aspect ratio closely matches that produced by the same lens on a regular film camera. The camera also has a high ISO rating, 12-frame burst mode and an SCSI interface for transferring images. Nikon's website can be found at *www.nikon.com*.

## CHAPTER 23
# TROUBLESHOOTING

Camera not working? There are hundreds of reasons why your camera may be having problems and this chapter is designed to provide you with some suggestions for diagnosing camera problems. As with other electronic equipment, a little care can go a long way to avoiding problems. The first section of this chapter outlines some tips for the care of the camera, while the later sections cover diagnosing actual problems.

## TIPS FOR CAMERA CARE

The following suggestions will help extend your camera life. Wherever possible, suggestions for avoiding these problems are provided:

**1.** Water, condensation, dirt and sand are the enemies of all electronic and mechanical equipment.

*Plastic zip-lock bags can be excellent impromptu protection against anything from major to minor accidents; and they aren't bulky like regular camera bags. Of course, they don't provide much padding.*

**2.** Avoid extremes in temperature.

*Both high temperatures and extreme cold can be hazardous to your camera. In particular, leaving it on a car dashboard in the middle of summer should be avoided at all costs.*

**3.** Two words: Fresh batteries!

*Camera operation can become erratic when the batteries get low. Using rechargeable batteries and recharging as soon as they get low should avoid many problems. Similarly, avoid leaving batteries in a camera for long extended periods without use; they can leak, causing damage to the camera.*

**4.** Protect the lens.

*If the camera doesn't have a lens cover, then get one, or make one (the plastic top from a 35mm film cartridge can be adapted for some small lenses). Many 35mm SLR owners use a UV filter as an extra protection against scratches, but the reality is that most damage to the lens occurs during storage, handling or in the camera bag. A lens cap or cover should protect against most of these problems. Clean the lens with a lens cloth only.*

**5.** Don't drop it!

*Using the camera strap (or adding one if the camera doesn't come with it) can prevent some accidents. Not leaving the camera on the edge of a desk or other precarious position is also recommended. Finally, if connected to AC power or the computer, make sure that the cord(s) aren't in a position where you can trip over them and send the unit flying across the room.*

## DIAGNOSING PROBLEMS

There are several things that can cause camera problems. Interestingly, as cameras become more intelligent they will become more susceptible to the problems computer users are familiar with; random crashes and sudden freezes. The Digita operating system (see Chapter 20) makes it possible for users to write their own scripts—and really mess up their cameras!

Obviously, if the camera isn't working when you get it, then there's a very good chance that it never worked—though you should always do the usual things (read the manual, check the batteries) before you assume it's defective. But if the camera worked yesterday and doesn't work today, identifying the problem becomes more tricky.

### STEPS IN DIAGNOSING A PROBLEM

**1.** Has anything changed since the last time the camera worked?

*Did you update the firmware or upload a new program? Did you drop the camera?*

**2.** Is there power?

*Are the batteries new, inserted correctly? (see below) Is the AC adapter plugged in and the power outlet live?*

**3.** Is the memory full?

**4.** Is the camera in the right mode for picture taking?

**5.** Is the camera connected to the computer?

*Many cameras go into a specific mode when a serial cable is inserted into the camera jack.*

**6.** Who used the camera last and what did they do with it?

**7.** Did you read the manual?

**8.** Firmware or OS problems?

*Cameras can suffer from firmware problems (see below).*

**9.** Is the memory card inserted?

*Some cameras won't work unless a memory card is inserted.*

**10.** Does the problem only occur with one memory card?

*There could be a format error on the card. See section "Operating System Problems" below.*

## DIAGNOSING INTERMITTENT PROBLEMS

Not all problems are user fixable, but even so there's a lot you can do to help someone else fix a problem with your camera. Hard problems—for example, the camera doesn't turn on—are relatively easy to diagnose and fix. It's the intermittent problems that can be really annoying and difficult to fix. For this reason, if your camera operates strangely every now and again, try to take notes. When did it happen? What is different? Look for things that have changed; memory card, mode of the camera, things like that. The more you know about the problem, the easier it will be to diagnose.

## POWER PROBLEMS

If the camera is malfunctioning, the most likely cause is the batteries. Make sure that the batteries are inserted in the correct orientation, are firmly positioned, and that the batteries are fresh. Do not mix old and new batteries. Also, don't assume that batteries included with the camera are high quality and that they will necessarily work.

Trying a fresh set of alkaline batteries just to eliminate any question about the batteries you are currently using (whether rechargeable or not) is the best way of eliminating the batteries as a possible cause of problems.

If it's extremely cold, the batteries will perform poorly. Some cameras, particularly if you use the flash and LCD panel, draw a lot of power. A cheap set of batteries can be sucked dry after only three pictures! Sometimes they don't even work at all! So make sure you use a good quality battery; at least if you are having problems, once you know how the camera works, you can experiment with different types of batteries to find out which ones work best.

If you've been using the same rechargeable batteries for several years, it might be time to replace them. Rechargeables, while they can be used many times, don't last indefinitely. Many are rated to about 500 recharges.

## OPERATING SYSTEM PROBLEMS

All cameras include software that operates the camera. In some cameras this software (called firmware) can be updated, and in cameras that run the Digita operating system you can even add programs and scripts that run on the computer.

You should check the site of the manufacturer of your camera occasionally because there might be updates released for the firmware for your camera. A digital camera news website can be a good way to keep on top of changes in digital cameras, such as:

*www.image-resource.com*

*www.dcresource.com*

For cameras that use CompactFlash, SmartMedia and PCMCIA cards, what's on the card can interact with the operation of the camera and actually cause problems. In many cases, switching to another card, or erasing the card, can fix the problem. Users of some cameras with CompactFlash cards have reported intermittent camera problems that have gone away after reformatting the memory card (rather than simply deleting the images on the card). Reformatting a card may not be possible using the camera; you may need to use a card reader to do this. How to know if that's your problem? If you have two cards and the problem only occurs with one, then that suggests a problem with the memory card rather than the camera.

## IF ALL ELSE FAILS

Resetting the camera—removing the batteries and waiting a few minutes—may be a fix for some camera problems. Ejecting the memory card and erasing that should also be considered. If the camera has a backup battery in addition to the main battery, you may try removing that as well (though check with the user manual, because the manual might say never to do that).

As always, checking the manual for specific recommendations about your camera is highly recommended.

## CONNECTION WITH COMPUTER NOT WORKING

The final area that can give you problems is connecting the camera to the computer. If the connection has worked in the past, then you are (theoretically at least) trying to solve a different problem than if this is the first time you have ever connected the camera to the computer. You at least know that it used to work, and now you only have to identify what has changed:

1. Many serial connections use comparatively fragile connections to the camera. While it takes a bit of doing, it's possible to bend one of the pins in the cable while inserting it into the plug. Check that the plug is not damaged. If the plug is damaged, try using needle-nose pliers to straighten the pin. You should be very careful when continuing to use that plug, as the pin will now be very likely to bend again, and is closer to breaking.

2. Other software can interfere with your serial port. Have you installed new hardware or software since the last time the camera was used? Try removing the hardware and/or software. Alternatively, you might have to try reinstalling the camera software. If your computer has more than one serial port, try switching to another serial port (you will have to reconfigure the computer software when you do this).

If you are having problems after installing the software, check that the serial port is configured correctly. Try switching to another port. If you can, try installing the software on another computer to see if the problem occurs there.

Also, make sure that you are using the right cable. The cable from your modem or another device attached to the serial port may look very much like the cable that came with the camera, but it might not have all of the individual lines in the cable active. Make sure you are using the camera cable. Alternatively, if the cable appears to have suffered some kind of damage, try another cable to see if it makes a difference.

## CONCLUSION

With care, a digital camera should last for a long time; certainly until a replacement costs a fraction of your camera's price! Most cameras require little maintenance; recharging or replacing the batteries being the biggest ongoing tasks. Treating them with care, avoiding heat, condensation and dirt, and protecting the camera from knocks, are the things you should do to extend camera life. Using the camera regularly, rather than leaving it in a bag for months on end, will also extend the life of the camera.

# CHAPTER 24
# CAMERA SPECIFICATIONS

This chapter provides information about cameras available at the time of printing, along with some recommendations for digital cameras. For more information about how to go about buying a camera, read Chapter 2.

## CAMERAS UNDER $500

Thinking about a digital camera that costs less than $500? The good news is that the cameras are coming down in price and you can get a camera that produces excellent results for less than $500. The bad news is that generally these cameras either lack the resolution or the optical zooms available in their more expensive brothers. However, over the next couple of years this should change too. So if resolutions of 1200 x 1000 are good enough, here are some cameras to consider:

Fuji MX-500

Toshiba PDR-M1

HP PhotoSmart C30

Nikon CoolPix 700

There are many other cameras in this category, but the above cameras have proved popular with users and have received favorable reviews in comparisons published by magazines.

## CAMERAS UNDER $1,000

In this category you'll find cameras that have resolutions of 1600 x 1200; which is just about good enough for 8 x 10 prints. Which camera is best? At the moment I still recommend putting resolution at the top of the list.

As one example, I recently was looking at buying a new camera, and I was comparing the CoolPix 950 and the Kodak DC265, which are both nice cameras. The ergonomics of the two cameras is very different, and when considering which camera to buy I found that the color accuracy of the DC265 images were rated slightly better than the earlier Nikon CoolPix 900s model in a magazine review. Despite this, the 950 offered the highest resolution in its price category (when originally released). For that reason I went with the CoolPix 950 sight unseen, despite having used the DC260 (the earlier model of the DC265). I think I made the right choice; the images are outstanding in detail reproduction and the color reproduction looks good too. But I can see why others might prefer the user interface of other cameras. I'm not sure that I really like the rotating body design of the Nikon, and I'm annoyed that when you start it up it is in telephoto mode, not wide-angle or something in-between. But these are minor gripes, and I'm happy. You should definitely try before you buy; if you can!

**Nikon CoolPix 950**. Great resolution, small package. Some annoying UI features.

**Kodak DC265**. Nice user interface, excellent images and Digita operating system. Resolution not as high as other cameras.

**Canon PowerShot Pro70**. Described as a "Pro" model, though reviews suggest it's really a high-end consumer model.

**Olympus C-2000Z**. Olympus digital cameras are very popular with their owners despite the fact that they have generally been more expensive and with a lower resolution than some other cameras. The C-2000Z offers the highest resolution in their line, though in a different form factor from the 500-620L models.

**Sony DSC-F55**. For those who want floppy disk. Some of Sony's high-end cameras look very good, though they seem to be expensive compared to other cameras with similar resolution.

# CAMERAS BY RESOLUTION

The following chart lists cameras by resolution. External Storage (CF=CompactFlash, SM=SmartMedia, PC=PCMCIA, FD=Floppy Disk, ST=Sony Memory Stick) and connection options other than serial (USB=USB, IEEE=IEEE-1394 Firewire, SCSI=SCSI, IrDA=InfraRad) are also listed. For other features, see the feature list below.

| | | | |
|---|---|---|---|
| Kodak DCS460 | 3060 x 2036 | PC | SCSI |
| Kodak DCS560 | 3040 x 2008 | PC | IEEE |
| Nikon D1 | 2000 x 1312 | CF | IEEE |
| Fuji MX-2700 | 1800 x 1200 | SM | |
| Fuji MX-2900 | 1800 x 1200 | SM | |
| Olympus C-2500L | 1800 x 1200 [1] | SM | |
| Ricoh RDC-5000 | 1792 x 1200 | SM | |
| Kodak DCS520 | 1728 x 1152 | PC | IEEE |
| Kodak DCS620 | 1728 x 1152 | PC | IEEE |
| **1600 x 1200** | | | |
| Agfa ePhoto 1680 | (1343 x 973) [2] | SM | |
| Agfa ePhoto CL50 | (1280 x 960) [2] | SM | |
| Konica Q-M200 | 1600 x 1200 | CF | |
| Nikon CoolPix 700 | 1600 x 1200 | CF | |
| Nikon CoolPix 950 | 1600 x 1200 | CF | |
| Olympus C-2000Z | 1600 x 1200 | SM | |
| Polaroid PDC-2000 | 1600 x 1200 | | SCSI |
| Polaroid PDC-3000 | 1600 x 1200 | CF | SCSI |
| Sony DSC-F55 | 1600 x 1200 | FD | |
| Toshiba PDR-M4 | 1600 x 1200 | SM | |

(continued next page)

| | | | |
|---|---|---|---|
| Canon PowerShot Pro70 | 1536 x 1024 | CF, PC | |
| Kodak DC260 | 1536 x 1024 | CF | IrDA, USB |
| Kodak DC265 | 1536 x 1024 | CF | IrDA, USB |
| Minolta RD-175 | 1528 x 1146 | PC | SCSI |
| Kodak DCS420 | 1524 x 1012 | PC | SCSI |
| Kodak DCS315 | 1520 x 1008 | PC | IEEE |
| Agfa ePhoto CL30 | (1024 x 768)[2] | CF | USB |
| Sony DSC-D700 | 1344 x 1024 | PC, ST | |
| Minolta Dimâge EX1500 | 1344 x 1008 | CF | IrDA |
| Fuji DS-300 | 1280 x 1024 | PC | |
| Fuji DS-330 | 1280 x 1024 | PC | |
| Fuji DS-560/565 | 1280 x 1000 | PC | |
| Fuji MX-500 | 1280 x 1024 | SM | |
| Fuji MX-600 | 1280 x 1024 | SM | |
| Fuji MX-700 | 1280 x 1024 | SM | |
| Olympus D-600L | 1280 x 1024 | SM | |
| Olympus D-620L | 1280 x 1024 | SM | |
| Toshiba PDR-M1 | 1280 x 1024 | SM | |
| Toshiba PDR-M3 | 1280 x 1024 | SM | |
| **1280 x 960** | | | |
| Agfa ePhoto 1280 | (1024 x 768)[2] | SM | |
| Casio QV-500SX | 1280 x 960 | CF | |
| Casio QV-7000SX | 1280 x 960 | CF | |
| Epson 700 | 1280 x 960 | CF | |
| Epson 700Z | 1280 x 960 | CF | |
| Kodak DC120 | (984 x 850)[2] | CF | |
| Kodak DC240 | 1280 x 960 | CF | USB |
| Nikon CoolPix 900 | 1280 x 960 | CF | |

(continued next page)

| | | | |
|---|---|---|---|
| Nikon CoolPix 900s | 1280 x 960 | CF | |
| Olympus D-340L | 1280 x 960 | SM | |
| Olympus D-340R | 1280 x 960 | SM | |
| Olympus D-400Z | 1280 x 960 | SM | |
| Ricoh RDC-4200 | 1280 x 960 | SM | |
| Ricoh RDC-4300 | 1280 x 960 | SM | |
| Sanyo VPC-Z400 | 1280 x 960 | SM | |
| Sony MVC-FD88 | 1280 x 960 | FD | |
| Sony MVC-FD83 | 1216 x 912 | FD | |
| HP PhotoSmart C-20 | 1152 x 872 | CF | |
| HP PhotoSmart C-30 | 1152 x 872 | CF | |
| Konica Q-M100 | 1152 x 872 | CF | |
| Kodak DC200 | 1152 x 864 | CF | |
| Kodak DC200 plus | 1152 x 864 | CF | |
| Kodak DC210 | 1152 x 864 | CF | |
| Kodak DC210 plus | 1152 x 864 | CF | |
| Kodak DC220 | 1152 x 864 | CF | IrDA, USB |
| **1024 x 768** | | | |
| Agfa ePhoto 780 | 1024 x 768 | SM | |
| Agfa ePhoto 780c | 1024 x 768 | SM | |
| Canon PowerShot A5 | 1024 x 768 | CF | |
| PowerShot A5 Zoom | 1024 x 768 | CF | |
| Olympus D-320L | 1024 x 768 | SM | |
| Olympus D-500L | 1024 x 768 | SM | |
| Polaroid PDC-700 | 1024 x 768 | CF | |
| Sanyo VPC-X300 | 1024 x 768 | SM | |
| Sanyo VPC-X350 | 1024 x 768 | SM | |
| Sony MVC-FD81 | 1024 x 768 | FD | |

(continued next page)

| | | |
|---|---|---|
| Sony MVC-FD91 | 1024 x 768 | FD |
| Vivtar ViviCam 2800 | 1024 x 768 | CF |
| Vivtar ViviCam 3000 | 960 x 800 | PC |
| Vivtar ViviCam 3100 | 960 x 800 | PC |
| Canon PowerShot 600 | 832 x 608 | PC |
| Ricoh RDC-2 | 768 x 576 | PC |
| Ricoh RDC-2E | 768 x 576 | PC |
| Kodak DC40 | 756 x 504 | |
| Kodak DC50 | 756 x 504 | PC |
| **640 x 480** | | |
| Apple QuickTake 100 | 640 x 480 | |
| Apple QuickTake 150 | 640 x 480 | |
| Apple QuickTake 200 | 640 x 480 | SM |
| Canon PowerShot 350 | 640 x 480 | CF |
| Casio QV-300 | 640 x 480 | |
| Casio QV-700 | 640 x 480 | CF |
| Casio QV-770 | 640 x 480 | |
| Casio QV-780 | 640 x 480 | CF |
| Fuji DS-7 | 640 x 480 | SM |
| Fuji DX-5 | 640 x 480 | SM |
| Fuji DX-7 | 640 x 480 | SM |
| Fuji DX-9 | 640 x 480 | SM |
| HP PhotoSmart | 640 x 480 | |
| Konica Q-mini | 640 x 480 | CF |
| Minolta Dimâge V | 640 x 480 | CF |
| Olympus D-220L | 640 x 480 | SM |
| Polaroid PDC-640 | 640 x 480 | SM |
| Ricoh RDC-300 | 640 x 480 | |

(continued next page)

| | | |
|---|---|---|
| Ricoh RDC-300Z | 640 x 480 | SM |
| Sanyo VPC-G200 | 640 x 480 | |
| Sanyo VPC-G250 | 640 x 480 | SM |
| Sony MVC-FD51 | 640 x 480 | FD |
| Sony MVC-FD71 | 640 x 480 | FD |
| Sony MVC-FD73 | 640 x 480 | FD |
| Toshiba PDR-5 | 640 x 480 | SM |
| Vivtar ViviCam 2700 | 640 x 480 | CF |
| Vivtar ViviCam 2750 | 640 x 480 | CF |
| Kodak DC20 | 493 x 373 | |
| Kodak DC25 | 493 x 373 | |
| **320 x 240** | | |
| Polaroid PDC-300 | 320 x 240 | |

1. Unknown at time of printing.

2. Resolution is interpolated. Actual sensor resolution is in brackets.

# FEATURE LIST

The table starting on the next page lists most digital cameras that have been released in the US in the last five years. Professional level digital backs have not been included.

The table spans both facing pages, so the features are continued from the left to right page for each model listed.

| Mfr. | Model | Intro.[5] | Description | Resolution[7] | Viewfinder/ LCD[1] | Timer | Focus | Focus Range | Zoom[2] | Sensi-tivity[6] |
|---|---|---|---|---|---|---|---|---|---|---|
| **Agfa** | **ePhoto 780 & 780c** | Jan 98 | Resembles a traditional point-and-shoot camera's form factor (unlike the 1280 and 1680). Upgraded to 780c in June 1999. The 780c features a metallic blue housing (instead of black) and is priced $199 ($100 lower than the 780.) It does not include Kai's PhotoSoap SE which is included with the 780. | 1024 x 768, 640 x 480, 320 x 240 at 24 bits (350,000 pixel CCD) | Optical plus 1.8-inch, color LCD (290 x 220 pixels) | Yes | Focus-free, all glass lens | Macro mode | No | 100 |
| **Agfa** | **ePhoto 1280** | Oct 97 | Lets you choose aperture or shutter. The largest resolution is interpolated by software. Includes charger and rechargeable batteries. Reasonably small form factor with an LCD that rotates up from base. Fairly long delay before shutter fires and it takes a long time to cycle between images. | 1280 x 960 (interpolated), 1024 x 768 High, 1024 x 768 Std, 640 x 480 High, 640 x 480 Std | 2" color low-temp poly-silicon TFT LCD, 110,000 pixels | 5 and 10 second | Automatic and manual focus settings | 10cm - infinity (wide) 80cm - infinity (tele.) Macro | 3x optical zoom with macro | 100 |
| **Agfa** | **ePhoto 1680** | Jun 98 | Same form factor as the 1280, but w/ higher resolution. Highest res is interpolated. No optical viewfinder, though LCD is sharp and performs well in bright light (compared to other LCDs). Black & White mode (1-bit color depth). Agfa originally bundled this with an Iomega Clik! Drive (Agfa plans to produce a camera with a Clik! drive built-in). | 1600 x 1200 pixels (1680 mode with PhotoGenie) 1280 x 960 pixels, 640 x 480 pixels (1.3M (1343 x 972) pixels CCD) | 2" color low-temp poly-silicon TFT LCD, 110,000 pixels | 5 and 10 sec self-timer (w/ on-screen count-down) | Automatic and manual focus settings | 20 cm - infinity (wide) 50 cm - infinity (tele.) Macro | 3x optical zoom w/ macro 2x digital (VGA only) | 60 |

| Mfr. | Model | Exposure | Flash Modes | Burst[3] Mode | Lens Aperture | Focal[4] Length | Shutter Speed | Storage | Interface | Wt. |
|---|---|---|---|---|---|---|---|---|---|---|
| **Agfa** | **ePhoto 780 & 780c** | Auto | Auto, Fill-in, Red eye reduction, Off (3-10' range) | No (1-2 sec. Per image) | Fixed aperture f4.0 | 50 mm | 1/30 – 1/10000 | 2 MB SSFDC SmartMedia card holds 8 to 96 images | Serial (Mac & PC) Video out (switchable NTSC/PAL) | 9.6 oz |
| **Agfa** | **ePhoto 1280** | Automatic with manual exposure and white point override | Auto, Fill-in, Red eye reduction, Off (7.8 Flash guide number) | No | Wide: f2.8, f5.6, f8.0<br>Tele: f3.5 , f6.4, f9.1 | 38-114 mm | Automatic: 1/4 sec. - 1/500 sec. Manual: 1/8 sec. - 1/500 sec. | 4 MB SSFDC SmartMedia card holds 6 to 60 images | Serial cable for PC. Serial cable adapter for Macintosh, Video cable for TV (PAL and NTSC) | 13.3 oz |
| **Agfa** | **ePhoto 1680** | Automatic with manual exposure and white point override | Auto, Fill-in, Red eye reduction, Off | No Up to 6.5 sec. in 1680 mode | F2.8 (wide), 3.5 (tele.) | 38-114 mm | 1/2 sec. - 1/500 sec. | 4 MB SSFDC SmartMedia card holds 6 to 48 images | Serial cable for PC. Serial cable adapter for Macintosh, Video cable for TV (PAL and NTSC) | 13.3 oz |

| Mfr. | Model | Intro.[5] | Description | Resolution[7] | Viewfinder/ LCD[1] | Timer | Focus | Focus Range | Zoom[2] | Sensi-tivity[6] |
|---|---|---|---|---|---|---|---|---|---|---|
| **Agfa** | **ePhoto CL30** | Feb. 99 | Simple, easy-to-use point-and-shoot form factor. Fixed lens focal length. Features USB. | 1440 x 1080 (PhotoGenie) 1152 x 864 pixels 1024 x 768 pixels Low 640 x 480 pixels | Optical plus 1.8" color LCD display | 10 sec | Auto and four manual focus positions: (1M, 2M, 5M, infinity) | macro: 10 cm - 1m. normal: 50 cm - infinity | No | 100 |
| **Agfa** | **ePhoto CL50** | Mar. 99 | While the nomenclature suggests the CL50 is an upgraded CL30, they are really two very different cameras. The CL50 features optical and digital zoom, but does not include USB (like the CL30). | 1600 x 1200 (PhotoGenie), 1280 x 960 pixels, 640 x 480 pixels | Optical plus 2" LCD display (110,000 pixels) | 5 and 10 sec self timer | Auto and four manual focus | macro: 20 cm - 80cm. normal: 80 cm - infinity | 3x optical zoom w/ macro. 2x digital (VGA only) | |
| **Agfa** | **ePhoto Smile** | Sum 99? | Very cute (and small) camera. Not available at press time, and few details known. | 640 x 480 | | | | | | |
| **Apple** | **Quick Take 100** | 93? (D) | One of the first low-priced (under $1,000) digitals. Unusual shape; long and flat, almost like binoculars. Images were really only acceptable at half their actual size (i.e.,320 x 240) but its ease of use, particularly in transferring images to the computer, set the standard for cameras to follow. | 640 x 480 320 x 240 | Optical | Yes | fixed | 4 feet to infinity | No | 85 |
| **Apple** | **Quick Take 150** | 95 (D) | An upgrade to the 100, offered marginally better image quality and more images can be captured (though it's still limited in number). Included plastic clip-on macro lens of questionable usefulness. | 640 x 480 | Optical | Yes | fixed | 4 feet to infinity | No | 85 |

| Mfr. | Model | Exposure | Flash Modes | Burst[3] Mode | Lens Aperture | Focal[4] Length | Shutter Speed | Storage | Interface | Wt. |
|---|---|---|---|---|---|---|---|---|---|---|
| **Agfa** | **ePhoto CL30** | Automatic with manual white point override | Auto, Fill-in, Red eye reduction, Off | No (6 - 11 sec.) | f/2.8 - 8.0 | 43mm | 1/30 - 1/700 second | 4 MB CompactFlash card holds 6 - 36 images | Serial, USB (serial and USB cables included) Video cable for TV (PAL and NTSC) | 9.4 oz |
| **Agfa** | **ePhoto CL50** | | Auto, Fill-in, Red eye reduction, Off | No (4 - 8.5 sec.) | f/2.8 (wide) - 4.7 (telephoto ) | 34 -102 mm | 1/2 - 1/500 of a second | 8 MB +(3.3V) SSFDC SmartMedia holds 12 - 96 images | Serial cable for PC, Mac adapter cable, A/V cable for TV output | 10.6 oz |
| **Agfa** | **ePhoto Smile** | | | | | | | 2MB internal memory holds 16 images | Serial (PC only) Video out | |
| **Apple** | **Quick Take 100** | Auto | automatic, on or off Range: 4 to 9 feet | No | f2.8 to f6 | 50mm | 1/30 to 1/175 | 1MB of FLASH EPROM holds 8 –16 images | Serial (Mac & PC) | 1 lb |
| **Apple** | **Quick Take 150** | Auto | automatic, on or off Range: 4 to 9 feet | No | f2.8 to f16 | 50mm | 1/30 to 1/175 | 1MB of FLASH EPROM holds 16 - 32 images | Serial (Mac & PC) | 1 lb |

| Mfr. | Model | Intro.[5] | Description | Resolution[7] | Viewfinder/ LCD[1] | Timer | Focus | Focus Range | Zoom[2] | Sensi-tivity[6] |
|---|---|---|---|---|---|---|---|---|---|---|
| **Apple** | **Quick Take 200** | Mar 97 (D) | A completely different camera from the QuickTake 100 & 150, the QuickTake 200 is based on the Fuji DS-7. Software only works with Macs (while Fuji's software only works with PCs). | 640 x 480 350,000 CCD | 1.8 inch LCD live viewing at 30 fps | 10 sec | Three zone | Macro 3.5 to 5"; close-up 1.5 to 3'; distant 3' to infinity | No | 100 |
| **Canon** | **Power Shot 350** | May 97 (D) | Includes Ni-CD batteries and charger. Good color images when used in natural light, but artificial lighting may pose a problem. LCD can be tilted for above-crowd shooting. Slow refresh rate, which can make picture framing of fast changing scenes difficult. No optical viewfinder, so battery consumption is high. | 640 x 480 pixels Economy, Normal and Fine compression (350,000 pixel CCD) | 1.8" color LCD with adjustable viewing angle | 10-sec self timer | Fixed Focus | Macro 1.2 to 26" Normal 26" to infinity | No | 120 |
| **Canon** | **Power Shot 600** | 96 (D) | Microphone used to record WAV files. Wide-angle lens adapter also adjusts optical view-finder. Color in images subdued compared to other cameras. Text mode captures as gray-scale rather than color. Fairly long delay between pressing the shutter and picture capture. 28mm wide-angle lens available. Incl Ni-Cads, charger and docking station. | 832 x 608, 640 x 480, 320 x 240 JPEG: Fine, Normal, Economy. Optional uncompressed mode | Optical | 2 sec or 10 | Fixed plus macro | Macro 4" to 16" Normal 16" to infinity | No | 100 |

| Mfr. | Model | Exposure | Flash Modes | Burst[3] Mode | Lens Aperture | Focal[4] Length | Shutter Speed | Storage | Interface | Wt. |
|---|---|---|---|---|---|---|---|---|---|---|
| **Apple** | **Quick Take 200** | Auto white balance | No | No | 2 step: indoor f2.2 and outdoor f8 | 38mm | 1/4 to 1/5000 | 2MB SmartMedia holds 20-30 images | Serial (Macintosh) Video-out | 8.5 oz. (241 g) |
| **Canon** | **Power Shot 350** | TTL (through-the-lens) programmed auto exposure. Optional expos. compensation of +/-4 steps | On, off, auto and red-eye reduction | No | f/2.8 | 42mm | 1/4 sec. to 1/1,000 | 2MB CompactFlash holds 11 – 47 images | Serial PC interface cable and Mac adapter Video-out cable | 10.2 oz |
| **Canon** | **Power Shot 600** | TTL (through-the-lens) programmed automatic exposure. Exposure compensation of +/- 1/3 steps | Flash Guide: 9 Coverage: 8" to 12 feet | No | f/2.5 | 50mm | 1/30 sec. to 1/500 sec | 1MB RAM (holds up to 18 images at 832 x 608) Type II & Type III PCMCIA Cards | Parallel port for PC | 14 oz. |

| Mfr. | Model | Intro.[5] | Description | Resolution[7] | Viewfinder/ LCD[1] | Timer | Focus | Focus Range | Zoom[2] | Sensi-tivity[6] |
|---|---|---|---|---|---|---|---|---|---|---|
| **Canon** | **Power Shot A5** | Jul 98 | Small; size of Canon Elph camera. Anodized aluminum body with metal tripod socket (rare in digital cameras). For taking panoramas, Stitch Assist displays previous photo on right half of LCD screen and live image on left. LCD shows about 90% of actual image. Illuminator for the autofocus system is used in low light. Includes Ni-MH battery and charger. | 1024 x 768, 512 x 384 Fine, normal, CCD raw (810,000 pixel CCD) JPEG and CRW (Canon proprietary raw data format) | Optical/ LCD 2" TFT (active color matrix) LCD can be used as viewfinder. Display Speed: 25 fps | ?? | TTL Auto-focus | 3.5" to infinity | No | ISO 100-400 (ISO 400 in Small Mode and low light only) |
| **Canon** | **Power Shot A5 Zoom** | Mar 99 | Zoom lens version of A5. Similar features and options but with zoom lens. | 1024 x 768, 512 x 384 Fine, normal, CCD raw (810,000 pixel CCD) JPEG and CRW (Canon proprietary raw data format) | Optical with approx. 84% coverage. LCD 2" TFT (active color matrix). | Yes | TTL Auto-focus | 19.7" to infinity (Normal), 6.7 to 19.7" (Macro). | 2.5x | ISO 100-400 (ISO400 in Small Mode and low light only) |
| **Canon** | **Power Shot Pro70** | Mar 99 | Semi-pro model. Camera body resembles SLR camera. Excellent exposure controls. The telephoto lens is comparatively wide (28mm). LCD can be used as viewfinder. No built-in flash, but will work with Canon's external flash units and some third-party strobes. Microphone. Includes Ni-MH battery and charger. | 1536 x 1024, 768 x 512 Fine, normal, CCD raw (1.68 million pixel 1/2-inch CCD sensor) JPEG and CRW (Canon proprietary raw data format) | Optical, LCD 2" TFT (active color matrix). Display Speed: 25 fps | Yes | TTL Auto-focus | 2.3' (normal) 4.7" (macro mode) | 2.5x | ISO 100-400 (ISO400 in Small Mode and low light only) |

| Mfr. | Model | Exposure | Flash Modes | Burst[3] Mode | Lens Aperture | Focal[4] Length | Shutter Speed | Storage | Interface | Wt. |
|---|---|---|---|---|---|---|---|---|---|---|
| **Canon** | **Power Shot A5** | Automatic Plus +/- 2 EV in 1/3 step increment s | Auto, On, Off, Red-eye Reduction | 1 fps up to 15 frames (small mode only) | f/2.5 (- f/11?) | 35mm | 1/6 - 1/750 sec | 8MB CompactFlash holds 44 - 89 images | Serial (Mac & PC cables) video-out (NTSC) | 8.1 oz |
| **Canon** | **Power Shot A5 Zoom** | TTL eval meter-ing using CCD output. Range: EV2-16.5; Com- pensation: +/- 2 EV in 1/3 step incr. Manual & Auto White bal. | Auto, On, Red-Eye Reduction, Off. Range: Wide 0.6 to 11.5'; Tele: 0.6 to 8.2' | 1 fps up to 15 frames (small mode only) | f/2.6-f/4.0 | 28- 70mm | 1/6-1/750 sec. Slow Shutter OFF; 2-1/750 sec. Slow Shutter ON) | 8MB CompactFlash holds 44 - 125 images | Serial (Mac & PC cables) video-out (NTSC) | 9.2 oz |
| **Canon** | **Power Shot Pro70** | Auto-exp, aperture priority and exp compensa tion. Exp Compens ation: ± 2 EV in 1/3 step incr | Accessory shoe | 4 fps up to 20 frames (small mode only) | f/2.0 (wide), f/2.4 (telelphoto ) - f/8.0 plus f/16 | 28- 70mm | 1/2 - 1/500 sec. (large mode) 1/2 - 1/8,000 sec. (small mode) | 8MB CompactFlash . (holds 20 – 40 images.) Two card slots including one CF Type II | Serial (Mac & PC cables) video-out (NTSC) | 21 oz |

| Mfr. | Model | Intro.[5] | Description | Resolution[7] | Viewfinder/ LCD[1] | Timer | Focus | Focus Range | Zoom[2] | Sensi-tivity[6] |
|---|---|---|---|---|---|---|---|---|---|---|
| **Casio** | **QV-300** | 97 | No flash or optical viewfinder, but LCD is large and bright. Fixed focus swivel lens manually switchable between telephoto and wide-angle. LCD viewfinder updates at 4 fps. High power consumption. Manual aperture adj (up to 6 stops). Lens subject to flare. | 640 x 480, 320 x 240 1/4-inch CCD (360,000 pixels) | 2.5" TFT, low-glare color LCD (61,380 pixels) | 10 sec | Dual focal point with macro | | Two position | 100-1600 |
| **Casio** | **QV-700** | 98 | Large 2.5" LCD. Swivel lens. Exp range of EV 5 to 18. HIgh amplification in low light which can create a lot of noise. Very fast cycle time; little as 1 sec per image (w/o flash). Includes buffer memory for up to 4 images captured rapidly. "Past" mode saves 3 frames prior to pressing shutter. Captures images constantly into buffer until shutter press, then saves. Time-lapse mode. Panorama mode displays part of previous frame in viewfinder. Some limited image editing and titling included. Color accuracy good, but saturation lower. | 640 x 480, 320 x 240 1/4-inch CCD (350,000 pixels) | 2.5"TFT, low-glare color LCD (84,480 pixels) | 10 or 2 sec | Fixed with macro | Normal: 0.7m to infinity Macro: 14cm to 16cm | No | ~3200 |
| **Casio** | **QV-770** | Jan 98 | A lower-end version of the QV-700. Smaller LCD, no support for removable memory. Can capture up to 32 frames in a movie mode. | 640 x 480 1/4-inch CCD (350,000 pixels) | 1.8" TFT, hyper amorphous silicon TFT low-glare color LCD (122,100 pixels) | 10 sec | Fixed with macro | Normal: 0.7m to infinity Macro: 14cm to 16cm | No | ~3200 |

| Mfr. | Model | Exposure | Flash Modes | Burst[3] Mode | Lens Aperture | Focal[4] Length | Shutter Speed | Storage | Interface | Wt. |
|---|---|---|---|---|---|---|---|---|---|---|
| **Casio** | **QV-300** | Aperture priority exposure | No | No | F/2.6 – f/11 | 47mm 106mm | 1/8 to 1/4,000 second | 32Mbit internal flash memory holds 64 - 192 images. | Serial (Mac & PC cables) video-out (NTSC) | 250g |
| **Casio** | **QV-700** | Program AE/Aperture priority AE. Exp Correction : -2EV to +2EV (1/2 EV steps). Auto/Fixed White bal | On ("forced on" or "fill"), Off, and Automatic Range: 2.3-9.8 ft (0.7-3.0m) | Up to 1 fps | f/2, f/8 | 38mm | 1/8 to 1/4,000 second | 2MB CompactFlash holds 14 – 47 images | Serial (Mac & PC cables) video-out (NTSC) | 10 oz/ 290g |
| **Casio** | **QV-770** | Program AE, aperture priority AE. Compensation: -2 to +2EV (1/4EV steps) Auto/Fixed White bal | AUTO, ON, OFF | No | f/2, f/8 | 38mm | 1/8 to 1/4000 second | 4MB internal memory holds 30 – 120 images | Serial (mac & PC cables) IR-TranP video-out (NTSC) | 290g |

| Mfr. | Model | Intro.[5] | Description | Resolution[7] | Viewfinder/ LCD[1] | Timer | Focus | Focus Range | Zoom[2] | Sensi-tivity[6] |
|---|---|---|---|---|---|---|---|---|---|---|
| **Casio** | **QV-780** | Sep 98 | Mini-movie Capability 3.2 second (32 Motion Frames). Panorama Capability Capture and View on built-in screen or TV without PC. | 640 x 480<br>1/4-inch CCD (350,000 pixels) | 2.5" TFT Screen | 2, 10 sec | Fixed with macro | Normal: 0.7m to infinity<br>Macro: 14cm to 16cm | No | not known |
| **Casio** | **QV-5000SX** | May 98 | Captures stills, movies (6.4 sec video) and panoramas. Titles. 8MB internal memory; no removable memory support. Includes optical viewfinder (LCD consumes power fast!), and built-in UV filter. Includes built-in sharpening option. Average image quality. Upcoming QV-5500SX will have 16MB memory. | 1280 x 960<br>640 x 480 | Optical/ 1.8" TFT, hyper amorphous silicon TFT low-glare color LCD (122,100 pixels) | 10 sec | Fixed with macro | NORMAL: 0.3m to infinity<br>MACRO: 0.1m | No | not known |
| **Casio** | **QV-7000SX** | Nov 98 | Swivel lens, zoom up to 8x, and HTML file generator software. Image color and resolution slightly better than QV-5000SX. Movie mode: 3.2 sec--320 x 240 at 5 fps 6.4 sec -- 160 x 120 at 10 fps 12.8 sec --160 x 320 at 5 fps. Panorama Capture Guide – prev. image remains on screen. Capture up to 360 deg. | 1,280 x 960<br>640 x 480<br>1/3 inch CCD<br>1,320,000 pixels<br>3 compression levels | Optical/ 2.5" HAST screen (122,100 pixels -- 555 x 220) | 2, 10 sec | Auto, Manual and Macro | Auto Focus -- 0.23cm<br>Manual Focus -- 0.10cm<br>Macro -- 10cm | 2x optical<br>4x digital | not known |

| Mfr. | Model | Exposure | Flash Modes | Burst[3] Mode | Lens Aperture | Focal[4] Length | Shutter Speed | Storage | Interface | Wt. |
|---|---|---|---|---|---|---|---|---|---|---|
| **Casio** | **QV-780** | Program AE. Correction 2 to +2EV (in 1/4 EV Units). White Bal: Auto, WB1, WB2 and WB3 | Automatic, On and Off | Mini-movie | f/2, f/8 | 38mm | 1/8 to 1/4000 second | 2MB CompactFlash holds 14 – 47 images | Serial (Mac & PC cables) video-out (NTSC) | 290g |
| **Casio** | **QV-5000SX** | Program AE. Exp Correction 2EV to +2EV (in 1/2 EV Units) White Bal: Automatic, fixed | AUTO, ON, OFF, Red-eye Reduction | No | f/2.8 | 35mm | 1/8 to 1/500 second | 8MN internal flash memory plus CompactFlash ?? | Serial (Mac & PC cables) video-out (NTSC) | 9.1 oz |
| **Casio** | **QV-7000SX** | Automatic Manual -- 3 mode selectable | Automatic, Off Fill— Forced Red Eye Reduction | Movie mode | Auto/ Manual Selectable F2.8, F5.6 and F11 | 32mm to 64mm (optical zoom) | 1/4 second to 1/1,000th second | 8MB CompactFlash included holds 14 - 33 images | Serial (Mac & PC cables) video-out (NTSC) | |

| Mfr. | Model | Intro.[5] | Description | Resolution[7] | Viewfinder/ LCD[1] | Timer | Focus | Focus Range | Zoom[2] | Sensi-tivity[6] |
|---|---|---|---|---|---|---|---|---|---|---|
| **Epson** | **700** | 98 | Upgrade from PhotoPC 600. 4MB internal memory; supports CompactFlash. Digital zoom. Upload images as well as download. Can print to Epson printer, but utility takes up camera memory (it can be removed). Burst mode can capture 10 images at 2 fps at 640 x 480. Four AA Ni-MH batteries and compact charger incl. Print directly to Epson printers | 1280 x 960, 640 x 480 | Optical/2 inch LCD can be used as viewfinder | No | Auto and Macro | 4" to infinity | 2x digital (640 x 480) | 60/12 0 |
| **Epson** | **750Z** | Feb 99 | Can print directly to Epson printers. | 1600 x 1200 (interpolated) 1280 x 960, 640 x 480 1/2.7" CCD with 1.25 million pixels | Optical/2 inch solar assist LCD can be used as viewfinder | ?? | Auto and Macro | Normal: 32"-infinity (0.8m- inf) Macro: 8" to 32" (0.2 - 0.8m) | 3x optical/2 x digital | 90/18 0/360 |
| **Fuji** | **DS-7** | | No zoom or optical viewfinder. Camera also available as QuickTake 200. Can print directly to Fujifilm's Digital Video Printer NC-2. Time and date stamp images. | 640 x 480 350,000 CCD | 1.8 inch LCD | Yes | Three zone | Macro 3.5 to 5"; Close-up 1.5 - 3'; Distant 3' to infinity | No | 100 |
| **Fuji** | **DS-300** | Jul 97 | Body based on GA645 camera. Optical viewfinder, no LCD. Camcorder LCD can be used with video out. Can output directly to Fuji printers. Manual and auto focus. Manual focus: hold down a button and then rotate a knob to select focus. Images can be saved in monochrome. Optional Extension Unit allows continuous shooting at 4.5 fps. | 1,280 x 1,024 640 x 480 2/3inch CCD 1.4 million pixels JPEG (Fine, Normal, Basic) and TIFF | Optical | | Auto and Macro | 7.8" to infinity | 3x opt./ 2x dig. | 100/4 00 |

| Mfr. | Model | Exposure | Flash Modes | Burst[3] Mode | Lens Aperture | Focal[4] Length | Shutter Speed | Storage | Interface | Wt. |
|---|---|---|---|---|---|---|---|---|---|---|
| **Epson** | **700** | Programmed AE or +/- 2 step. White balance Auto/Fixed | Auto, Forced and Off Range: 8" to 8' (0.2m - 2.4m) | 2 fps up to 10 sec. | f/2.8, f/5.6, f/11 | 36mm | 1/4 to 1/500 seconds | 4MB internal. CompactFlash slot holds 6 – 42 images | Serial (Mac & PC) Video-out | |
| **Epson** | **750Z** | Programmed AE or +/- 2 step. White bal Auto/Fixed /User preset | Standard and slow synchro | 2 fps up to 16 sec. | f2.8 and f8 | 34mm to 102mm | 1/4 to 1/750 flash: 1/30 to 1/750 sec | 4MB internal. CompactFlash slot and 8MB card included | Serial (Mac & PC) Video-out | |
| **Fuji** | **DS-7** | Auto white balance | No | | 2 step: indoor f2.2 and outdoor f8 | 38mm | 1/4 - 1/5000 | 2MB SmartMedia | Serial (Windows 3.1+) Video-out | 5.6oz |
| **Fuji** | **DS-300** | Programmed AE, Aperture-Priority AE, Shutter Speed-Priority AE & Manual 0.3 EV steps (-0.9+1.8) White bal fixed/auto | Auto flash plus hot shoe | 4.5 fps with add on | F3.5, 5.6, 8 and 11 | 35-105mm | º - 1/1000 | ATA Type I or II PCMCIA SmartMedia or CompactFlash with PC Adapter | Serial Video SCSI (with extra adapter) | 620g/ 21.9oz |

| Mfr. | Model | Intro.[5] | Description | Resolution[7] | Viewfinder/ LCD[1] | Timer | Focus | Focus Range | Zoom[2] | Sensi-tivity[6] |
|---|---|---|---|---|---|---|---|---|---|---|
| **Fuji** | **DS-330** | Jun 98 | Upgrade to DS-330 Camcorder style LCD viewfinder available as accessory. Extension Unit increases burst-mode to 4.5 fps for up to 12 frames. CCD improved slightly and cost was lowered. | 1,280 x 1,024<br>640 x 480<br>2/3inch CCD<br>1.4 million pixels<br>JPEG (Fine, Normal, Basic) and TIFF | Optical | | Auto and Macro | 7.8" to infinity | 3x opt./ 2x dig. | 100/400 |
| **Fuji** | **DS-560/565** | 98 | SLR (Nikon) body. Accepts Nikon F-Mount lenses. Condenser optics system reduces light from lens to CCD (so the viewfinder matches what's captured). 560 burst mode 1 fps. 565 3 fps for 12 frames. The 565 adds burst rate up to 3 fps. | 1,280 x 1,000 | Optical viewfinder with approximately 98% frame coverage | | Auto / Manual | NA | NA | 800, 1600, 3200 |
| **Fuji** | **DX-5** | Jun 97 (D) | Same as DS-10. Pop-up flash. Optical viewfinder only. Manual light-level selection (indoor and outdoor) can cause problems if you forget to change setting (and lack of LCD means you won't notice). No macro or video out. | 640 x 480<br>1/3" CCD with 350,000<br>JPEG | optical | No | Fixed | 28" to infinity | No | 150 (200 with flash) |
| **Fuji** | **DX-7** | (D) | Comparatively large unit with optical viewfinder and LCD. Slider to select between two apertures. Fixed focus lens with macro. Takes more than 5 seconds to recycle. Acceptable image quality. | 640 x 480<br>1/3" CCD with 350,000<br>JPEG | Optical/ 1.8" LCD monitor | Yes | Fixed and macro | 8" to infinity, Macro 3-8 in at f8 | No | 150 (200 w/ flash) |

| Mfr. | Model | Exposure | Flash Modes | Burst[3] Mode | Lens Aperture | Focal[4] Length | Shutter Speed | Storage | Interface | Wt. |
|---|---|---|---|---|---|---|---|---|---|---|
| **Fuji** | **DS-330** | Programmed AE, Aperture-Priority AE, Shutter Speed-Priority AE & Manual 0.3 EV steps (-0.9- + 1.8 EV). White bal: fixed/auto | Auto flash plus hot shoe | 4.5 fps with add on | F3.5, 5.6, 8 and 11 | 35-105mm | º - 1/1000 | ATA Type I or II PCMCIA SmartMedia or CompactFlash with PC Adapter | Serial Video SCSI (w th extra adapter) | 620g/ 21.9oz |
| **Fuji** | **DS-560/565** | Programmed AE, Aperture Priority AE, Shutter Priority AE, and Manual Modes | | 560: 1 fps 565:3 fps for 12 frames | 4.8 - 38 (set by condenser optics system) | NA | Auto/Man1 /2000 - 1/8sec. Synchro-flash: 1/250sec. max. | ATA compliant Type I or Type II PC Card PCMCIA | SCSI NTSC / PAL | 4.08 lb |
| **Fuji** | **DX-5** | TTL Apterture-Priority AE | Auto (range 3.6m) | No | F3.1/8 | 38 | º - 1/5,000 | 2MB SmartMedia holds 22 – 30 images | Serial (Mac & Windows) | 160 g / 5.6 oz |
| **Fuji** | **DX-7** | TTL Apterture-Priority AE | Auto Red-Ey, Fill-in and Off (range 3.1m) | | F3.1/8 | 38 | º - 1/5,000 | 2MB SmartMedia holds 22 – 30 images | Serial (Mac & Windows) Video out (U model NTSC, E model PAL) | 250 g / 8.8 oz |

| Mfr. | Model | Intro.[5] | Description | Resolution[7] | Viewfinder/ LCD [1] | Timer | Focus | Focus Range | Zoom [2] | Sensi-tivity [6] |
|---|---|---|---|---|---|---|---|---|---|---|
| **Fuji** | **DX-9** | Jan 98 | Includes manual focus as well as automatic. Accepts both 5 and 3.3-volt SmartMedia cards. Images not rated as well as competing cameras in some reviews. | 640 x 480 1/4 inch CCD with 350,000 JPEG | Optical/ 2" LCD monitor | Yes | Auto | 20cm/8in. to infinity | 3x | 100 |
| **Fuji** | **MX-500** | Sep 98 | High resolution, though images may not match those produced by other cameras in same resolution. Digital zoom. Very long delay between taking an image. | 1,280 x 1,024 640 x 480 JPEG (Fine, Normal, Basic) ½" CCD with 1.5 million pixels | Optical/ 1.8" LCD monitor | | Fixed and macro | Macro: 3.5" | No | 100 |
| **Fuji** | **MX-600** | Mar 99 | Same CCD as MX-700 but with 35- to 105-mm optical lens. Soft Focus Filter and Cross Filter to add special effects. Same as the Toshiba PDR-M3. Uses Lithium-ion battery and includes charger. | 1,280 x 1,024 640 x 480 JPEG (Fine, Normal, Basic). ½" CCD with 1.5 million pixels | Optical/ 1.8 inch LCD monitor | Approx. 10 sec | Auto | Normal 90 cm/3' to infinity; Macro 25 cm/9.8" to 90 cm/3' | 3x | 100 |
| **Fuji** | **MX-700** | | Compact autofocus, autoflash with digital zoom. Takes about 5 seconds to store images, but multi-record mode saves 4, 9 or 16 images within 2 seconds. Lithium-ion battery. | 1,280 x 1,024 640 x 480 JPEG (Fine, Normal, Basic). ½ inch CCD with 1.5million pixels | Optical/2 inch LCD monitor | Yes | Auto | | 2x digital | 100 |
| **Fuji** | **MX-2700** | Feb 99 | 3:2 aspect ratio. Pocket-sized with digital telephoto. Styling resembles MX-700. Can print direct to Fujifilm MX-70 printer. Includes Lithium-ion battery and charger. | 1800 x 1200 640 x 480 JPEG (Fine, Normal, Basic). CCD 2.3 million pixels | Optical/ 2 inch polysilicon LCD with 130,000 pixels | Yes | Auto | 9 cm (3.5") to infinity | 2.5x digital | 120 |

| Mfr. | Model | Exposure | Flash Modes | Burst[3] Mode | Lens Aperture | Focal[4] Length | Shutter Speed | Storage | Interface | Wt. |
|---|---|---|---|---|---|---|---|---|---|---|
| **Fuji** | **DX-9** | TTL Aperture-Priority AE | Auto, Red-Eye Red, Fill-in flash, Off Range: 2.4m/7.9ft | Up to 8 images per second | f2.4 to f3.3 | 32mm to 96mm | º - 1/5,000 | 2MB SmartMedia | Serial (Mac & Windows) video | 10oz / 290 g |
| **Fuji** | **MX-500** | Programmed AE | Auto flash, 8.2 ft/2.5m range | No | f3.2 | 35mm | º - 1/1,000 | 4MB SmartMedia | Serial (Mac & Windows) video | 8.6 oz / 245 g |
| **Fuji** | **MX-600** | Program AE/Aperture-Priority AE White bal: fix & 4 man. modes | Auto Flash (range 3m/9.8 ft) | | F3.8/F7.6(wide) to F5.5/F11(tele) | 35mm-105mm | 1/4-1/1,000 sec | 4MB SmartMedia | Serial (Mac & Windows) video | 330 g/11.6 oz |
| **Fuji** | **MX-700** | Programmed AE White balance: fix and 4 manual modes | Auto (range 2.5m) | No 5 sec per image | F3.2/8 | 35mm | º-1/1000 | SmartMedia 2MB card Fine:2, Normal 5, Basic 11 images | Serial video | 245g / 8.6 oz |
| **Fuji** | **MX-2700** | White balance, exposure control, flash intensity control | Auto (range of approx. 2.5 m/8.2 ft.) | | 2 steps (É3.2/8) | 35mm | 1/4 -1/1000 | SmartMedia 8MB card Fine:8, Normal 17, Basic 35 images | Serial (Mac/PC) video | 240 g / 8.5 oz |

| Mfr. | Model | Intro.[5] | Description | Resolution[7] | Viewfinder/ LCD [1] | Timer | Focus | Focus Range | Zoom [2] | Sensi-tivity [6] |
|---|---|---|---|---|---|---|---|---|---|---|
| **Fuji** | **MX-2900** | Aug 99 | aka FinePix 2900Z. Int flash plus hot shoe mounting for ext flash. Optional 43mm filter and 28mm wide-angle attachments. Manual adjust shutter and aperture settings, as well as white balance, exposure compensation, manual exposure, flash brightness adjustment, light metering mode and continuous shooting. Uncompressed TIFF format plus JPEG. Lithium-ion battery and charger. | 1800 x 1200<br>640 x 480<br>JPEG (Fine, Normal, Basic)<br>CCD 2.3 million pixels | Optical (with diopter)/<br>2" polysilicon LCD with 130,000 pixels | | | | 3x optical and 2.5x digital | |
| **HP** | **Photo-Smart Digital Camera** | (D) | Familiar point-and-shoot body design. No LCD display. Software is Windows TWAIN driver. Recycle time is 2 seconds. Similar or same as Konica Q-ez? | 640 x 480<br>Normal, Fine, Superfine | optical | 10 sec | Auto and macro | 2.5 to infinity | No | 100/200 |
| **HP** | **Photo-Smart C-20** | Feb 98 | Plastic body w/ rubber grip. Autofocus may have trouble in low light, but can be turned off; however will then only work with objects outside macro range. Lens cover acts as power switch. Metal tripod socket. Optical and LCD viewfinders. LCD can be used as view-finder, shows 100% of captured image. LCD frame rate about 5 fps. Focus and exposure lock reduces shutter release delay. Exposure compensation with flash. Incl AC adapter | 1152 x 872<br>576 x 436<br>Normal, Fine, SuperFine<br>1/3" CCD | Optical/<br>Color 1.8" LCD | 10 sec | Auto focus, plus macro | 2' to infinity - Macro focus from 9.5" to 2' | No | 100 |

| Mfr. | Model | Exposure | Flash Modes | Burst[3] Mode | Lens Aperture | Focal[4] Length | Shutter Speed | Storage | Interface | Wt. |
|---|---|---|---|---|---|---|---|---|---|---|
| **Fuji** | **MX-2900** | Average, Spot and Multi White balance. | | | | 35-105mm | 3s – 1/500 | SmartMedia 8MB card Fine:8, Normal 17, Basic 35 images | Serial (Mac/PC) USB Video | 12 oz |
| **HP** | **Photo-Smart Digital Camera** | Auto | Auto, On, OFF | | F2.8 | 43mm | | 2MB Card holds 4 – 32 images | Serial (PC) | .55kg / 1.4lb |
| **HP** | **Photo-Smart C-20** | Automatic +/- 2EV, 0.5 EV steps | On, Off, Red-eye, Fill | | f2.8, 5.6, 11 | 39 mm | 1/500 to 2 seconds | 4 MB CompactFlash holds 8 – 40 images | Serial (PC) video | .68 lbs |

| Mfr. | Model | Intro.[5] | Description | Resolution[7] | Viewfinder/ LCD[1] | Timer | Focus | Focus Range | Zoom[2] | Sensi-tivity[6] |
|---|---|---|---|---|---|---|---|---|---|---|
| **HP** | **Photo-Smart C-30** | Oct 98 | C20 with digital zoom and some tweaks. Low price. LCD has low refresh rate and will drain batteries. Video output is switchable. Includes AC adapter. No camera cover included. | 1152 x 872 Normal, Fine, SuperFine | Optical/ Color 1.8-inch LCD | | Auto focus, plus macro | 2' to infinity - Macro focus from 9.5" to 2' | 2x digital | 100 |
| **Kodak** | **DC20** | 96 (D) | Very small & light, but very low-res images. Simple; no focus or exposure adjustments or flash. Image quality mediocre and 493 x 373 is highest resolution. Max 16 image capture. Considerable delay between shutter press & image capture. Switching image resolution can only be done connected to computer. | 493 x 373, 320 x 240 at 24 bits | Rangefinder | No | Focus-free | 0.5m to infinity | No | 800/1 600 |
| **Kodak** | **DC25** | 96 (D) | Adds built-in flash, LCD for review of images & more memory than DC20. Can switch resolution; accepts flash memory card. | 493 x 373, 320 x 240 at 24 bits | 1.6" LCD | 10 sec | Focus-free | 0.5m to infinity | No | 800/1 600 |
| **Kodak** | **DC40** | 95 (D) | One of the earliest low-cost digital cameras (after the QuickTake). Kodak helped Apple develop QuickTake. DC40 & QuickTake 150 are similar in features, though DC40 resolu-tion is higher. Limited functionality, no LCD display. Image quality marginally better than QuickTake, but still not great. Flash tends to wash out close subjects and backgrounds too dark (a problem with many early digitals). | 756 x 504 | optical | | Focus-free | 4 ft to infinity | No | 85 |

| Mfr. | Model | Exposure | Flash Modes | Burst[3] Mode | Lens Aperture | Focal[4] Length | Shutter Speed | Storage | Interface | Wt. |
|---|---|---|---|---|---|---|---|---|---|---|
| **HP** | **Photo-Smart C-30** | automatic | On, Off, Red-eye, Fill | | f2.8, 5.6, 11 | Unknown | 1/500 to 2 seconds | 4 (later models 8) MB CompactFlash holds 8 - 40 images | Serial (PC) Video (NTSC & PAL switchable) | 0.7 lb |
| **Kodak** | **DC20** | automatic | No | No | f/4 to f/11 | 47 mm | 1/30 to 1/4000 | 1 MB internal holds 8 - 16 image | Serial (Mac & PC) | 4.2 oz/133 g |
| **Kodak** | **DC25** | automatic | Auto, fill, off | No | F/4 to f/11 | 47 mm | 1/30 to 1/4000 | 2 MB internal. Holds 14 - 29 images. Supports CompactFlash cards | Serial (Mac & PC) | 9.5 oz/ 270g |
| **Kodak** | **DC40** | Auto plus +/- 1 f-stop override | Auto, On, off. (8ft range.) | No | F/2.8 to f.16 | 42mm | 1/30 to 1/175 | 1 MB internal holds 48 - 99 images | Serial (Mac & PC) | 1 lb (0.5 kg) |

| Mfr. | Model | Intro.[5] | Description | Resolution[7] | Viewfinder/ LCD[1] | Timer | Focus | Focus Range | Zoom[2] | Sensi-tivity[6] |
|---|---|---|---|---|---|---|---|---|---|---|
| **Kodak** | **DC50** | 96 (D) | Similar to DC40 in features and image performance, but with a zoom lens. Optical range-finder zooms with lens. As one of the first zoomable low-priced digitals, it was great at the time, but image quality is not as good as newer models. Also available, with slightly different features, as the Chinon ES-3000 and Dycam 10-C. | 756 x 504 compression: good, better, best | optical | 10 sec | Focus-free | 29" to infinity. Close-up mode: 19 in | 3x optical | 84 |
| **Kodak** | **DC120** | Apr 97 | Similar body to DC50, with similar zoom lens, but camera itself very different. Much higher-res CCD. Image size is larger than resolution of the CCD (images enlarged by software). Images look best when scaled down to 640 x 480; at highest resolution you see effects caused by interpolation and CCD quality. | 1280 x 960 CCD resolution: 850 x 984 | Optical 1.6" LCD for review only. | 10 sec | Auto | Normal: 19" to infinity Macro 7.9" – 19" | 3x optical | 160 |
| **Kodak** | **DC200 & DC200 plus** | Feb98/ Mar 99 | Similar to, and released after the DC210 (see below) but without the zoom lens. DC200 plus adds: Faster operation, 20% increase in battery performance, Auto Exposure lock, Date stamp, Template options, User interface supports multiple languages | 1152 x 864 640 x 480 Image compression: Good, Better, Best CCD: 1160 x 872 | Optical plus 1.8" LCD for review | 10 sec | Focus-free | 27" to infinity | No | 140 |

| Mfr. | Model | Exposure | Flash Modes | Burst[3] Mode | Lens Aperture | Focal[4] Length | Shutter Speed | Storage | Interface | Wt. |
|---|---|---|---|---|---|---|---|---|---|---|
| **Kodak** | **DC50** | Auto plus +/- 1 f-stop override | Auto, on, off (range 14ft) | No | F/2.5 to f/24 | 38mm to 114mm | 1/16 to 1/500 | 1 MB internal holds 7 - 22 images. Supports PCMCIA storage cards | Serial (Mac & PC) | 22 oz |
| **Kodak** | **DC120** | Auto or manual with auto white balance | Auto, fill, off. Optional flash sync cable for external flash | No | Wide: f/2.5 – f/16 Tele: f/3.8 – f/24 | 38-114 | 1/500 – 16 seconds | 2 MB internal (20 pictures) CompactFlash slot. | Serial (Mac & PC) | 1.15 lb (520 g) |
| **Kodak** | **DC200 & DC200plus** | Auto, manual override (+/-2 EV) auto white balance | Auto, fill, red-eye, fill/red eye, off Range up to 9.8' | | f/4.0 to f/13.5 | 39mm | ½ to 1/362 | 4MB CompactFlash card (13 to 60 pictures) | Serial (PC) Macintosh connection kit available separately | .75lb (340g) |

| Mfr. | Model | Intro.[5] | Description | Resolution[7] | Viewfinder/ LCD[1] | Timer | Focus | Focus Range | Zoom[2] | Sensi-tivity[6] |
|---|---|---|---|---|---|---|---|---|---|---|
| **Kodak** | **DC210 & DC210 plus** | Oct 97/ Aug 98 | Comparatively large size, noticeable parallax on close images (use LCD for close shots). Lens has noticeable barrel distortion. Max shutter speed comparatively slow. Image quality and color not as accurate as compar-able cameras. DC210 plus is upgrade. | 1152 x 864<br>640 x 480<br>CCD: 1174 x 884 | Optical plus 1.8" TFT LCD for review | 10 sec | Focus-free | 19.8" to infinity: close-up 8" | 2x | 140 |
| **Kodak** | **DC220** | Jun 98 | Same body and similar features to DC260, but with lower resolution and shorter zoom lens. Does includes Digita operating system and USB. | 1152 x 864<br>640 x 480<br>Image compression: Good, Better, Best<br>CCD: 1174 x 884 | Optical/2" TFT LCD for review | 10 sec/ time lapse | Focus-free/ macro | Tele: 39" to infinity<br>Wide: 19.8"<br>Macro: 8" | 2x/2x digital | 140 |
| **Kodak** | **DC240** | Mar 99 | Falls between "low-end" 1 megapixel DC200 etc, and 1.6 megapixel DC265, the DC240 offers 1.31 megapixel sensor. 3x optical zoom, macro to 9.75 inches. Ni-MH batteries and charger. Sepia and B&W modes. | 1280 x 960<br>640 x 480<br>Image compression: Good, Better, Best<br>CCD: 1344 x 971 | Optical/1.8" TFT LCD for review | 10 sec | Auto focus | 19.6" to infinity<br>Macro 9.8" | 3x zoom/ 2x digital | 140 |
| **Kodak** | **DC260 & DC265** | Jun 98/ Mar 99 | DC260 provides burst mode, high res, USB and Digita OS. Good ergonomics & software. Image quality & color accuracy very good. A great camera, though others now offer higher res. DC265 is upgrade to DC260; has faster start-up and processing. Faster preview rate. 1.6 million (1536 x 1024) pixel tripled multi-picture burst capability. Longer battery life. | 1536 x 1024<br>1152 x 768<br>768 x 512<br>Image compression: Good, Better, Best<br>CCD: 1548 x 1032 | Optical/2" TFT LCD for review | 10 sec | Auto focus | 12" to infinity | 3x zoom/ 2x digital | 100 |

| Mfr. | Model | Exposure | Flash Modes | Burst[3] Mode | Lens Aperture | Focal[4] Length | Shutter Speed | Storage | Interface | Wt. |
|---|---|---|---|---|---|---|---|---|---|---|
| **Kodak** | **DC210 & DC210plus** | Auto, manual override (+/-2 EV) auto white balance | Auto, fill, red-eye, fill/red eye, off Range up to 9.8' | | Wide: f/4.0 to f/13.5 Tele: f/4.7 to f/16.0 | 29 to 58mm | ½ to 1/362 | 8MB CompactFlash card (up to 120 images) | Serial (PC) Macintosh connection kit available separately | .75lb (340g) |
| **Kodak** | **DC220** | Auto, manual override (+/-2 EV) auto white balance | Auto, red-eye, fill, off Range up to 9.8' | | Wide:f/4.0 - f/13.5 Tele: f/4.7- f/16 | 29 to 58mm | ½ to 1/360 | 8MB CompactFlash card (up to 120 images) | Serial, USB, IrDA (PC) USB (Macintosh) video | 1.2lb (525g) |
| **Kodak** | **DC240** | Auto, manual override (+/-2 EV) auto white balance | Auto, red-eye, fill, off Range up to 9.8' | | Wide:f/2.8 to f/16 Tele: f/4.5 to f/25.7 | 39 to 117mm | ½ to 1/755 | 8MB CompactFlash card (21 to 56 pictures) | Serial, USB, (PC) USB (Macintosh) video | .7lb (328 g) |
| **Kodak** | **DC260 & DC265** | Auto, manual override (+/-2 EV) auto white balance | Auto, red-eye, fill, off Range up to 9.8' Optional sync cable for external flash | 2-8 pictures (0.1 to 3 fps) | Wide: f/3.0 to f/14.0 Tele:f/4.7 to f/22 | 38mm to 115mm | º to 1/400 | 8MB CompactFlash card (up to 95 pictures) | Serial, USB, IrDA (PC) USB (Macintosh) video | 1.2lb (525g) |

| Mfr. | Model | Intro.[5] | Description | Resolution[7] | Viewfinder/ LCD [1] | Timer | Focus | Focus Range | Zoom [2] | Sensi-tivity [6] |
|---|---|---|---|---|---|---|---|---|---|---|
| **Kodak** | **DCS 315** | Jun 98 | Professional Product: based on Nikon Pronea 6I SLR body. Use standard Nikon lenses, but multiply by focal length multiplier to find comparable focal length. | 1520 x 1008 10/bits color | Optical/LCD | Yes | Auto/ manual | NA | NA | 100-400 |
| **Kodak** | **DCS 420** | 94 | Professional Product: based on Nikon N90s. Use standard Nikon lenses, but multiply by focal length multiplier to find comparable focal length. | 1524 x 1012 12/bits color | Optical | Yes | Auto/ manual | NA | NA | 100-400 |
| **Kodak** | **DCS 460** | 94 | Professional Product: based on Nikon N90s. Use standard Nikon lenses, but multiply by focal length multiplier to find comparable focal length. | 3060 x 2036 12/bits color | Optical | Yes | Auto/ manual | NA | NA | 80 |
| **Kodak** | **DCS 520** | May 98 | Professional Product: based on Canon EOS-IN. Use standard Nikon lenses, but multiply by focal length multiplier to find comparable focal length. | 1728 x 1152 12/bits color | Optical/LCD | Yes | Auto/ manual | NA | NA | 200-1600 |
| **Kodak** | **DCS 560** | Sep 98 | Professional Product: based on Canon EOS-IN. Use standard Nikon lenses, but multiply by focal length multiplier to find comparable focal length. | 3040 x 2008 12/bits color | Optical/LCD | Yes | Auto/ manual | NA | NA | 80-200 |
| **Kodak** | **DCS 620** | Feb 99 | Professional Product: based on Nikon F5. Use standard Nikon lenses, but multiply by focal length multiplier to find comparable focal length. | 1728 x 1152 12/bits color | Optical/LCD | Yes | Auto/ manual | NA | NA | 200-1600 |

| Mfr. | Model | Exposure | Flash Modes | Burst[3] Mode | Lens Aperture | Focal[4] Length | Shutter Speed | Storage | Interface | Wt. |
|---|---|---|---|---|---|---|---|---|---|---|
| **Kodak** | **DCS 315** | White balance | Pop-up flash and hot shoe | 2 fps max 3 frames | NA | 2.6 focal length mult. | 30 sec - 1/4000 | Type II, III (300 images) | IEEE 1394/ PC Card | 1.77 kg |
| **Kodak** | **DCS 420** | No White balance | | 2 fps max 5 frames | NA | 2.6 focal length mult. | | Type III (346 images) | SCSI/PC Card | 1.7 kg |
| **Kodak** | **DCS 460** | No White balance | | 1 fps max 2 frames/8 sec | NA | 1.3 focal length mult. | | Type III (85 images) | SCSI/PC Card | 1.7 kg |
| **Kodak** | **DCS 520** | White balance | | 3.5 fps/ 12 images | NA | 1.6 focal length mult. | | Type II, III (250 images) | IEEE 1394/ PC Card | 1.7 kg |
| **Kodak** | **DCS 560** | White balance | | 1 fps/ 3 frames | NA | 1.3 focal length mult. | | Type II, III (150 images) | IEEE 1394/ PC Card | 1.7 kg |
| **Kodak** | **DCS620** | White balance | | 3.5 fps/ 12 frames | NA | 1.6 focal length mult. | | Type II, III (250 images) | IEEE 1394/ PC Card | 1.58 kg |

| Mfr. | Model | Intro.[5] | Description | Resolution[7] | Viewfinder/ LCD[1] | Timer | Focus | Focus Range | Zoom[2] | Sensi-tivity[6] |
|---|---|---|---|---|---|---|---|---|---|---|
| **Konica** | **Q-mini** | 97 | Small and compact, CompactFlash, LCD and video out. No optical viewfinder. Includes Ni-Cd batteries and charger. Manual white balance. Good color images when used outside. | 640 x 480 | 1.8" LCD (312 x 230) | Yes | Focus free | Macro to 1.4" | | 120 |
| **Konica** | **Q-M100** | 97 | Available in gold and silver. Compact size. Integrated lens cover acts as power switch. Optical and LCD viewfinder which is bright, though frame rate only 2 fps. Digital zoom. Good image color. Can capture a sequence of four images that is saved as one frame. | 1,152 x 872<br>576 x 436<br>(Economy, Fine and SuperFine) | Optical/1.8" LCD | Yes | Fixed | 0.2m to infinity | 2x digital | 100 |
| **Konica** | **Q-M200** | Jun 99 | Small point-and-shoot with high-resolution. | 1600 x 1200<br>1152 x 872<br>640 x 480<br>2.11million pix CCD | Optical/1.8" LCD | Yes | Auto | 4" to infinity | 2x digital | |
| **Minolta** | **Dimâge V** | 97 (D) | Zoom lens contained in a tube-like attach-ment, detachable from the body of the camera using 3-foot cable. Flash on camera. LCD viewfinder (no optical) poor in strong light. Good auto white balance. No manual setting, but has exposure compensation. | 640 x 480 | 1.8 inches LCD | | Pan focus | | 2.5x optical | 100 |
| **Minolta** | **Dimâge EX1500** | 98 | Available as Zoom 1500, Wide 1500 and 3D 1500 which has same camera body but different lens module. Has detachable lens assembly. | 1344 x 1008<br>640 x 480<br>CCD 1,500,000 pixels | Optical/ 2" TFT LCD | 10 sec | External passive AF system | | Zoom: 3x optical | 125 |

| Mfr. | Model | Exposure | Flash Modes | Burst[3] Mode | Lens Aperture | Focal[4] Length | Shutter Speed | Storage | Interface | Wt. |
|---|---|---|---|---|---|---|---|---|---|---|
| **Konica** | **Q-mini** | Auto white balance/ Exposure compensation | Light adjustable, red & white reduction and auto | No | f/2.8 | 43mm | 1/4-1/2,000 | 2MB CompactFlash (up to 47 images) | Serial (Mac & PC) video | 10.5 oz/ 290g |
| **Konica** | **Q-M100** | Auto white balance/ Exposure compensation | Light adjustable, red & white reduction and auto | Speed-shot 4 images | f/2.8 | 39mm | 1/8 to 1/500 | 4MB CompactFlash 10 to 50 images | Serial (Mac & PC) video | 290g |
| **Konica** | **Q-M200** | White balance | Auto, red eye | 4 frame capt. mode | f/3.2/f6.9 | 38mm to 76mm | 1/8 to 1/1000 | CompactFlash | Serial (Mac & PC) video | 9.5 oz |
| **Minolta** | **Dimâge V** | Auto with Manual compensation in +/-5 increments | Automatic flash, Daylight flash, non-flash option | | f/5-5.6 | 34-92 mm | 1/30 to 1/10,000 | 4MB CompactFlash | Serial (Mac & PC) | |
| **Minolta** | **Dimâge EX1500** | Programmed AE. +/- 2 EV in 1/3 incr | Auto, red-eye auto, fill, red-eye fill, off | 3.5 fps Timelapse: 1-60 sec | Zoom: f/3.5-5.6 Wide: f/1.9 | Zoom:38 - 115mm Wide: 28mm | 2 sec – 1/4000 | 4MB CompactFlash | Serial (Mac & PC) IrDA video | 11 oz (310 g) |

| Mfr. | Model | Intro.[5] | Description | Resolution[7] | Viewfinder/ LCD [1] | Timer | Focus | Focus Range | Zoom [2] | Sensi-tivity [6] |
|---|---|---|---|---|---|---|---|---|---|---|
| **Minolta** | **RD-175** | | SLR body, accepts Minolta A-type lenses | 1528 x 1146 | Eye-level fixed roof mirror. Field of view: 90% | | Auto/ manual | NA | NA | 800 |
| **Nikon** | **Cool Pix 700** | 99 | Similar in specifications to CoolPix 950, but with a digital zoom rather than optical. | 1,600 x 1,200 640 x 480 Saves files in TIFF and JPEG format: Hi = TIFF, Fine 1:4 JPEG, Normal 1:8 JPEG, Basic 1:16 JPEG ½ CCD: 2.11 million pixels | Optical/1.8" 112,000 pixel TFT LCD Adjustable back light. Update rate: 30 fps | 10 and 3 sec | Continu ous AF, Single AF mode, Manual | 12 inches to infinity. Macro: 3.5 inches to infinity | 4 step digital: 1.25x/1.6 x/2x/2.5x | 80 |
| **Nikon** | **Cool Pix 900 & 900s** | 98 | Extremely popular camera which offered high resolution at a reasonable price. Image quality very good, though some early models had a weakness in the swivel assembly that splits the body in half. 900s is updated version of 900 (and included a lens cover!). | 1,280 x 960 640 x 480 Fine 1:4 JPEG, Normal 1:8 JPEG, Basic 1:16 JPEG CCD: 1.3 million pixels | Optical/2" 114,000 dot TFT LCD. Update 30 fps. | 10 sec | Continu ous AF, Single AF mode | 20 in – infinity Macro: 3.1 – 20 in | 3x optical | 64 |

| Mfr. | Model | Exposure | Flash Modes | Burst[3] Mode | Lens Aperture | Focal[4] Length | Shutter Speed | Storage | Interface | Wt. |
|---|---|---|---|---|---|---|---|---|---|---|
| **Minolta** | **RD-175** | TTL phase detection. Pro-grammed AE, Aperture-prior-ity, shutter-prior-ity. 3 EV (.5 EV incr) | Red-eye reduction, range 39ft. | | NA | NA | ½ to 1/2000 | PCMCIA Type III | SCSI, PCMCIA Type III Mac & PC software | 2.4 lbs |
| **Nikon** | **CoolPix 700** | Programmed AE, Manual exp compensation (+/- 2EV in 1/3 EV steps.) Auto white bal, manual & preset. | Auto, off, force-fill, slow-synch and red-eye reduction | 1.5 fps VGA mode 2 fps | 3 steps: f/2.6, f/3.8, f/8.0 | 38mm | 1 sec to 1/750 | 8MB CompactFlash | Serial (Mac & PC) Video out | 9.5 oz |
| **Nikon** | **CoolPix 900 & 900s** | 3-mode TTL metering; Matrix, Spot and Center-weighted. Auto white bal: man & preset | Auto, flash cancel, anytime, slow sync, red-eye. Range: 9.8' | 2 fps in VGA | f/2.4 – f/6.6 | 38-115mm | º to 1/750 | 4MB CompactFlash | Serial (Mac & PC) Video out | 12.7 oz 360 g |

| Mfr. | Model | Intro.[5] | Description | Resolution[7] | Viewfinder/ LCD[1] | Timer | Focus | Focus Range | Zoom[2] | Sensi-tivity[6] |
|---|---|---|---|---|---|---|---|---|---|---|
| **Nikon** | **Cool Pix 950** | 99 | Same form factor as the 900, but with higher resolution CCD and other improvements. Body more substantial and strengthening in the swivel mechanism. Reasonably small body (though not the smallest). Lens is at maximum zoom when turned on. Offers mixture of manual and automatic features, though some find the user interface a little complicated. | 1,600 x 1,200<br>1,024 x 768<br>640 x 480<br>Hi = TIFF, Fine 1:4 JPEG, Normal 1:8 JPEG, Basic 1:16 JPEG<br>CCD: ½inch 2.11 million pixels | Optical/2 inch 130,000 TFT LCD. Update 30 fps | 10 sec and 3 sec | Contin. AF, Sngl AF mode, Manual Focus; 10 steps 4" to infinity | 12 inches to infinity Macro: 0.8 inches to infinity | 3x optical 4 step digital: 1.25x/1.6 x/2x/2.5x | 80 |
| **Nikon** | **D1** | | Professional Digital Camera. Compatible with virtually any F-mount Nikkor /lens (Picture angle with D1 is equivalent to 1.5x focal length in 35mm [135] format). | 2000x1312<br>2.74 million pixel CCD | Optical viewfinder/ 2-in., 120,000-dot low-temperature polysilicon TFT LCD | 2-20 sec | NA | NA | NA | 200, 400, 800 and 1,600 |
| **Olympus** | **D-220L** | 98 | All glass lens. Capable of printing directly to Olympus Photo Printer. | 640 x 480<br>2 levels of compression | Optical/2-inch LCD | 12 sec | auto | 29.5" (0.75m) to infinity Macro: 7.9" (20cm) - 29.5" (0.75m) | No | 130 |
| **Olympus** | **D-320L** | 97 (D) | Similar to 220L but with higher resolution. Recycle time can be many seconds. Capable of printing directly to Olympus Photo Printer. | 1024 x 768<br>640 x 480<br>CCD: 1/3 inch 810,000 pixels | Optical/2" LCD | 12 sec | auto | 19 inch to infinity macro: 7.9" to 27" | No | 100 |

| Mfr. | Model | Exposure | Flash Modes | Burst[3] Mode | Lens Aperture | Focal[4] Length | Shutter Speed | Storage | Interface | Wt. |
|---|---|---|---|---|---|---|---|---|---|---|
| **Nikon** | **CoolPix 950** | Programmed AE, Manual exposure compensation (+/- 2EV in 1/3 EV steps.) Auto white balance, manual & preeset | Auto, off, force-fill, slow-synch and red-eye reduction Range: 30' | 1 fps (2 fps for VGA) | 3 steps: f/2.6, f/4, f/7.0 | 38 to 115mm | 8 sec to 1/750 | 8MB CompactFlash | Serial (Mac & PC) Video out | 12.3 oz |
| **Nikon** | **D1** | 3D Color Matrix, Center-Weighted, Spot Metering. Programmed, Shutter-Priority & Aperture-Priority | Works with Nikon Flashes. | 4.5 fps up to 21 images | NA | 1.5x focal length in 35mm | 30 to 1/16,000 sec. and Bulb | CompactFlash | Firewire video | |
| **Olympus** | **D-220L** | Auto with manual override +/- 1 full f-stop | 4-mode flash w/ redeye reduction, force-fill flash, auto low-light and back-light mode | 9 shots in one second (captured to one frame) | 2.8 | 36mm | 1/4 to 1/10,000 | 2 MB SmartMedia | Serial (Mac & PC) Video out | 9.2 oz / 261g |
| **Olympus** | **D-320L** | Auto with manual override +/- 1 full f-stop | Auto, red-eye reduction, fill-in, Off | 9 shots in 1sec capt to 1frame | F/2.8 - 11 | 35mm | | 2 MB SmartMedia | Serial (Mac & PC) Video out | |

| Mfr. | Model | Intro.[5] | Description | Resolution[7] | Viewfinder/ LCD [1] | Timer | Focus | Focus Range | Zoom [2] | Sensi-tivity [6] |
|---|---|---|---|---|---|---|---|---|---|---|
| **Olympus** | **D-340L D-340R** | 98 | Further upgraded camera similar to D-320 and D0220 but with even higher resolution. Capable of printing directly to Olympus Photo Printer. Calendar works up to year 2030! | 1280 x 960 640 x 480 CCD: 1,310,000 pixels | 320L: Optical/2 inch LCD 320R: Optical/1.8 inch LCD | 12 sec | auto | 20 inch to infinity. Macro: 4 inches to 20 inches | Digital Tele 2x | 320L: 60 320R: 100/2 00 |
| **Olympus** | **D-400Z** | 98 | (C-900 in Europe) Plastic body, 3X zoom. Dioptic viewfinder. Autofocus lens. Can store uncompressed images in TIFF format. Panorama mode helps take sequence of over-lapping images. Spot metering. High quality images. Saving to TIFF takes 30 seconds. Can print directly to Olympus Photo Printer. | 1280 x 960 640 x 480 CCD: 1/3 inch 1,310,000 pixel | Optical with dioptic correction/ 1.8 inch TFT LCD with approx 72,000 pixels | 12 sec | Auto | 8” to infinity | 3x | 100 |
| **Olympus** | **D-500L** | Oct 97 | D-500L and D-600L essentially the same, but 600L has higher-res CCD. SLR design, very lightweight for size. Similar to Olympus IS-series film camera. Very good optics. LCD for review only. D600L focal length is “wider” and rated slower than 550L (differences between CCD used). Separate macro lens available. Can print direct to Olympus Photo Printer. | 1024 x 768 640 x 480 Progressive scan 850K pixels | Optical through the lens with dioptic correction/ LCD for review | | Auto with lock | 1.97’ to infinity. Macro: 11.8” to 2’ | 3x | 180 |
| **Olympus** | **D-600L** | Oct 97 | See 500L. Capable of printing directly to Olympus Photo Printer. | 1280 x 1024 640 x 512 CCD: 2/3” 1,410,000 pixels | Optical through the lens, 95% or image/1.8” LCD | 12 sec | Auto | 24” to infinity Macro: 12 to 24” | 3x optical | 100 |

| Mfr. | Model | Exposure | Flash Modes | Burst[3] Mode | Lens Aperture | Focal[4] Length | Shutter Speed | Storage | Interface | Wt. |
|---|---|---|---|---|---|---|---|---|---|---|
| **Olympus** | **D-340L D-340R** | Auto. Over/Under exposure of 1.3 steps 320R: Auto. +/- 1 step | Auto, red-eye reduction, fill-in, Off Range: 9.8ft | 6-10 shots in 640 x 480 SQ mode at 2 fps | 320L:f/2.8 320R: f/2.8, f/5.6, f/11 | Standard 36mm. Digital Tele: 72mm | ½ to 1/500 | 4 MB SmartMedia | Serial (Mac & PC) Video out | 9.4 oz |
| **Olympus** | **D-400Z** | Auto plus presets: daylight, overcast, tungsten, fluorescent | Auto, red-eye reduction, fill-in, Off Range: 13 ft | 10 shots at 2 fps | f/2.8, f/5.6, f/11 | 35 to 105mm | ½ to 1/1000 | 8 MB SmartMedia | Serial (Mac & PC) Video out | 9.5 oz |
| **Olympus** | **D-500L** | TTL auto white bal & exp control. Exposure compensation with +/- three f-stop incr | Four mode with red-eye reduction Range: 15.7 feet | | f/2.8 | 50mm to 150mm | º to 1/10,000 | 2 MB SmartMedia | Serial (Mac & PC) | 16.5 ounces |
| **Olympus** | **D-600L** | Auto spot meter. +/- 3 steps | Four mode with red-eye reduction Range: 15.7' | | F/2.8 | 36 to 110mm | º to 1/10,000 | 4 MB SmartMedia | Serial (Mac & PC) | |

| Mfr. | Model | Intro.[5] | Description | Resolution[7] | Viewfinder/ LCD[1] | Timer | Focus | Focus Range | Zoom[2] | Sensi- tivity[6] |
|---|---|---|---|---|---|---|---|---|---|---|
| **Olympus** | **D-620L** | 98 | Further upgrade to the 500/600L design, with even higher resolution CCD. Capable of printing directly to Olympus Photo Printer. | 1280 x 1024<br>640 x 512<br>CCD: 2/3"<br>1,410,000 pixels | Optical through the lens, 95% or image/1.8" LCD approx: 61,000 pixels | 12 sec | Auto | 1' to infinity | 3x optical | 100 |
| **Olympus** | **C-2000Z** | 99 | Whole new camera design, first Olympus to offer 2 megapixel resolution. Surprisingly small (though thick) body design. Aperture and Shutter priority modes. Optical viewfinder with diopter correction. 3x zoom. Includes infrared remote control. Works with ext. flash. | 1600 x 1200<br>1024 x 768<br>640 x 480<br>½" CCD<br>2,110,000 pixels | Optical viewfinder/1.8 inch TFT LCD with manual brightness | 12 sec | | 16" to infinity Macro: 8 in to 16 in | 3x optical; 1.6x, 2.0x, 2.5x digital | 100, 200, 400 |
| **Olympus** | **C-2500L** | Jul 99 | Not yet released; few details available. | 1800 x 1200?<br>2/3 inch progressive scan CCD<br>2,510,000 | Optical/LCD | | | | 3 x | 100,2 00,40 0 |
| **Polaroid** | **PDC-300** | 98 | Good combination of features, but surprisingly low resolution and image quality not that good. | 320 x 240<br>JPEG | LCD 1.8", adjustable brightness | 10 sec | | 15" - infinity | | 200 |
| **Polaroid** | **PDC-640** | Jul 98 | Lens cover acts as power switch. LCD can be used as viewfinder. Reasonable image quality. | 640 x 480<br>320 x 240 | Optical/LCD 1.8" | Yes | fixed focus | 3 ranges: macro, close-up and normal | No | Unkno wn |

| Mfr. | Model | Exposure | Flash Modes | Burst[3] Mode | Lens Aperture | Focal[4] Length | Shutter Speed | Storage | Interface | Wt. |
|---|---|---|---|---|---|---|---|---|---|---|
| **Olympus** | **D-620L** | Auto spot meter. +/- 3 steps | Auto, red-eye, fill-in, flash off, fill-in flash with ext flash and ext flash only | 5 shots at 3 fps | W: f/2.8, f/5.6 T: f/3.9, f7.8 | 36 to 110 mm | º - 1/10,000 | 8 MB SmartMedia | Serial (Mac & PC) | |
| **Olympus** | **C-2000Z** | Aperture pre-ferred, Shutter-preferred, w/manual control +/- 2 EV steps in 1/3 incr. | Auto, red-eye, fill-in and off | 2 fps | f/2.0 – f/11 | 35 – 105mm | ½ - 1/800 | SmartMedia | Serial (Mac & PC) | |
| **Olympus** | **C-2500L** | Aperture preferred, Shutter-preferred, with manual control +/- 2 EV steps in 1/3 incr. | Auto, red-eye, fill-in and off | | | 36 – 110mm | | SmartMedia | Serial (Mac & PC) | |
| **Polaroid** | **PDC-300** | | Range 15" – 8' | | f/5/6 | 52mm | 1/30 – 1/2000 | 2MB capacity: 64 images | Serial Video (NTSC) | 10 oz |
| **Polaroid** | **PDC-640** | Auto | Auto, fill-in, red-eye, off | No | f5.6 | Unknow n | 1/30 to 1/10000 | 2MB SmartMedia capacity: 36 images | Serial (PC only software) video | 2 lbs. |

| Mfr. | Model | Intro.[5] | Description | Resolution[7] | Viewfinder/ LCD[1] | Timer | Focus | Focus Range | Zoom[2] | Sensi-tivity[6] |
|---|---|---|---|---|---|---|---|---|---|---|
| **Polaroid** | **PDC-700** | | Low priced unit, good image quality. LCD has manual brightness adjustment. Internal memory with optional CompactFlash card. Takes over 5 seconds to recycle. | 1024 x 768 | 2.0" LCD | 10 sec | auto | Normal: 5' - infinity Close-up: 1.5 - 5' Macro: 6" - 1.5' | | 200 |
| **Polaroid** | **PDC-2000** | Mar 96 | Unusual shape: looks like alien space-craft! Can be controlled from computer. Optical viewfinder poor; no correction for parallax error. 3 models: PDC-2000/40 stores 40 images on internal hard disk. PDC-2000/60 stores images in solid-state memory. PDC-2000/T is tethered to host computer. Autofocus sonar based, not infrared. 60mm-equivalent lens available. | 1600 x 1200<br>800 x 600<br>1 million pixel CCD | Optical | | Auto | 10" to infinity | No | 100 |
| **Polaroid** | **PDC-3000** | Jan 98 | Upgraded version of PDC-2000. Viewfinder improved. | 1600 x 1200<br>800 x 600<br>1 million pixel CCD<br>No compression & lossless modes: 2:1, 5:1, 10:1 | Optical | | Auto | 10" to infinity | No | 100, 50, 25 |
| **Ricoh** | **RDC-2**<br>**RDC-2E** | | RDC-2 features wide and normal lens. RDC-2E has wide setting only. Includes remote control for camera operation. Records sound using ITU-T G.721 ADPCM method at 8 KHz (10 seconds in length). | 768 x 576<br>1/4", 410,000 pixels (effective 380,000 pixels) | Optical with diopter correction | | Auto | 0.01m (0.03') -- Infinity | 1.5 x Bi-focal, | |

| Mfr. | Model | Exposure | Flash Modes | Burst[3] Mode | Lens Aperture | Focal[4] Length | Shutter Speed | Storage | Interface | Wt. |
|---|---|---|---|---|---|---|---|---|---|---|
| **Polaroid** | **PDC-700** | Unknown | Auto, fill-in, red-eye Flash range: 3'-8' | | f/5.6 | 52mm | Unknown | 4MB internal (up to 50 images) plus CompactFlash | Serial (PC only) | 10 oz |
| **Polaroid** | **PDC-2000** | Backlight compensa tion White bal for incan & fluorescen t | Range to 15' | | f/2.8 | 38mm | 1/25 to 1/500 | | SCSI (Mac & PC TWAIN driver) | 0.9 kg, 2.0 lbs. |
| **Polaroid** | **PDC-3000** | Backlight compensa tion White bal for incan & fluorescen t | Range to 15 feet | | f/2.8 - f/11 | 38mm | 1/25 to 1/500 | 20 MB card CompactFlash | SCSI (Mac & PC TWAIN driver) | 0.9 kg, 2.0 lbs. |
| **Ricoh** | **RDC-2 RDC-2E** | Auto. Manual exp adjust range +- 1.5 EV (in 0.3 EV incr) | Auto | Approx. 1 image/sec ond | f/3.7, f/5.6 | Normal: 55mm Wide: 35mm | 1/8-1/1000 | 2MB internal memory Type I and Type II, ATA-compliant: PCMCIA2.1 | NTSC serial | 285g 10.1 oz |

| Mfr. | Model | Intro.[5] | Description | Resolution[7] | Viewfinder/ LCD[1] | Timer | Focus | Focus Range | Zoom[2] | Sensi-tivity[6] |
|---|---|---|---|---|---|---|---|---|---|---|
| **Ricoh** | **RDC-300** | Jul 97 | Tilt-up LED screen with brightness adjustment. Computer cable not included. Exposure compensation control +/- .5EV per step. | 640 x 480<br>320 x 240<br>160 x 120<br>1/4", 350,000 pixels | 1.8" LCD | Yes | Fixed | 60cm to infinity. Manual macro - 1 to 60cm | No | 80 |
| **Ricoh** | **RDC-300Z** | Nov 97 | Zoom version of the RDC-300. | 640 x 480<br>CCD: 350,000 pixels | 1.8" LCD | 10 sec | Auto | Macro: 1cm (<1") to infinity | 3x | 80 |
| **Ricoh** | **RDC-4200** | Sep 98 | Lens mounted at end of camera and rotates. LCD display difficult to see in sunlight (no optical viewfinder) and consumes power. Time-lapse mode takes pic every 30 secs. Optional remote. Support up to 16MB SmartMedia. | 1280 x 960<br>640 x 480<br>1.32 million square pixel CCD | 1.8" TFT color LCD display | | Auto and manual plus macro | 35cm to infinity Macro 8-35cm (3.2" - 13.8") | 3x optical 2x digital (VGA only) | 80-100 |
| **Ricoh** | **RDC-4300** | Jun 98 | Similar design and functionality as 4200, but has larger LCD and can also record audio. | 1280 x 960<br>640 x 480<br>1.32 million square pixel CCD | 2 inch LCD | 10 sec | Auto | 40cm to infinity Macro: 8 cm to 40 cm | 3x | 40-80 |
| **Ricoh** | **RDC-5000** | May 99 | Built-in lens cover, AC adapter, optional remote. Plastic cover for LCD. Supports up to 32MB SmartMedia. LCD brightness adjustment. Auto macro mode. Interval mode 30 sec to 3 hours. USB connection. | 1792 x 1200<br>896 x 600<br>2,300,000 pixel CCD | 1.8" TFT LCD (110,000 pixels) | 10 sec | Auto | 4 cm to infinity | 2.5x | 100 |
| **Sanyo** | **VPC-G200** | | Also captures 6 sec audio per image. Saves JPEG, WAV and AVI file format. Built-in earphone jack. | 640 x 480<br>320 x 240<br>1/3 inch CCD<br>350,000 pixels | Optical/ 2 inch LCD (110K pixels) | 10 sec | Fixed | 20 cm to infinity | No | 80 |
| **Sanyo** | **VPC-G250** | | Capture to AVI at 320 x 240 10 fps max of 5 seconds. | 640 x 480<br>320 x 240<br>1/3 inch CCD<br>350,000 pixels | Optical/1-inch 110,000 pixel LCD | | Fixed | 60cm to infinity. Macro: 10cm – 50 cm | No | 80 |

| Mfr. | Model | Exposure | Flash Modes | Burst[3] Mode | Lens Aperture | Focal[4] Length | Shutter Speed | Storage | Interface | Wt. |
|---|---|---|---|---|---|---|---|---|---|---|
| **Ricoh** | **RDC-300** | TTL Programmed AE | Range: 60cm-3m | No | f/3.8 | 38mm | 1/5 - 1/8,000 | 4 MB internal | Serial (Mac & PC) NTSC | |
| **Ricoh** | **RDC-300Z** | Automatic Manual Adjust | Auto | | f/3.8 | 45-135mm | 1/5 - 1/8000 | 2 MB SmartMedia | Serial (Mac & PC) | 9.2 oz (260g) |
| **Ricoh** | **RDC-4200** | Auto | | | f2.8-4.8 (wide) f2.8-14 (tele) | 35-105mm | 1-1/500 | 4 MB SmartMedia holds 6 - 26 images | Serial (Mac & PC) Video | 10.05 oz |
| **Ricoh** | **RDC-4300** | Programmed AE +/- 2 EV in .5 EV incr. Auto/Manual White Bal | Auto, Fill-in, Off Range: 3m | | f/5.6 – f/16.8 | 35-105mm | 1/8-1/500 | 4MB SmartMedia | Serial (Mac & PC) Video | 10 oz 285 g |
| **Ricoh** | **RDC-5000** | Programmed AE +/- 2 EV in .5 EV incr | Auto, ON, Slow synchro, red-eye, off | | f/8.0 | 38-86mm | 1-1/500 | 8MB internal plus SmartMedia | Serial (Mac & PC) USB Video | 315 g |
| **Sanyo** | **VPC-G200** | Auto | Auto, Forced, Off Range 24 ft | 16 images at .1 or .2 sec apart | f/2.8, f/5.6, f/11 | 36mm | º - 1/10,000 | 4MB built-in flash memory 60 images at high resolution. | Serial (Mac & PC) Video | 260 g |
| **Sanyo** | **VPC-G250** | Programmed AE with manual override (7 settings) | Auto, Forced, Off | 15 images at .1 or .2 sec apart | f/2.8/8 | 43mm | º to 1/10,000 | 4MB SmartMedia | Serial (Mac & PC) Video | 200 g |

| Mfr. | Model | Intro.[5] | Description | Resolution[7] | Viewfinder/ LCD [1] | Timer | Focus | Focus Range | Zoom [2] | Sensi-tivity [6] |
|---|---|---|---|---|---|---|---|---|---|---|
| **Sanyo** | **VPC-X300** | | Audio annotation (four secs per image). | 1024 x 768<br>640 x 480<br>CCD: 1016 x 794 (962 x 774 effective) | Optical/ 2 inch LCD (110K pixels) | 10 sec | Auto | 19.7" to infinity. Macro: 7.8" to 19.7" | Digital: 1.3/1.6/2.0/2.5/3.0 (VGA Mode) | 100 |
| **Sanyo** | **VPC-X350** | Dec 98 | Optical viewfinder and LCD. Burst mode captures up to 20 sequential shots in either one-and-a-half or three-second intervals. Can record up to fifteen seconds of full-motion video with audio. 320 x 240 and 160 x 120 for video clips. | 1024 x 768<br>640 x 480<br>320 x 240<br>160 x 120<br>1/3" CCD 850K pixels | Optical/ 2 inch LCD 110,000 pixels | 10 sec | Fixed | Macro 10 cm to 50 cm (4 – 20 inches) | Digital 3x | 100 |
| **Sanyo** | **VPC-Z400** | Dec 98 | | 1280 x 960<br>640 x 480 | Optical/ 2 inch LCD | Yes | Fixed | Macro 20 cm to 80 cm ( 8-40 inch) | 3x optical and 2x digital | |
| **Sony** | **MVC-FD51** | Jul 98 | Uses double-speed floppy. Stores 40 JPEGs in standard mode. 410,000-pixel interlaced CCD. Email mode saves images at 320 x 240. Lithium batteries. | 640 x 480<br>º inch 410K interlaced CCD | 2.5" LCD | 10 sec | Fixed | | 10x optical | 100 |
| **Sony** | **MVC-FD71** | Jul 98 (D) | Similar to FD51 but with progressive scan CCD. 10x optical zoom with auto and manual focus. | 640 x 480<br>º" 350K progressive scan CCD | 2.5" LCD | 10 sec | Auto with Manual and Macro | 1" Macro Capability | 10 x optical | 100 |
| **Sony** | **MVC-FD73** | May 99 | 10x optical zoom and double speed floppy disk drive, E-Mail Mode and a 2.5-inch LCD screen. | 640 x 480 progressive scan CCD | 2.5" LCD | 10 sec | Macro | 1 cm to infinity | 10x optical | 100 |
| **Sony** | **MVC-FD81** | Oct 98 | Similar to earlier Sony models but with larger resolution images. Still limited to floppy disk (which many find attractive). MPEG movie mode and voice annotation. Optical zoom is only 3x. | 1024 x 768<br>640 x 480<br>320 x 240<br>160 x 112<br>progressive scan CCD w/ 850K pixels | 2.5" LCD | 10 sec | Auto with Manual | 1" Macro Capability | 3x optical | 100 |

| Mfr. | Model | Exposure | Flash Modes | Burst[3] Mode | Lens Aperture | Focal[4] Length | Shutter Speed | Storage | Interface | Wt. |
|---|---|---|---|---|---|---|---|---|---|---|
| **Sanyo** | **VPC-X300** | Auto | Auto, Forced, Off | 9 pics on one image, .1 – .2 sec | f/2.8, f/5.6, f/11 | 36 | º to 1/500 | 4M3 SmartMedia | Serial (Mac & PC) Video | 9.6 oz 270 g |
| **Sanyo** | **VPC-X350** | Auto/Man (7 steps) | Auto, On, Off | Yes + Video | f/2.8/8 | 36 | 1/4 - 1/10,000 | 8M3 SmartMedia holds 30 - 120 images | Serial (Mac & PC) Video | 200 g |
| **Sanyo** | **VPC-Z400** | Auto/Man (7 steps)/white balance | Auto, red-eye, On, Off | 2 images per sec up to 10 images | f/2.8-4.7 | 34-102mm | | 8M3 SmartMedia | Serial (Mac & PC) Video | 300g |
| **Sony** | **MVC-FD51** | Auto with Auto white balance. 7 settings (+/- 1.5 EV) | Auto, On, Off | | Max: f/2.0 | 40-400mm | 1/60 to 1/2000 | Up to 40 images per floppy | floppy | 1.16 lbs |
| **Sony** | **MVC-FD71** | Auto with 6 pre-programmed modes and White balance | Auto/On/Off | Multi-image 9 pics in 2 sec | Max: f/1.8-2.9 | 40 - 400mm | 1/60 – 1/4000 | Up to 40 images per floppy | floppy | |
| **Sony** | **MVC-FD73** | Auto and manual | flash on/off, auto flash. range:2-4m | | f/1.8 - 2.9 | 40 - 400 mm | 1/60 - 1/2,000 | floppy | floppy | 590 g |
| **Sony** | **MVC-FD81** | Auto with 6 pre-programmed modes. | Auto, On, Off | | Max: f/2.0-2.1 | 37-111mm | 1/60-1/4000 | floppy | floppy | |

| Mfr. | Model | Intro.[5] | Description | Resolution[7] | Viewfinder/ LCD [1] | Timer | Focus | Focus Range | Zoom [2] | Sensi-tivity [6] |
|---|---|---|---|---|---|---|---|---|---|---|
| **Sony** | **MVC-FD83** | Jun 99 | Sound recording up to 5 seconds per image, video recording up to 60 seconds in MPG 160 x 112 pixels; 15 seconds with 320 x 240 pixels. Stores images on floppy. | 1,216 x 912 (interpolated) 1,024 x 768 640 x 480 1/3" CCD chip with 850,000 pixels | 2.5" LCD | 10 sec | Auto with Macro | 1cm to infinity | 3x optical/6 x digital | 100 |
| **Sony** | **MVC-FD88** | Jun 99 | Higher resolution CCD. | 1,280 x 690 1,024 x 768 640 x 480 1/3.6" CCD chip w/ 1,300,000 pixels | 2.5" LCD | 10 sec | Auto with Macro | 1cm to infinity | 8x optical/1 6x digital | 100 |
| **Sony** | **MVC-FD91** | Nov 98 | MPEG mode captures up to 60 secs of video and audio to a floppy. Store 5 seconds of audio per image. Optical SteadyShot system. Uses standard 3.5" floppy disk as storage. | 1024 x 768 640 x 480 1/3" Progressive CCD, 850K pixels | Color viewfinder 180K (like camcorder) 2.5" LCD | 10 sec | Auto & Manual with Macro | 1" Macro Capability | 14x optical | 100 |
| **Sony** | **DSC-F55** | May 99 | 10 oz w/battery. 2.1million megapixel. Hybrid LCD absorbs light to generate a bright LCD screen, Carl Zeiss optics, MPEG Movie Mode for recording full-motion video and audio, "crop & save," 180-deg rotational lens. Built-in mic. Sup-ports MPEG video: 160 x 112 and 320 x 240. Uses Sony's Memory Stick storage media. | 1600x 1200 1280 x 768 1024 x 768 640 x 480 320 x 240 160 x 112 ½ inch CCD with >2 million pixels | 2" LCD (122K pixels) | 10 sec | High speed Auto | Macro to 1" | 2.5x digital | 100 |

| Mfr. | Model | Exposure | Flash Modes | Burst[3] Mode | Lens Aperture | Focal[4] Length | Shutter Speed | Storage | Interface | Wt. |
|---|---|---|---|---|---|---|---|---|---|---|
| **Sony** | **MVC-FD83** | Auto with 9 pre-programmed exposure modes | fon, off, auto flash) Range: 2 m to 4 m | Movie mode | f/2 - 2.1 | 37 to 111 mm | 1/30 - 1/4,000 | floppy | Video floppy | 660 g |
| **Sony** | **MVC-FD88** | Auto and Manual | Auto, On, Off | Movie Mode | f/2.8 - 3 | 41 to 328 mm | 1/30 - 1/4,000 | floppy | Video floppy | |
| **Sony** | **MVC-FD91** | Auto & Manual. White Balance. 7 settings (+/- 1.5 EV) | Auto, On, Off | Movie Mode | Max: f/1.8-3.2 | 37 - 518mm | 1/60 – 1/4000 | floppy | floppy | |
| **Sony** | **DSC-F55** | Auto/Manual f/-1.5 to f/+1.5 | Auto, On, Off | Movie mode | Max: f/2.8 | 37 mm | 1/60 – 1/750 | floppy | Serial video | 10 oz |

| Mfr. | Model | Intro.[5] | Description | Resolution[7] | Viewfinder/ LCD[1] | Timer | Focus | Focus Range | Zoom[2] | Sensi- tivity[6] |
|---|---|---|---|---|---|---|---|---|---|---|
| **Sony** | **DSC-D700** | Dec 98 | Sophisticated camera that isn't limited by floppy; however, resolution is lower than less expensive models. Sony will replace the DSC-D700 with the DSC-D770 in fall '99. Will have faster focus, better sharpening algorithms, histogram functions, and a better manual focus. | 1344 x 1024 Progressive CCD with 1.5 million pixels | Optical Through The Lens viewfinder/2.5 " LCD (180K pixels) | Yes | Auto and manual | Auto macro. 20 cm to infinity | 5x optical | 100, 200, 400 |
| **Toshiba** | **PDR-5** | Jul 98 | Small design. Upgrade of PDR-2 that adds flash and LCD. Uses CMOS chip. Images rated well. Uses Lithium battery. | 640 x 480<br>320 x 240<br>º inch progressive scan CCD | 2.5" LCD | Yes | Fixed plus macro | Macro .1m – 2m | 2x | 100 |
| **Toshiba** | **PDR-M1** | Jul 98 | High resolution at a low price. Circular built-in lens cap. 5 levels of flash, plus you can get the white balance. Takes over 5 seconds to power up and recycle between images. LCD goes blank while camera focuses. | 1280 x 1024<br>640 x 480<br>½ inch Million pixel CCD | 1.8" LCD | 10 sec | Auto | Macro (0.09m – 0.5m) | 2x | 100 |
| **Toshiba** | **PDR-M3** | Feb 99 | Same as the Fuji MX-600. | 1280 x 1024<br>640 x 480<br>½ inch CCD with 1.5 million pixels | Optical/1.8 inch LCD | Yes | Auto with Macro | 0.9m to infinity | 3x optical | 100 |
| **Toshiba** | **PDR-M4** | Jun 99 | Toshiba's own design. | 1600 x 1200<br>1024 x 768<br>640 x 480<br>CCD - 1/2" with 2.14 million pixels | 1.8" LCD | yes | | | 2x digital | |
| **Vivitar** | **Vivi Cam 2700** | | Inexpensive camera. Has received mixed reviews for image quality. | 640 x 480<br>320 x 240<br>410,000 pixel CCD | Optical/1.8" color LCD | 10 sec | fixed | 10" to infinity | | 200 |

| Mfr. | Model | Exposure | Flash Modes | Burst[3] Mode | Lens Aperture | Focal[4] Length | Shutter Speed | Storage | Interface | Wt. |
|---|---|---|---|---|---|---|---|---|---|---|
| **Sony** | **DSC-D700** | Program AE, Aperture Priority AE, Shutter Priority AE & Manual. Man & Auto White Bal | Automatic, On, Off plus Hotshoe | | Wide: f/2-f/l. 1; Tele: f/2.4-f/3.2 | 28-140mm | 4 seconds to 1/2000 | ATA Type PC Card or Sony Memory Stick (supplied) | | |
| **Toshiba** | **PDR-5** | Auto and Manual +/- 2 Auto and Fixed White Bal | Auto, On, Off, Red-eye | 4 fps (1/4 VGA) | f/2.5, f/5.6 | 39mm | 1/8 – 1/2000 | 2MB SmartMedia | Serial (Mac & PC) video | 6.77 oz |
| **Toshiba** | **PDR-M1** | Auto | Auto, On, red-eye, Off | | f/3.2 f/8 | 35mm | º - 1/1000 | 4MB SmartMedia | Serial (Mac & PC) video | 8 oz |
| **Toshiba** | **PDR-M3** | Auto, red-eye, fill-in, off. Range 10 ft | on/off/auto, auto redeye red., slow & ext sync. Range 1.3 - 6.5' | | f/3.8, f/5.5, f/7.4 | 35 – 105mm | 1/4 - 1/1000 | SmartMedia | Serial (Mac & PC) video | 350 g |
| **Toshiba** | **PDR-M4** | | Auto, on, off, red eye reduction | | f3.2/f8 | 35mm | 1/4 - 1/1000 | 16 MB internal plus SmartMedia | Serial USB | 290 g |
| **Vivitar** | **ViviCam 2700** | auto | Auto, On/Fill, Off | | f/2.8 & f/8 | | 1/60 – 1/1000 | 2MB internal memory 50 images plus CompactFlash | Serial (PC only) NTSC video | 8.5 oz |

| Mfr. | Model | Intro.[5] | Description | Resolution[7] | Viewfinder/ LCD[1] | Timer | Focus | Focus Range | Zoom[2] | Sensi-tivity[6] |
|---|---|---|---|---|---|---|---|---|---|---|
| **Vivitar** | **Vivi Cam 2750** | Jan 99 | Upgraded version of 2700. Includes Ni-MH batteries. | 640 x 480 progressive scan CCD | Optical/1.8" color LCD | | Focus-free with Macro | 16" to infinity. Macro: 6 - 12" | | 200 |
| **Vivitar** | **Vivi Cam 2800** | | Improved resolution, and USB support. Audio recording up to 10 seconds. Unfortunately the zoom is digital. | 1024 x 768<br>640 x 480<br>512 x 384<br>1/3" CCD<br>800,000 pixel CCD | Optical/1.8" color LCD | 10 sec | Auto with Macro | | 2x digital | |
| **Vivitar** | **Vivi Cam 3000** | Nov 97 | Also available as IMAX MDX 8000. CMOS sensor. Audio recording 4 secs per image. | 960 x 800<br>480 x 400<br>800,000 pixels CCD | Optical | 12 sec | fixed | 1m to infinity | No | 50 |
| **Vivitar** | **Vivi Cam 3100** | Nov 97 | Requires six AA batteries! No LCD display. Resolution is interpolated. CMOS. Image quality ranked so-so. Audio up to 20 seconds per image | 960 x 800 (1920 x 1600 interpolated)<br>800,000 pixels CCD | Optical | 12 sec | fixed | | No | 50 |
| **Vivitar** | **E-cam** | Late 99 | Point-and-shoot camera for Kids! Selling for less than $80. | | Optical | 10 sec | Fixed | | | |

1: LCD: some LCD's can be used as a viewfinder, while others can only be used to review a picture after it is taken. Where known the functionality is indicated.
2: Zoom: Digital or optical zooming is indicated where known
3: Burst Mode: minimum time between pictures
4: Focal Length: equivalent to a 35mm camera.
5: Introduction Date: (D) = Discontinued. Models may still be available through the retail channel.
6: Sensitivity: ISO. As quoted by manufacturer.
7: Resolution: Manufacturers sometimes indicate the resolution based on computer screen resolutions: Most noticeably 1280 x 960 (SXGA) 640 x 480 (VGA)

| Mfr. | Model | Exposure | Flash Modes | Burst[3] Mode | Lens Aperture | Focal[4] Length | Shutter Speed | Storage | Interface | Wt. |
|---|---|---|---|---|---|---|---|---|---|---|
| **Vivitar** | **ViviCam 2750** | auto exp matrix weighting system | Auto, On/Fill, Off | | f/5.6 | 52mm | 1/30 – 1/10,000 | 2MB internal memory 16 images plus CompactFlash | Serial (PC only) NTSC video | 10.6 oz |
| **Vivitar** | **ViviCam 2800** | Auto exp w/ EV override. Select weighting criteria: centr point matrix. White bal Auto or Daylight | On/Fill, Off, Auto, and Redeye modes | | | | 2 sec to 1/8000 | 4 MB CompactFlash | Serial & USB (PC only) NTSC video | 10.8 oz./3 07 g |
| **Vivitar** | **ViviCam 3000** | Auto | Yes | | f/4 | 48 | 1 sec to 1/2000 | Internal Memory Plus PCMCIA type I card | Serial | 6 oz |
| **Vivitar** | **ViviCam 3100** | Auto | Range: 9 feet | | f/4 | 48mm | 1 sec to 1/2000 | 200Kb (up to 3 images) Plus PCMCIA type I card (2MB included) | Serial | 8 oz |
| **Vivitar** | **E-cam** | Auto | | | | | | | Serial (PC software only) | |

# CHAPTER 25
# FUTURE

Better. Faster. Cheaper. It's the story of computers, and it's the story of the digital camera. The cameras that will be available next year will be better and cheaper than the camera available this year. And most importantly, digital cameras have jumped an important hurdle. While digital cameras have been available for several years, it was only in 1998 that the quality of the sub-$1,000 cameras could be considered "good enough" for the average user. Now more people are buying the cameras. With increased sales, the cost of manufacturing is going down—and the quality is going up—attracting more people. And so the cycle continues: not only does this book cover digital cameras, it's an introduction to economics too!

Many people wonder what the digital camera will do to film cameras. At the moment film has many advantages over digital cameras; a high-quality film camera can be had for a lot less money than a digital camera. And if you only want snapshots, going to the photo processor might be a little easier than using your computer to print them all. And you could get a lot of pictures printed for the cost of the computer, printer and paper and ink.

Still, it's a safe bet that digital camera sales will grow, and eventually replace most of the film market.

Kodak, a company very dependent upon film sales, has been in the vanguard of digital camera development. They've also worked to make film fit better with computers through the development of Photo CD and Picture CD.

Interestingly, so far digital cameras haven't seemed to impact the sales of film. A Kodak representative said that so far, as they sell more digital cameras, the sales of film increases too. He went on to relate how he had recently gone on vacation, taking a digital camera along with a film camera. While he shot a lot of digital images, he found that he was shooting more regular pictures too; the immediacy of the digital camera actually encouraged taking pictures; and increased the use of the film camera.

## THE SHAPE OF CAMERAS TO COME

While all of the low-cost cameras are fairly limited affairs, in the high end we already see cameras that look very like today's SLR cameras. In most cases they are SLR film cameras that have been converted into a digital form. Unfortunately, these cameras are still far too expensive for most people, but that will change and soon there should be cameras that work with 35mm film camera lenses that cost about $1,000. When that happens the digital camera market will really be here to stay.

Just as with current film cameras there are the point-and-shoot and SLR cameras, so too with digital cameras there will be the two types of cameras; one for those who just want to take snapshots and one for those who want to have complete control of the picture taking process.

Whether it will be possible to retrofit older cameras is a question for debate. There is at least one company that claims to be developing an insert that converts a traditional film camera into a digital camera. The insert looks like a film cartridge with a plastic panel containing the image sensor sticking out the side like a piece of film. Just drop the device into an SLR camera an away you go. Unfortunately, there seem to be a lot of technical hurdles to solve; for example, cameras are based on focusing on a piece of film which is very thin. Replacing it will a CCD or CMOS chip seems to present several problems. Then there's the problem of how the device senses that the shutter has been fired. Such a device would seem to be limited in the number of images it could hold, and once full, how do you download the images?

Which is not to say that it might not be possible to produce such a device. Some cameras let you change the back of the camera. This seems the most plausible way of upgrading existing cameras, though obviously it limits the number of cameras that the unit would work with.

Just remember to see the images before you buy any digital camera.

## SMARTER CAMERAS

Digital cameras include a computer chip that handles the image processing and downloading tasks. Most film cameras include a computer processor. But digital cameras also include an LCD display, which makes possible much more complicated interfaces. The Digita operating system adds more intelligence to the camera, and this is a trend that will probably continue. Hopefully, smarter cameras will be easier to use, not more complicated.

## EASIER PRINTING, DISPLAY AND STORAGE

Now you can buy printers that print from your camera or memory card without the need for a computer. You could become a digital photographer without even buying a computer. Of course, you'd miss out on some of the possibilities for image improvement (and manipulation), but many people don't care about that. While some of these printers offer image editing (Epson Print-On) and image archiving (Lexmark), users need a really good way to archive and then find their images. A printer with an LCD display, removable disk storage, and an easy-to-use interface could easily make it possible to do away with the computer. It's the archiving and storage that's probably the biggest problem.

### META DATA

Many computers already save Meta Data in the EXIF JPEG files. Making it simple and easy to add Meta Data—information about where a picture was taken and who is in it—will enable users to make better use of their photo collections. While there are things that can be done in software to help users do this, the truth is, like cleaning your office, no software can do it for you; the user has to want to do it.

## MERGING WITH VIDEO?

Digital cameras grew out of video cameras. Digital video cameras offer many of the features of digital still cameras, but at a lower resolution. Will the two join together?

The next step in the evolution of video is HDTV (High Definition TeleVision) which promises much higher resolution than current video. Unfortunately, even at its highest resolution, HDTV has a resolution of 1920 x 1080. While that's as good as or maybe a little better than current sub-$1,000 digital cameras, it's not as good as the high-end digital cameras can produce. For the enthusiast and professional, cameras must offer the resolution of 35mm film; which is about 3072 by 2048 pixels. While many people will be happy with half that resolution, the enthusiasts and pros won't be.

For these reasons, while it's likely that there will be consumer HDTV camcorders sometime in the future, and they will offer good quality resolution stills that many people will be happy with, they won't be good enough for everyone. And for that reason, as well as because of ergonomics and other issues, digital and video cameras will remain a separate category for many people.

## REALLY SMART CAMERAS

As camera and computer processing power increases, more options for image processing start to develop. Software that can identify parts of a photograph become possible. An application that can identify which parts of an image are people (and which are faces) could make searching images much simpler, particularly for family albums where the number of people appearing in the pictures is fairly small. The software might be able to, most of the time, determine who the people are based on previous photographs.

Video cameras already feature image stabilization technologies that could find their way into still cameras. Motion tracking software could theoretically make it possible for the camera to focus on a fast moving subject, and assuming the capture device had enough resolution, frame a portion of the image and follow the object within that frame.

## CAMERAS IN OTHER DEVICES

Adding cameras to other devices—computers and phones for example—becomes more possible as their cost decreases. Already Sony has released a notebook with a built-in camera, and another company has displayed a cellular phone with a built-in camera.

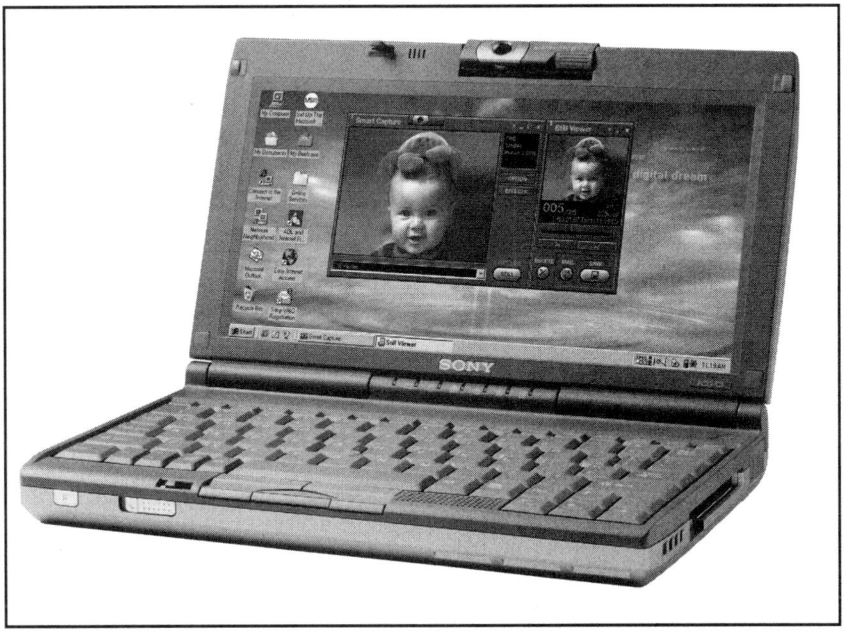


*Figure 25-1. Vaio C1X PictureBook (photo courtesy Sony Electronics, Inc).*

Sony's VAIO C1X PictureBook is a notebook computer that includes a CCD camera at the edge of the case, above the computer screen. The camera is rotatable, and the camera's PictureGear software can be used to create 360-degree panoramas. An Auto Shutter feature automatically captures images when the system detects motion. It can capture in JPEG format, or up to 60 seconds

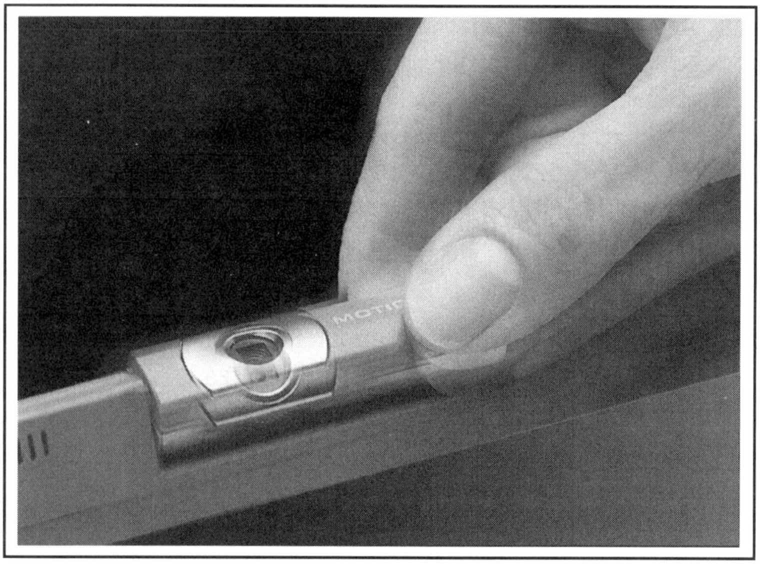

*Figure 25-2. Rotatable camera of the Vaio C1X PictureBook (photo courtesy Sony Electronics, Inc).*

in AVI format. Software is included that makes emailing files easy, and can turn documents into HTML format. The camera also has an IEEE-1394 (FireWire) interface. Sony calls it i.Link. The only downside—the camera is a 1/6" CCD with only 270,000 resolution.

## REAL SOON

Just around the corner, MetaCreations is developing MetaFlash, a 3D camera that uses a special flash and converts digital photos into 3D wireframe models. The technology is supposed to be released soon, but there are no example images or good descriptions of what it is or how it does what it does. It sounds fascinating. We just have to wait a little bit longer to see what it really is. Maybe by the time you read this book it will be out. If so, you can read about it at:

*www.m2w.net/digitalcamera*

# IS NOW THE RIGHT TIME TO BUY?

If cameras are going to get better and cheaper, then why buy one today? Good question. And just like with computers, the answer is a complicated one, but it boils down to "Can you use it for the task you want?" Sure, if you wait a year the camera you buy then will be better than the camera you bought this year. But the same will be true the year afterwards, and if you follow that to its logical conclusion you'll never buy a camera. At the same time, if you're not going to be able to use the images the camera produces, if you're not even going to be using the camera much, then maybe you should wait.

But if, like me, you find the resolution acceptable for many purposes, and that the benefits outweigh the liabilities, then why wait? Sure, next year's camera will be better, but you'll have missed a whole year of pictures by then.

# GLOSSARY

**AE:** see Auto Exposure

**AF:** see Auto Focus

**Anti-aliasing:** Inserting pixels of intermediate color at the border of high-contrast color regions to remove the "jaggy" effect sometimes seen in bitmapped graphics.

**APS:** Advanced Photo System. Consumer film format that's slightly smaller than 35mm film, and is packaged in a cartridge for easier film loading. Information about the images (shutter speed, etc.) is stored along with the images. APS supports different aspect ratios, though these are created by masking the negative. One nice thing about APS is that films are returned in the cartridge; and newer film scanners have been developed that support these cartridges.

**Aperture:** The size of the opening that lets light through the lens. It's measured in f-stops.

**Artifact:** Pixels that are visibly incorrect in color or brightness. This can be caused during capture (low light produces noise, etc.) or by heavy compression of the image. (see *JPEG*)

**ASA:** American Standards Association (also see *ISO*). Film speed rating. Indicates the sensitivity of the film to light.

**Auto Exposure:** Camera uses internal light meter to measure the amount of light entering the camera and adjusts aperture and shutter speed to produce a correctly exposed scene. (see also *Center Weighted*, *Bottom Weighted*, *Spot* and *Matrix*)

**Auto Focus:** Method of focusing the lens, typically done with some kind of rangefinding mechanism such as an infrared light beam. Not to be confused with Focus Free lens, which is a fixed focus lens with a fairly wide depth of field.

**Barrel Distortion:** Distortion in image caused by lens. Edges appear to curve outward. This is most visible in straight lines in the image that can appear to bend outward rather than run straight. (see also *pincushion*)

**Bayer Pattern:** A pattern of red, green and blue filters that are arranged in front of a CCD and are used to create the final color image. This is required in a single-chip CCD because a CCD is not color-sensitive and each pixel returns only a single brightness level. Because the human eye is more sensitive to green, the Bayer pattern contains twice as many green filters than blue and red ones.

**Bit:** The smallest unit of digital memory. A bit can contain two values: 1 and 0. Bits are usually used in groups of eight called a *byte*. A byte can be used to represent 256 different values.

**Bit Depth:** Indicates the number of bits used to store the color value of a single pixel (either for a monitor display, or in a digital image file). An 8-bit depth (one byte) can store one of 256 different values or colors. For 16-bit, thousands of different values (colors) are possible, and with 24-bit millions of values are possible.

**Bitmap:** Storing a digital image as individual pixel values; essentially an array that contains the color values for each pixel. (see also *vector graphics*)

**Blooming:** Bleeding of colors from one pixel to another.

**BMP:** Common bitmapped graphic format used by the Windows operating system.

**Bottom Weighted:** Automatic exposure system that is biased towards the bottom of the picture. (see also *Center Weighted*, *Spot* and *Matrix*)

**Burst Mode:** Capturing multiple images in rapid succession (more than 1 frame per second).

**Byte:** Unit of digital memory consisting of 8 bits. A byte can be used to represent any one of 256 different values.

**Card Reader:** Device that attaches to a computer and can be used to read data from a memory card. There are three types of card (and reader) available: CompactFlash, SmartMedia and PCMCIA.

**CCD:** Charged-Coupled Device. Light-sensitive chip used to capture an image. These are not color-sensitive, merely sensitive to light. To create color either three chips are used, each with a single color filter, or a color mask called the *Bayer pattern* is overlaid in a red-green-blue pattern. More green is used because the eye is more sensitive to this color. Some manufacturers do use a CMYK and other filter patterns. CCDs are analog devices, so the measurement must go through an analog-to-digital converter.

**CD:** CompactDisc. (see also *Picture CD* and *Photo* CD)

**CDR:** Compact Disc Recordable. A device and special media that can be used to write a CD (either a "Data" disc or an audio disc). CDR discs cannot be overwritten, though it is possible to add data to a multisession disc. (see also *CDRW*)

**CDRW:** Compact Disc ReWriteable. A device and special media that can be used to write a CD (either a "Data" disc or an audio disc). CDRW discs can be erased and then rewritten, and can usually also be used to write CDR discs. CDRW media is currently more expensive than CDR blanks, and the drives are slightly more expensive than a CDR drive.

**Center-Weighted:** Auto exposure system that measures light at the center of the image. (see also *Spot Metering*, *Bottom Weighted*, *Matrix* and *Auto Exposure*)

**CF:** CompactFlash.

**CMD:** Charge Modulated Device. A variation of CCD and CMOS technology. CMDs are noisier than CCDs.

**CMOS:** Complementary Metal-Oxide Semiconductor. A light-sensitive chip manufactured using a different process from CCDs. CMOS chips are supposed to be cheaper to manufacture, require less power, and be able to produce images as good as those produced by CCDs. So far, CCDs remain the most popular devices in digital cameras.

**CMYK:** Cyan, Magenta, Yellow and blacK. Subtractive (reflective) color process used in printing. (see also *RGB*)

**Color Depth:** Number of colors (or bit-depth) supported by a monitor or display device. Also used to indicate the number of bits used to store color information in a bitmapped data file. The most common bit-depths are 24, 16, and 8-bit.

**Color Filter Array:** Filter dyes placed over the light-sensitive chip (CCD, CMOS or CMD) to enable the capture of color information. These devices are not color-sensitive; they only measure the brightness of the light that strikes them. Color filters are used to measure particular colors; usually red, green and blue.

**CompactFlash:** Small memory card that is used to store digital information. Available in a wide range of sizes (measured in Megabytes), CompactFlash cards retain the digital information even after being removed from a camera. A Card Reader can be used to transfer information from the card to a computer.

CompactFlash cards use a subset of the ATA format standard used by PCMCIA cards. Using an inexpensive adapter, a CompactFlash card can be read using a PCMCIA card reader. (see also *SmartMedia*)

**COM Port:** The serial port on Windows computers. Often used to connect digital cameras to the computer.

**Compression:** Reducing the size of a digital file by using different methods of encoding that take advantage of duplication (or similar) data values within a file: Two types of compression—*lossy* and *lossless*—are used. Lossy compression produces the smallest file sizes, but may physically alter the appearance of the image.

**Compression Ratio:** Size of compressed file to original size. 1:4 (1 to 4) produces a compressed file that is a quarter the size of the original.

**Contrast:** Visible difference in brightness between one pixel and another. By adjusting contrast this difference can be expanded (creating a vivid, posterized image) or contracted (creating a muted image).

**Dark Current:** The signal noise generated by the CCD even when no light is present. This varies over time.

**Depth-of-Field:** A lens is in focus for a single distance from the lens, but points closer and further from the lens will also appear to be in focus. The depth of field (the region in the scene which is in focus) depends upon the lens and the aperture. The wider the aperture, the smaller the depth of field.

**Digital Zoom:** Zoom effect (apparent changing of the focal length of the lens) created by taking the image and enlarging it as though a zoom lens had been used. The quality of the produced image will not be as good as that produced by an optical zoom lens.

**Diopter Correction:** Focus adjustment that adjusts the viewfinder to match problems in the vision of the user. Diopter correction adjusts the focus of the viewfinder and enables the user to see the image in the viewfinder without using glasses.

**DPI:** Dots per Inch. Used to indicate the resolution of a printer, and is a measurement of the size of the dots that the printer places on a page. The higher the number, the smaller the size of the dots.

**DRAM:** Dynamic RAM. Inexpensive memory that is slower than other types of RAM but is inexpensive.

**Dye Sub:** Dye Sublimation. Printing process that takes a solid, turns it into a gas (sublimate) and applies it to the paper. A thermal print head is used to heat a dye ribbon to transfer the dye. The print head can control the size and amount of dye that is transferred, making it possible to create a continuous tone image. The problem is that this method creates soft edges, so text is not as sharp.

**Dynamic Range:** Range of light values over which differences between values are detectable by the recording device. The measurement device has a lowest and highest value that it can record. Values that are higher or lower than those values will appear to be the same.

**EPS:** Encapsulated PostScript. A file format used for **PostScript** files that includes a low-resolution bitmap preview of the PostScript data.

**EPP:** Enhanced Parallel Port. An updated version of the directional printer port that features higher transmission speeds. Some digital cameras and scanners can use the EPP port.

**EXIF:** A variation of the **JFIF** file format used for storing JPEG compressed images. EXIF adds a series of tags for storing information such as shutter speed, aperture, etc.

**Exposure:** Allowing light to strike the light sensitive measurement device (i.e. CCD) or light sensitive film. The term "Correct Exposure" refers to allowing enough light to strike the measurement device so that an acceptable image is created. While this amount can be estimated by a light meter, due to limitations of the dynamic range of most recording devices, photographers often to decide which part of the image they want to be "correctly exposed" i.e. do they want to capture detail in the bright regions of the image, or in the darker regions.

**Exposure Compensation:** Lets the user adjust the exposure that the camera's light meter suggests by slightly over or under exposing the image.

**f-stop:** A measurement of the aperture for the lens; the size of the opening through which light can enter the lens. The larger the f-stop value the smaller the size of the opening. Most camera lenses provide a range of apertures (for example f2.8 - f16) though some low priced cameras use a fixed aperture and only adjust the shutter speed to adjust the exposure.

Generally, the smaller the size of the opening, the greater the depth of field.

**Film Speed:** A measurement of the sensitivity of film to light. Standards for measuring film speed include **ASA** and **ISO**.

**Filter:** Traditionally, a filter is an add-on attachment to a camera lens that removes (filters) certain colored light from the scene. Often used to adjust artificial light (such as that produced by interior lighting) so that the images appear more natural.

Filters can also include special effects, and is used in software editing applications to refer to routines that perform special effects.

**FireWire:** Apple's product name for the IEEE-1394 high-speed serial bus. It can be used for attaching peripherals such as hard disks, printers and cameras to a computer, but has been most widely adopted by DV (digital video) camcorders.

**Firmware:** Software in a camera or computer that controls the basic operation of the hardware. While this is infrequently done, it is possible in many cases for manufacturers to develop new versions of the firmware and provide them to the owners so that they can upgrade their equipment.

**Fixed Focus:** A camera lens that has a fixed focus point. Usually the lens will be "in focus" (produce sharp images) over a very wide range (from a couple of feet to infinity.)

**Flash Memory:** Nonvolatile computer memory that retains information when power is not being supplied to the chips. Nearly all digital cameras use Flash memory in the form of CompactFlash and SmartMedia cards.

**Flash RAM:** see Flash Memory

**Flash Range:** Distance over which the flash is effective. The longer the range, the stronger the flash.

**Flash Sync:** Connection on a camera that can be used to trigger an external flash. (see also *PC*)

**Floppy Disk Adapter:** A device with the same form factor as a 3.5" floppy disk that accepts a SmartMedia card and can be inserted into a floppy disk drive. Using special driver software, the contents of the SmartMedia card can then be transferred to the computer. Also referred to as SSFDC: Solid State Floppy Disc Card.

**Focal Length:** Measurement of the angle of view of the camera lens, usually expressed in mm. Since 35mm cameras (that's the film size, not the focal length) have been so dominant, the focal length of digital cameras is often expressed as the equivalent value for a 35mm lens.

**GIF, GIF89a:** Compressed file format used for Internet graphics. Limited to a maximum of 8-bit color per pixel. Also supports animation (sequence of frames) and transparency. GIF makes use of a compression algorithm that is patented and owned by UNISYS.

**IEEE-1394:** see FireWire

**Image Sensor:** The CCD or CMOS chip used to measure light.

**Index Color:** Method of storing color information for bitmapped graphics. Used for 8-bit and lower graphics, each pixel is stored as a value that points to a color table containing the actual RGB values for the pixel. For an 8-bit file, the table can contain 256 different colors because an 8-bit value can hold a number between 0 and 255 (256 values.)

**Interlaced:** Method of capturing and displaying images that is used by video cameras and television monitors. The image is drawn by the cathode ray tube as a sequence of horizontal lines. The odd lines are drawn first, followed by the even lines. (see also *Progressive Scan*)

**Interpolation:** When scaling up an image, extra pixels must be added to the image. Interpolation estimates the values for the new pixels based on the pixels that surround them.

**IR:** InfraRed (aka IRda)

**ISO (International Standards Organization):** Film speed rating. Indicates the sensitivity of the film to light.

**Jaggies:** Stair-step effect visible at the intersection of high-contrast color regions that is sometimes visible in bitmapped graphics. **Anti-aliasing** can be used to reduce this effect.

**JFIF:** The file format used to save JPEG compressed images. JPEG is actually the compression algorithm, not the file format, however most people refer to a JFIF file as a JPEG file.

**JPEG:** Highly efficient compression algorithm used to reduce the size of photographic images. The JPEG algorithm works on 8 x 8 pixel areas. JPEG is lossy; some information is lost in compression, so the decompressed image is not exactly the same as the original. In many cases, the differences are not detectable by the average viewer. However, as compression level is increased the image can degrade to the point where the 8 x 8 blocks start to become visible.

**JPG:** File extension used for JPEG files (i.e. filename.jpg)

**Kilobyte (KB):** Unit of measurement for computer memory. 1024 bytes. Note: It's not 1000 because memory is a binary system, not base-10.

**Lab Color:** CIElab. A color model developed by the Commission Internationale de l'Eclairage (International Committee on Illumination, or CIE). Unlike other color models (RGB, CMYK) it is based upon how the eye perceives color, and can be used to reproduce any color. It is often used as an intermediate format when dealing with multiple color spaces.

**LCD (Liquid Crystal Display):** Used as a display device (typically about 1.5 – 2 inches in width.) Most cameras also use a small black and white LCD display to display information about the number of pictures remaining and the mode of the camera.

Note that the color LCD's consume large amounts of power, so turning them off will extend battery life. Also, they are often hard to view in bright sunlight.

**Lithium-Ion:** Rechargeable battery technology (see also: Ni-Cd and Ni-MH.) Using Lithium based chemicals (ions) these batteries are light, hold a large charge, and like NiMH batteries, have no memory effect. However they are currently more expensive than other batteries.

**Lines per Inch (lpi):** Measurement of resolution used for commercial printers and printing presses (most consumer printers use dots per inch). Indicates the resolution that the printing press is capable of producing. Newspapers have an lpi of 85, magazines are typically about 140.

**Lossless:** Compression algorithm that does not remove information to reduce the size of a file. LZW is an example of a lossless compression algorithm often used with TIFF files.

**Lossy:** Compression algorithm that removes some information when compressing a file. The compressed image can often not be distinguished from the original by the human eye. JPEG is an example of a lossy compression algorithm.

**Macro:** Lens that can focus on objects close to the lens (typically less than 2 feet.) This function is vital if you want to take close-ups of small objects.

**MAh:** Thousands of Amp hours. Measurement of the capacity of a battery, or the power consumption of a device.

**Matrix Metering:** Method of calculating correct exposure for a scene by measuring the exposure for multiple areas in a scene.

**Megabytes:** Millions of bytes (MB). Measurement of the size of computer memory. Used for disk drives and RAM memory. Also used for memory cards used in cameras.

**Memory Effect:** A drop in voltage and capacity of a battery that occurs due to recharging a battery that was not completely discharged before being recharged. Nickel-cadmium batteries are particularly vulnerable to this problem. Nickel-metal hydride and Lithium ion batteries suffer negligible such effects.

**MB:** see Megabytes

**MD:** MiniDisc. Developed by Sony as a digital medium for recording and rerecording audio at near CD quality (CD audio is at 44 kHz, 16 bit non-compressed, MiniDiscs use a compression algorithm.) MD Data is used for storing data, though it has not proven popular with computer users and manufacturers. Sharp manufactured a camera that used MD discs for storage (each holds about 120 MB) but did not release it in the US.

**Megapixel:** Used as a marketing term for cameras that either have a CCD with more than 1 million pixels, or produce an image with more than 1 million pixels (sometimes through interpolation.)

**MiniDisc:** see MD

**mm:** Millimeter. Metric measurement used as a measurement of the focal length of a lens even in non-metric countries!

**Multi-Capture:** Capturing several images in rapid succession. (see also *burst mode*)

**Multi-Spot Focusing:** Auto-focus system that determines correct focus by measuring focus distance for a number of different points in an image.

**NiCd:** Nickel Cadmium (aka Nicad). Rechargeable battery often used for camcorders and portable cameras. Has largely been replaced by Nickel-Metal Hydride and Lithium ion batteries because they do not suffer from the memory effect.

**NiMH:** Nickel-Metal Hydride. Rechargeable batteries that use a potassium hydroxide electrolyte and a nickel oxyhydroxide electrode. Along with Lithium ion is the most popular rechargeable battery for cameras, camcorders and portable computers.

**Noise:** Erroneous signal or pixels created by imperfections in electronics, CCD manufacture, or countless other causes.

**NTSC:** Television format used in the US and some other countries. Other common format is **PAL**.

**Optical Zoom:** Lens that changes its focal length by moving optical elements within the lens. See also Digital Zoom.

**Overexposure:** Too much light is allowed to strike the light sensitive material (film or CCD) resulting in a washed out image with little detail.

**PAL:** Television format used in Japan, Great Britain and other countries. Other common format is NTSC.

**Parallax Error:** Difference between what is seen in the viewfinder and what the lens sees; occurs in rangefinder cameras because the rangefinder and camera do not have the same viewpoint. SLR cameras do not have this problem.

Parallax is also used to describe what occurs when rotating the camera around its center point to take a panoramic image. Objects in the scene appear to move relative to other objects as the camera is rotated.

**PC:** External flash sync connector. (see also *Flash Sync*)

**PC Card:** see PCMCIA

**PCMCIA:** A standard for small cards that can be used for adding Flash RAM, hard disks, modems and other devices to portable computers. There are three types: Type I, II, and III, which have different thicknesses, and so can be used for different purposes. Some high-end cameras use PCMCIA cards. CompactFlash and SmartMedia cards can be placed in a PC Card Adapter that lets you read these card on a computer with a PCMCIA slot.

**PCX:** File format originally developed for the PC paint program PaintBrush.

**Photo CD:** Kodak Proprietary CD disk format for storing digital scans of slides and negatives. Originally developed for use with 35mm film, but now also used for larger format work. The 35mm scans have a resolution of 3072 x 2048 pixels.

**Photosite:** Area of the CCD that is light sensitive.

**Picture CD:** Film digitizing service developed by Kodak which scans film and places it on a CD-ROM. Differs from Photo CD in that it uses standard file formats (JPEG), the scans are at half the resolution (1536 x 1024) and the disc includes software for viewing and editing the images.

**Pincushion:** Distortion in image caused by lens. Edges appear to curve inward. Is most visible in straight lines in the image that can appear to bend inward rather than run straight. (see also *Barrel Distortion*)

**Pixel:** Picture element. The elements used to capture or display digital images.

**PPI:** Pixels per inch. Measurement of resolution. Usually used for displays (**dots per inch** is usually used for printers).

**Plug-In:** Add-on software that adds functionality to an existing application (such as special effects filters for image editing software).

**Point and Shoot:** Marketing term used for a camera that is fully automatic (automatic exposure, automatic focus.) User does not need to do anything other than press the shutter button.

**Posterize:** Reducing the number of tonal values in an image, often to much less than 20 tonal values (or colors.) Creates very stylized images.

**Progressive Scan:** CCD and video monitors that display or capture the horizontal lines of an image in sequence. As opposed to interlaced, which captures or displays the odd lines and then the even lines. Computer monitors use progressive scan, and many digital cameras now use progressive scan CCDs.

**Prosumer:** Marketing term used for equipment that's considered somewhere in between consumer and professional equipment.

**Rangefinder:** Viewfinder that is a separate optical piece to the lens, but provides a close approximation of what the lens sees (see also Parallax error.)

**Red-Eye:** The red color that appears in people's eyes when photographs are taken with a flash. It is the light bouncing off the retina of the eye. Can be reduced by moving the flash further away from the lens of the camera.

**Red-Eye Reduction:** A feature that reduces the red-eye effect by firing the flash briefly one or more times before the photograph is taken. This flashing causes the iris of the eye to close, and reduces the light being reflected and the red-eye effect.

**Resolution:** The number of pixels on a CCD, or the number of pixels per inch of a printer. In all cases, it indicates the size of the smallest light sensitive detector, or reproduction elements (monitor pixel or ink drop.) The higher the resolution, the more detail should be captured or reproduced.

**RGB:** Red Green Blue. Additive color (emitted color) process used by projectors, monitors and by digital cameras. It is also the most common method for encoding color information when saving it in a digital data file.

**ROM:** Read Only Memory. Type of computer memory that retains information permanently. Used for storing the basic operating functions of a camera or computer.

**RS-232:** Serial communications protocol used for connecting devices to a computer (often used for modems.) Is the most common interface for connecting digital cameras to computers. See also Com Port.

**SCSI:** Small Computer Standard Interface. A communication protocol for connecting devices such as hard disks and scanners to a computer. Also used by some high-end digital cameras.

**Serial Port:** Same as "RS-232" above.

**Shutter:** Mechanical or electronic device that controls the length of time that light passes through the lens and is captured by the light sensitive film or CCD.

**Shutter Speed:** Measurement of time that the shutter is opened. This speed (usually in fractions of a second) is dependent upon the subject being photographed, the sensitivity of the CCD, and the size of the **aperture**.

**Single Lens Reflex:** (SLR) Camera viewfinder system that reflects the light that comes through the lens into the viewfinder so that the photographer is seeing exactly what the lens will capture. This is accomplished with a mirror that reflects the light up to the viewfinder. Before the photograph is taken the lens rotates up and out of the path of the light.

**Slave Strobe:** Flash that has a light sensor that fires the flash when the flash attached to the camera is fired.

**SLR:** see Single Lens Reflex

**SmartMedia:** A flash memory card used by Agfa, Epson, Fuji, Olympus and others. There are two types: 3.3v and 5v. (see also *CompactFlash*)

**Spot Focus:** Method of focusing that measures focus at a single point in the center of the image.

**Spot Metering:** Method of measuring exposure that is based on a single point in the center of the image. See also Matrix metering.

**SSFDC:** Solid State Floppy Disc Card. Device that accepts a SmartMedia card and can be inserted and read in a Floppy Disk drive. (see also *Floppy Disk Adapter*)

**Teleconverter:** Add-on lens that increases the focal length of the lens.

**Telephoto Lens:** Lens with a focal length higher than a "standard" lens, and therefore a narrow field of view. For a 35mm camera a standard lens is 50mm, and a telephoto lens is considered anything greater i.e. 100mm.

**TIFF:** Tagged Image File Format . File format used for digital images.

**TTL:** Through the Lens

**TWAIN:** A software protocol method for communicating between software and scanners and digital cameras.

**Type I, II, III:** Different kinds of PCMCIA cards.

**Underexposure:** Too little light was available to the light sensitive material during exposure, resulting in a dark image. See also overexposure.

**USB:** Universal Serial Bus. New communications protocol for connecting devices including digital cameras to a computer.

**Vector Graphics:** Graphics programs and file formats that use lines and curves to create images. Programs that make extensive use of vector graphics are the illustration programs Adobe Illustrator and Macromedia FreeHand. Not usually suitable for digital photographs (note that many vector graphics programs also support bitmapped graphics, though they may not provide many editing functions).

**Video Out:** Ability to connect camera to television monitor or VCR. Typically available as an RCA (composite video) connection.

**Viewfinder:** Optical or LCD display that lets the photographer see the scene that the camera lens will capture.

**White Balance:** Adjusting the balance of red, green and blue spectrum so that bright objects appear white and not some other color.

**Wide-Angle Lens:** Lens with a focal length lower than a "standard" lens, and therefore a wider field of view. For a 35mm camera a standard lens is 50mm, and a wide-angle lens is considered anything less i.e. 28mm.

**WYSIWYG:** What You See Is What You Get. Term coined to describe computer displays which displayed a close approximation to the output produced by a printer.

**Zoom Lens:** Lens with a variable focal length.

# INDEX

## Symbols



## A



## B

## D



## E

## M

## N



## O



## P

## Q



## R

## U



## V



## W



## Z

PROMPT®
PUBLICATIONS

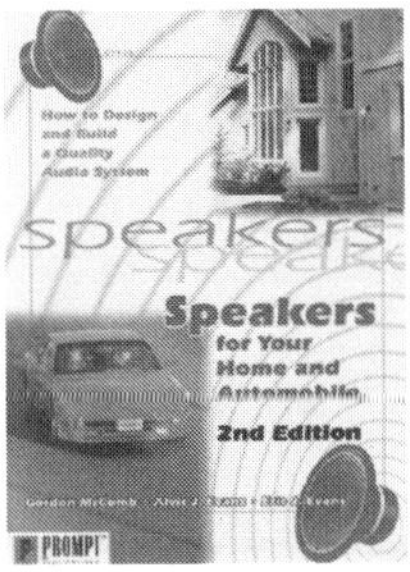


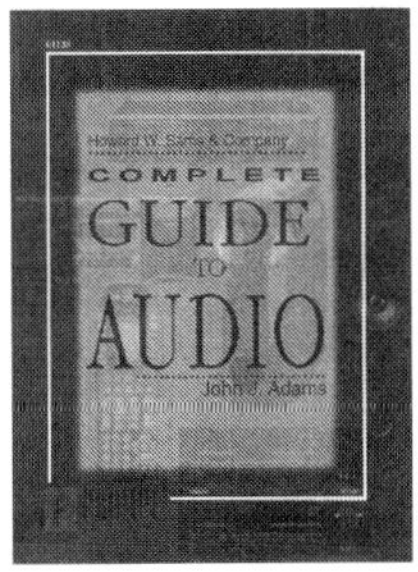


## Speakers For Your Home and Automobile, *Second Edition*

*Gordon McComb, Alvis Evans, Eric Evans*

The cleanest CD sound or the clearest FM signal are useless without a fine speaker system. *Speakers for Your Home and Automobile* will show you the hows and whys of building quality speaker systems for home or auto installation. With easy-to-understand instructions and clearly-illustrated examples, this book is a must-have for anyone interested in improving their sound systems.

The comprehensive coverage includes:

Speakers, Enclosures, and Finishing Touches
Construction Techniques
Wiring Speakers
Automotive Sound Systems and Installation
Home Theater Applications

Audio
192 pages • paperback • 6 x 9"
ISBN: 0-7906-1119-8 • Sams: 61119
$24.95

## Complete Guide to Audio

*John J. Adams*

*Complete Guide to Audio* was written for the consumer who wants to know more about sound systems. With comprehensive, simple explanations, it answers questions you may have asked salespeople in the past but were unable to get answers for.

In addition, this book explains some common problems that you may run into while setting up your home entertainment center. The information in the book will help you make successful purchasing decisions and demystify the jungle of wires and connections that come with your audio system.

Topics include: Audio Basics, Sound, Stereophonics, Home Theater, Amplifiers and Preamplifiers, Receivers and Surround-Sound, Cassette and CD Decks, DVD, MiniDisc and Phonographs, Speakers, Headphones and Microphones, Computer Sound, Brands and Choices, Hookups and Accessories.

Audio
163 pages • paperback • 7-1/4 x 9-3/8"
ISBN: 0-7906-1128-7 • Sams: 61128
$29.95

**To order your copy today or locate your nearest Prompt® Publications distributor : 1-800-428-7267 or www.hwsams.com**

Prices subject to change.

PROMPT®
PUBLICATIONS

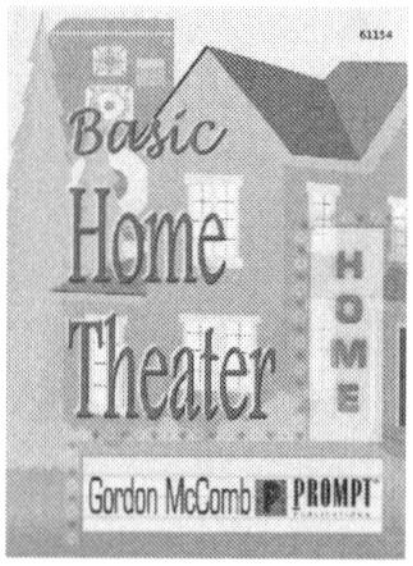


## Basic *Home Theater*

Gordon McComb

In today's electronic age, it is quite possible to have the experience of a movie theater right in your own home. Everything from surround-sound stereo and digital sound tracks to large-screen television sets have helped bring Hollywood to living rooms, family rooms and even specially designed home theater dens everywhere. This book will provide you everything you need to know to plan, design, install and maintain your very own home theater.

Topics covered include: What Is Home Theater?; Program Sources; TV: The Silver Screen of Home Theater; Sound: The Other Half of Home Theater; Hooking Up the Home Theater System; Satellite Dish & "Off-the-Air" Hookups; Wiring Your Home Theater; Room Design and Ergonomics; Home Theater Maintenance; and, Troubleshooting Common Problems.

Video Technology
140 pages • paperback • 6 x 9"
ISBN: 0-7906-1154-6 • Sams 61154
$24.95

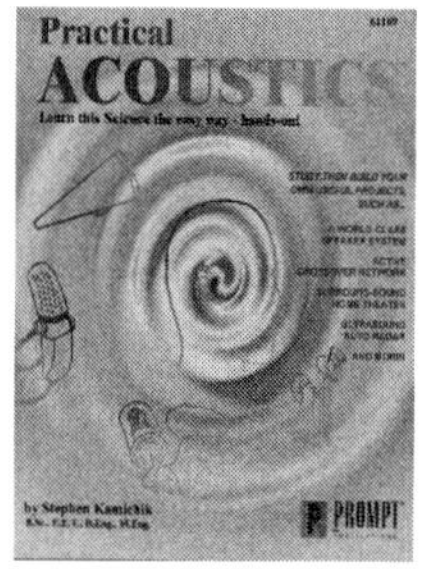


## *Practical Acoustics*

Joe Kamichik

Acoustics is the physics of sound. As simple as that may sound, though, it is actually a complex multidisciplinary science-one that is not taught in most universities and technical schools. A full understanding of acoustics, then, demands that the student grasp the many concepts and practical applications of this fascinating subject.

*Practical Acoustics* is a two-part in-depth study of this science. Part One contains eleven chapters of theory, such as acoustical waves, decibels, transmission through media, ultrasonics, human ear anatomy, underwater acoustics, microphones and speakers. Each chapter has test questions at the end. Part Two contains eight chapters of projects, such as a speaker enclosure, graphic equalizer, crossover network, ultrasonic auto radar, preamplifier circuits, surround-sound, tone control, and more. Problem solutions and a full glossary of related terms is also included.

Audio
259 pages • paperback • 7-3/8 x 9-1/4"
ISBN: 0-7906-1169-4 • Sams 61169
$29.95

**To order your copy today or locate your nearest Prompt® Publications distributor : 1-800-428-7267 or www.hwsams.com**

Prices subject to change.

PROMPT®
PUBLICATIONS

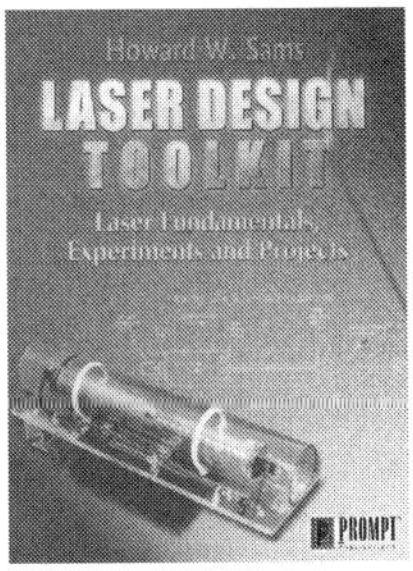


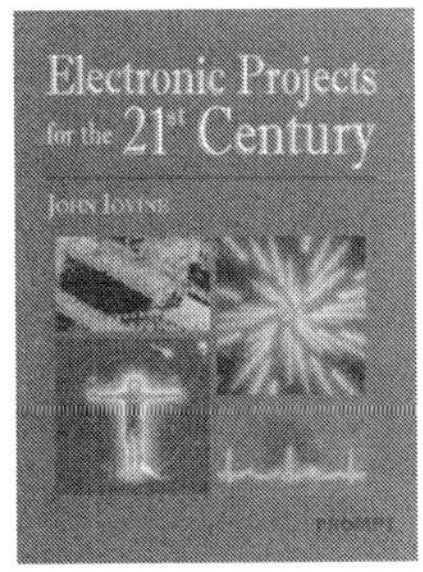


## Howard W. Sams Laser Design Toolkit

*Carl Bergquist*

For over 30 years, the laser has enriched our lives to an extent few can fully appreciate. Unfortunately for the hobbyist, working with lasers was too expensive, but today that is no longer the case! The costs involved in constructing home lasers is now no worse than many other areas of electronics.

There is so much to lasers, that it's hard not to like them once you get to know them, and you can't describe the thrill associated with watching that "red spot" on the wall, projected from a helium-neon laser you built with your own hands. Like most tubes, they have a life span of several thousand hours, and run at full performance for most of that span.

The *Laser Design Toolkit* will introduce you to this fascinating field, and help you complete exciting projects, like building your first HeNe laser, working with optics, constructing a laser light show, hologram processing, building a laser light meter, working with laser diodes, and more!

Projects
336 pages • paperback • 7-3/8 x 9-1/4"
ISBN: 0-7906-1183-X • Sams 61183
$29.95

## Electronic Projects for the 21st Century

*John Iovine*

If you are an electronics hobbyist with an interest in science, or are fascinated by the technologies of the future, you'll find *Electronic Projects for the 21st Century* a welcome addition to your electronics library. It's filled with nearly two dozen fun and useful electronics projects designed to let you use and experiment with the latest innovations in science and technology. This book contains the expert, hands-on guidance and detailed instructions you need to perform experiments that involve genetics, lasers, holography, Kirlian photography, alternative energy sources and more. You will obtain all the information necessary to create the following: biofeedback/lie detector device, ELF monitor, Geiger counter, MHD generator, expansion cloud chamber, air pollution monitor, laser power supply for holography, pinhole camera, synthetic fuel from coal, and much more.

Projects
256 pages • paperback • 7-3/8 x 9-1/4"
ISBN: 0-7906-1103-1 • Sams: 61103
$24.95

**To order your copy today or locate your nearest Prompt® Publications distributor : 1-800-428-7267 or www.hwsams.com**

Prices subject to change.

# AGREEMENT

**READ THIS AGREEMENT BEFORE OPENING THE SOFTWARE PACKAGE**

**BY OPENING THE SEALED PACKAGE YOU ACCEPT AND AGREE TO THE FOLLOWING TERMS AND CONDITIONS PRINTED BELOW. IF YOU DO NOT AGREE, DO NOT OPEN THE PACKAGE AND RETURN THE SEALED PACKAGE AND ALL MATERIALS YOU RECEIVED TO HOWARD W. SAMS & COMPANY, 2647 WATERFRONT PARKWAY EAST DRIVE SUITE 100 INDIANAPOLIS, IN 46214-2041 (HEREINAFTER "LICENSOR") WITHIN 30 DAYS OF RECEIPT ALONG WITH PROOF OF PAYMENT.**

Licensor retains the ownership of this copy and any subsequent copies of the Software. This copy is licensed to you for use under the following conditions:

**Permitted Uses. You may:** use the Software on any supported computer configuration, provided the Software is sued on only one such computer and by one user at a time; permanently transfer the Software and its documentation to another user, provided you retain no copies and the recipient agrees to the terms of this Agreement.

**Prohibited Uses. You may not:** transfer, distribute, rent, sublicense, or lease the Software or documentation, except as provided herein; alter, modify, or adapt the Software or documentation, or portions thereof including, but not limited to, translation, decompiling, disassembling, or creating derivative works; make copies of the documentation, the Software, or portions thereof; export the Software.

**LIMITED WARRANTY, DISCLAIMER OF WARRANTY**

Licensor warrants that the optical media on which the Software is distributed is free from defects in materials and workmanship. Licensor will replace defective media at no charge, provided you return the defective media with dated proof of payment to Licensor within ninety (90) days of the date of receipt. This is your sole and exclusive remedy for any breach of warranty. EXCEPT AS SPECIFICALLY PROVIDED ABOVE, THE SOFTWARE IS PROVIDED ON AN "AS IS" BASIS. LICENSOR, THE AUTHOR, THE SOFTWARE DEVELOPERS, PROMPT PUBLICATIONS, HOWARD W. SAMS & COMPANY, AND BELL ATLANTIC MAKE NO WARRANTY OR REPRESENTATION, EITHER EXPRESS OR IMPLIED, WITH RESPECT TO THE SOFTWARE, INCLUDING ITS QUALITY, ACCURACY, PERFORMANCE, MERCHANTABILITY, OR FITNESS FOR A PARTICULAR PURPOSE. IN NO EVENT WILL LICENSOR, THE AUTHOR, THE SOFTWARE DEVELOPERS, PROMPT PUBLICATIONS, HOWARD W. SAMS & COMPANY, AND BELL ATLANTIC BE LIABLE FOR DIRECT, INDIRECT, SPECIAL, INCIDENTAL, OR CONSEQUENTIAL DAMAGES (INCLUDING BUT IS NOT LIMITED TO, INTERRUPTION OF SERVICE, LOSS OF DATA, LOSS OF CLASSROOM TIME, LOSS OF CONSULTING TIME) OR LOST PROFITS ARISING OUT OF THE USE OR INABILITY TO USE THE SOFTWARE OR DOCUMENTATION, EVEN IF ADVISED OF THE POSSIBILITY OF SUCH DAMAGES. IN NO CASE SHALL LIABILITY EXCEED THE AMOUNT OF THE FEE PAID. THE WARRANTY AND REMEDIES SET FORTH ABOVE ARE EXCLUSIVE AND IN LIEU OF ALL OTHERS, ORAL OR WRITTEN, EXPRESSED OR IMPLIED. Some states do not allow the exclusion or limitation of implied warranties or limitation of liability for incidental or consequential damages, so that the above limitation or exclusion may not apply to you.

**GENERAL:**

Licensor retains all rights, not expressly granted herein. This Software is copyrighted; nothing in this Agreement constitutes a waiver of Licensor's rights under United States copyright law. This License is nonexclusive. This License and your right to use the Software automatically terminate without notice from Licensor if you fail to Comply with any provision of this Agreement. This Agreement is governed by the laws of the State of Indiana.